MODERN ANALYTICAL ULTRACENTRIFUGATION

EMERGING BIOCHEMICAL AND BIOPHYSICAL TECHNIQUES
Series Editor: Todd M. Schuster

MODERN ANALYTICAL ULTRACENTRIFUGATION

Acquisition and Interpretation of Data for
Biological and Synthetic Polymer Systems

TODD M. SCHUSTER
THOMAS M. LAUE
— *Editors* —

118 Illustrations

Boston · Basel · Berlin

Todd M. Schuster
Department of Molecular & Cell Biology
and Analytical Ultracentrifugation Facility
University of Connecticut
75 North Eagleville Road
Storrs, CT 06269-3125
USA

Thomas M. Laue
Department of Biochemistry
University of New Hampshire
Durham, NH 03824
USA

QP
519
.9
.U47
M63
1994

Library of Congress Cataloging-in-Publication Data

Modern analytical ultracentrifugation : acquisition and interpretation
    of data for biological and synthetic polymer systems / Todd M.
    Schuster, Thomas M. Laue, editors.
        p.   cm.
    Includes bibliographical references and index.
    ISBN 0-8176-3674-9
        1. Ultracentrifugation. 2. Biomolecules—Analysis. 3. Polymers—
    Analysis. I. Schuster, Todd M., 1933–   . II. Laue, Thomas M.,
    1950–   .
    QP519.9.U47M63   1994
    574.19'285—dc20   94-20182
                                                            CIP

Printed on acid-free paper.

Birkhäuser ®

©1994 Birkhäuser Boston

ISBN 0-8176-3674-9
ISBN 3-7643-3674-9

Camera-ready text prepared by authors.
Cover art by Maxine Marcy.
Printed and bound by Braun-Brumfield, Inc., Ann Arbor, MI
Printed in the United States of America

9 8 7 6 5 4 3 2 1

CONTENTS

PART IV: *SOME SPECIFIC EXAMPLES*

PREFACE

There are numerous examples in the history of science when the parallel developments of two or more disciplines, methodologies, technologies or theoretical insights have converged to produce significant scientific advances. The decades following the 1950s have produced several such significant advances, as a result of a convergence of developments in molecular biology and in solid state-based electronics instrumentation.

Since one of these areas of significant advancement, analytical ultracentrifugation, has been undergoing a renaissance, we thought it would be a useful activity to call upon a group of researchers who have been developing either the experimental or theoretical aspects of the methodology and gather in one place a group of articles summarizing the current status of the field. The success of recombinant DNA methodologies at producing biologically active macromolecules of commercial interest has evoked interests in mechanisms of function. Pursuit of the related questions has emphasized the importance of studies of macromolecular binding and interaction. Several contributions to this volume remind us that analytical ultracentrifugation is rigorously based on solid thermodynamic theory and, as such, is fully capable of providing comprehensive quantitative descriptions of molecular interactions in solution. Furthermore, a number of the chapters provide examples, along with innovative methods for carrying out these characterizations.

The past decade has seen several developments that reflect the rebirth of interest in analytical ultracentrifugation. One of these is the commercial development and availability of a new generation of instrumentation for analytical ultracentrifugation. Other important developments have been the establishment of National Analytical Ultracentrifugation Facilities Laboratories in the U.S. (Storrs, CT) and U.K. (Leicester).

A review of the chapter titles in this volume reveals that considerable effort has been devoted to developing computer assisted protocols for data analysis and interpretation. These approaches are both timely and necessary as researchers turn their attention to increasingly complex interacting and self associating systems. Examples are provided in the articles by Harding; Hensley, et al.; Cann; Shire; Philo; and by Holzman and Snyder.

One can expect the future to bring more examples from biotechnology and pharmaceutical laboratories as the chapters by Philo, Shire, Hensley and Holzman clearly presage.

An overview of this volume also reveals that the rebirth of interest in Analytical Ultracentrifugation will certainly call for a new generation of trained practitioners. With this potential teaching requirement in mind we have attempted to include chapters that will have didactic as well as expository value to students and trainees as well as to advanced researchers. It should be noted that fully half of the number of authors of this volume had not completed their graduate training at the time of publication of Professor Fujita's classic monograph, "Foundations of Ultracentrifugation Analysis." These statistics reflect a continuation of a scientific tradition extending back more than 50 years to the pioneering developments of The Svedberg.   The new computer controlled commercial analytical ultracentrifuge greatly simplifies the execution of experiments and the acquisition of data. Several of the chapters in this volume summarize novel approaches to the subsequent steps in ultracentrifuge investigation, data reduction, analysis and interpretation. The developments in Sedimentation Velocity analysis reported by Philo and by Stafford are particularly surprising since this oldest of Sedimentation methodologies was long thought to be of limited value for analysis of complex systems. But the ready availability of high speed and high capacity desk-top computers has made it possible to bring a new level of sophistication and rigor to the analysis of sedimenting boundaries. We look forward to these methods making Sedimentation Velocity a workhorse methodology of *Modern Analytical Ultracentrifugation*.

We wish to acknowledge the long term support for biological instrumentation and methodology development provided by the Biophysics and the Instrumentation Programs of the U.S. National Science Foundation.

Our thanks and acknowledgement to Ms. Alyson M. Blow for expert assistance with manuscript preparation for this volume.

T. M. Schuster
Storrs, Connecticut

T. M. Laue
Durham, New Hampshire

CONTRIBUTORS

Ian Brooks Department of Macromolecular Sciences, Smithkline Beecham Pharmaceuticals, King of Prussia, Pennsylvania 19406-0939, USA

John R. Cann Department of Biochemistry/Biophysics/Genetics, University of Colorado, Health Sciences Center, B-121, 4200 East 9th Avenue, Denver, Colorado 80262, USA

Winnie Chan Department of Macromolecular Sciences, Smithkline Beecham Pharmaceuticals, King of Prussia, Pennsylvania 19406-0939, USA

Hiroshi Fujita 35 Shimotakedono-Cho, Shichiku, Kita-Ku, Kyoto, Japan

Stephen E. Harding University of Nottingham, School of Agriculture, Sutton Bonington LE12 5RD, United Kingdom

David B. Hayes Department of Biochemistry, University of New Hampshire, Spaulding Life Science Building, Durham, New Hampshire 03824-3544, USA

Joe Hedges Beckman Instruments, Inc., 1050 Page Mill Road, Palo Alto, California 94304, USA

Preston Hensley Department of Macromolecular Sciences, Smithkline Beecham Pharmaceuticals, King of Prussia, Pennsylvania 19406-0939, USA

Thomas F. Holzman Abbott Laboratories, Protein Biochemistry, D-46Y Discovery Research, Pharmaceutical Products, Abbott Park, Illinois 60048, USA

Michael L. Johnson Departments of Pharmacology and Internal Medicine, Box 448, University of Virginia Health Sciences Center, Charlottesville, Verginia 22908, USA

Soon-Jong Kim   Laboratory of Biochemistry (LB), NCI, Building 37, Room 4C-09, National Institute of Health, Bethesda, Maryland 20892, USA

Jeffrey W. Lary   Department of Molecular & Cell Biology and Analytical Ultracentrifugation Facility, University of Connecticut, 75 North Eagleville Road, Storrs, Connecticut 06269-3125, USA

Thomas M. Laue   Department of Biochemistry, University of New Hampshire, Durham, New Hampshire 03824, USA

William R. Laws   Mount Sinai School of Medicine of the City University of New York, New York, New York 10029, USA

James D. Lear   DuPont Merck Pharmaceutical Company, P.O. Box 80328, Wilmington, Delaware 19880-0328, USA

Grace Lee   PerSeptive Biosystems, Department of Immunochemistry, Cambridge, Massachusetts 02139, USA

James C. Lee   Department of Human Biochemistry/Genetics, University of Texas Medical Branch, 617C Basic Science Building, F-47, Galveston, Texas 77555-0647, USA

Marc S. Lewis   Biomedical Engineering and Instrumentation Branch (BEIP), NCRR, Building 35, Room B101C, National Institute of Health, Bethesda, Maryland 20892, USA

Sen Liu   Department of Muscle Research, Boston Biomedical Research Institute, 20 Saniford Street, Boston, Massachusetts 02114-2500, USA

Daryl A. Lyons   Department of Biochemistry, University of New Hampshire, Durham, New Hampshire 03824, USA

Donald K. McRorie   Beckman Instruments, Inc., 1050 Page Mill Road, Palo Alto, California 94304, USA

Allen P. Minton   Laboratory of Biochemical Pharmacology, National Institute of Diabetes and Digestive and Kidney Diseases, National Institutes of Health, Bethesda, Maryland 20892, USA

Thomas P. Moody   Department of Biochemistry, University of New Hampshire, Durham, New Hampshire 03824, USA

Philip H. Olsen Department of Biochemistry, University of New Hampshire, Durham, New Hampshire 03824, USA

John S. Philo Protein Chemistry Department 14-2-A-223, Amgen, Inc., Amgen Center, Thousand Oaks, California 91320, USA

Surendran Rajendran Department of Human Biological Chemistry and Genetics, University of Texas Medical Branch, 617C Basic Science Buliding, F-47, Galveston, Texas 77555-0647, USA

Theresa M. Ridgeway Department of Biochemistry, University of New Hampshire, Durham, New Hampshire 03824, USA

J. B. Alexander Ross Department of Biochemistry, Mount Sinai School of Medicine of the City University of New York, New York, New York 10029

Shokoh Sarrafzadeh Beckman Instruments, Inc., 1050 Page Mill Road, Palo Alto, California 94304, USA

Todd M. Schuster Department of Molecular & Cell Biology and Analytical Ultracentrifugation Facility, University of Connecticut, 75 North Eagleville Road, Storrs, Connecticut 06269-3125, USA

Steven J. Shire Department of Pharmaceutical R&D, Genentech, Inc., 460 Point San Bruno Boulevard, South San Francisco, California 94080, USA

Richard I. Shrager Physical Sciences Laboratory (PSL), DCRT, Building 12A, Room 2007, National Institutes of Health, Bethesda, Maryland 20892, USA

Seth W. Snyder Abbott Laboratories, Protein Biochemistry, D-46Y, Discovery Research, Pharmaceutical Products, Abbott Park, Illinois 60048, USA

K. Karl Soneson KKS Software, 531 Westfield Drive, Exton, Pennsylvania 19341-1730, USA

Walter F. Stafford III Analytical Ultracentrifugation Research Laboratory, Boston Biomedical Research Institute, 20 Staniford Street, Boston, Massachusetts 02114-2500, USA

Martin Straume NSF Center for Biological Timing and Department of Internal Medicine, Gilmer Hall, University of Virginia, Charlottesville, Verginia 22903, USA

Donald G. Watts Department of Mathematics & Statistics, Queens University, Kingston, Ontario, K7L 3N6 Canada

Evan Waxman Department of Biochemistry, Mount Sinai School of Medicine of the City University of New York, New York, New York 10029, USA

Ronald Wetzel Department of Macromolecular Sciences, Smithkline Beecham Pharmaceuticals, King of Prussia, Pennsylvania 19406-0939, USA

Peter R. Wills Department of Physics, University of Auckland, Auckland, New Zealand

Donald J. Winzor Department of Biochemistry, University of Queensland, Queensland 4072, Australia

David A. Yphantis Department of Molecular & Cell Biology and Analytical Ultracentrifugation Facility, University of Connecticut, 75 North Eagleville Road, Storrs, Connecticut 06269-3125, USA

Part I
SEDIMENTATION EQUILIBRIUM

NOTES ON THE DERIVATION OF SEDIMENTATION EQUILIBRIUM EQUATIONS

Hiroshi Fujita

INTRODUCTION

In the traditional kinetic approach as often taught in physical chemistry courses, sedimentation equilibrium of a macromolecular solution in the ultracentrifuge cell is pictured as the state in which forward transport of a macromolecular solute by centrifugal force is exactly counterbalanced by backward diffusion transport of the same solute by its chemical potential gradient at any point in the solution column. For the formulation of sedimentation equilibrium of solutions containing more than one solute, however, this approach is of no use and has to be replaced by the one based on the thermodynamics of multicomponent systems (Fujita, 1975). Such formulation was initiated by Goldberg (1953), discussed by many authors, and finally almost completed by Casassa and Eisenberg (1964). Actually, there are few that can be added to their elegant theory, but many workers may not find simple to follow its sophisticated mathematical content.

This paper aims to add something that may help understand the basis of sedimentation equilibrium theory. Actually, we deal with two topics. One is the sedimentation equilibrium condition for compressible solutions. This subject is yet to come for a discussion, as far as the author is aware, though it may not be worth picking up because compressibility effects are considered negligible under conditions of usual sedimentation equilibrium measurements, especially on aqueous solutions. The other is the derivation of the differential equations for sedimentation equilibrium by application of the method described in Kurata's book (Kurata, 1982: pp. 251–254). This approach is interesting because it allows a simpler and more clear-cut derivation than does the Casassa–Eisenberg theory. The analysis is illustrated for a typical ternary system consisting of a liquid solvent, a low-molecular-weight solute, and a (monodisperse) macromolecular solute. The reader is advised to learn that the sedimentation equilibrium equations for a given solution are brought to different-looking form depending on the independent variables chosen.

EQUILIBRIUM CONDITIONS

In the experimental setup with an ultracentrifuge cell, a meniscus is formed between a test solution and an air bubble as illustrated in Figure 1. As the cell is spun at high speed about a fixed axis, a pressure gradient is produced extending from the meniscus to the cell bottom. If the solution is compressible, then the meniscus moves toward the cell bottom depending on the speed of rotation, and the volume of the solution decreases. This means that unless incompressible, the solution subject to ultracentrifugation cannot be treated as a system of constant volume. The question is how the sedimentation equilibrium behavior is affected by the compressibility of the solution.

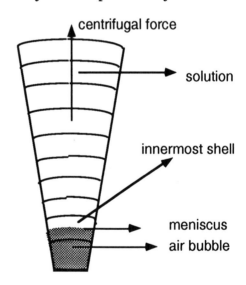

FIGURE 1. Ultracentrifuge sector cell, divided into a series of thin cylindrical shells. The air bubble remains when the cell is filled with a test solution.

We consider a compressible solution consisting of $q + 1$ thermodynamically well-defined components, one of which (say, component 0) is a liquid solvent. When spun at an angular velocity of ω, the unit mass of the solution at a radial distance r from the axis of revolution has a centrifugal potential $\phi(r)$ given by

$$\phi(r) = -(1/2)\omega^2 r^2 \tag{1}$$

We assume that the meniscus of the solution is impermeable, since otherwise the formulation would become complex. This assumption allows the solution to be treated as a *closed* system. When it is at sedimentation equilibrium, any part of the solution as well as the air bubble are in thermal equilibrium with the rotor which is maintained at a given temperature T_0 by an external device, while the composition and pressure in the cell vary with the radial distance r. We consider what relation governs this equilibrium state of the solution by applying the established criterion for thermodynamic equilibrium.

The criterion is for a closed system and described in standard textbooks of thermodynamics. Here we refer to Kurata's book (Kurata, 1982: p. 20), according to which it is given by

$$\delta U - T_e \delta S + p_e \delta V \geq 0 \tag{2}$$

where U, S, and V are the internal energy, entropy, and volume of the system, T_e and p_e are the temperature and pressure of the surroundings, and δ denotes a virtual infinitesimal variation. In applying eq. (2) to our closed system, we note that T_e can be replaced by the rotor temperature T_0, while p_e must be taken as the pressure p_b of the air bubble. Thus, we have

$$\delta U - T_0 \delta S + p_b \delta V \geq 0 \tag{3}$$

As illustrated in Figure 1, we divide the cell into a number of very thin cylindrical shells (their volumes may not be equal) and assume the centrifugal potential and all thermodynamic properties to be constant within each shell. The innermost shell, i.e., one closest to the axis of revolution, contains both solution and air bubble.

We denote the internal energy, entropy, pressure, and volume of the α shell by U^α, S^α, p^α, and V^α, respectively, with α as a running index to indicate the position of the shell. Equation (3) can then be written

$$\sum_\alpha \delta U^\alpha - T_0 \sum_\alpha \delta S^\alpha + p_b \sum \delta V^\alpha \geq 0 \tag{4}$$

The solution in each shell is an *open* system and has the same temperature T_0 (but the pressure varies with the shell). Hence, δU^α is expressed by (Kurata, 1982: p. 23)

$$\delta U^\alpha = T_0 \delta S^\alpha - p^\alpha \delta V^\alpha + \sum_{i=0}^{q} (\mu_i^\alpha + M_i \phi^\alpha) \delta n_i^\alpha \qquad (5)$$

where M_i is the molar mass of component i, and μ_i^α and n_i^α are the chemical potential and amount in moles of component i in the α shell, respectively. In writing this expression, we have considered that one mole of component i in the α shell has the centrifugal potential $M_i \phi^\alpha$. Substituting eq. (5) into eq. (4), we get

$$\sum_{\alpha} \sum_{i=0}^{q} (\mu_i^\alpha + M_i \phi^\alpha) \delta n_i^\alpha - \sum_{\alpha} (p^\alpha - p_b) \delta V^\alpha \geq 0 \qquad (6)$$

Each shell has a fixed volume and is filled completely with the solution except the innermost one. This condition means that nonzero δV^α is not permitted for the solution in any shell other than the innermost one. If the solution is incompressible, δV^α is zero even for the innermost shell. For a compressible solution this is not the case. However, if we neglect the interfacial tension at the meniscus, the pressure is continuous there so that p^α of the solution in the innermost shell can be equated to p_b. From these considerations we find that regardless of whether the solution is compressible or not, the last sum in eq. (6) vanishes, and obtain

$$\sum_{\alpha} \sum_{i=0}^{q} (\mu_i^\alpha + M_i \phi^\alpha) \delta n_i^\alpha \geq 0 \qquad (7)$$

Since δn_i^α can be made either positive or negative, it follows from this relation that the necessary and sufficient condition for the solution to be at sedimentation equilibrium is

$$\mu_i^\alpha + M_i \phi^\alpha = \text{constant} \quad (i = 0, 1, \ldots, q) \qquad (8)$$

Rigorously, this relation holds when each shell is taken infinitesimally thin. In this limit, eq. (8) yields, with eq.(1) substituted for $\phi(r)$,

$$d\mu_i(r)/dr - M_i \omega^2 r = 0 \quad (i = 0, 1, \ldots, q) \qquad (9)$$

where $\mu_i(r)$ and $\phi(r)$ denote the chemical potential of component i and the centrifugal potential at the point r, respectively. This equation agrees with what can be derived from the usual (but naive) consideration that the total potential of the solution must be uniform throughout the cell at sedimentation equilibrium. The above analysis shows that it applies regardless of whether the solution is compressible or not.

DIFFERENTIAL EQUATIONS FOR SEDIMENTATION EQUILIBRIUM

The chemical potential of each component in the solution under consideration is a function of $q + 2$ intensive variables, of which the temperature T of the solution is an obvious choice. The determination of these variables as functions of r at a given rotor speed requires $q + 2$ relations. They are given by $q + 1$ equations in eq. (9) for thermodynamic equilibrium along with the equation for mechanical equilibrium:

$$\frac{dp(r)}{dr} = -\rho(r)\frac{d\phi(r)}{dr} \qquad (10)$$

where $\rho(r)$ denotes the density of the solution at the point r. With eq. (1), eq. (10) can be written

$$\frac{dp(r)}{dr} = \rho(r)\omega^2 r \qquad (11)$$

These $q + 2$ differential equations can be transformed into various forms depending on the choice of $q + 2$ variables for the chemical potential. One of the basic problems relating to sedimentation equilibrium formulation is concerned with what choice is most adequate to approach the exact equations for a given system. Working equations for data analysis should be derived from them by introducing assumptions or approximations relevant to the experimental conditions as well as the system under study. In what follows, this well-known fact is illustrated on the ternary solution mentioned in the Introduction, i.e., one consisting of a pure solvent (component 0), a low-molecular-weight solute (component 1), and a macromolecule (component 2). Components 0 and 1 are sometimes called diffusible components in the sense that they can permeate through a membrane that prevents the macromecular component from passing through. Both components 1 and 2 may be electrolytes. No ionic species appear in the thermodynamic formulation of sedimentation equilibrium, because such species cannot be treated as thermodynamic components.

Incompressible Solutions

First, we deal with the case where the solution is incompressible so that it is a system of constant volume. Considerations of osmotic pressure show (Kurata, 1982: pp. 136–138) that the best choice of independent variables for this system is T, V, μ_0, μ_1, and n_2, where V is the volume of the solution, and n_2 is the amount in moles of component 2. With this choice the intensive variables governing the chemical potential of each component are given by T, μ_0, μ_1, and c_2, where c_2 is the molarity (or volume molality) of component i, i.e., $c_2 = n_2/V$.

Hence we have at constant temperature

$$d\mu_i = \sum_{k=0}^{1} \left(\frac{\partial \mu_i}{\partial \mu_k}\right)_{T,\mu_j,c_2} d\mu_k + \left(\frac{\partial \mu_i}{\partial c_2}\right)_{T,\{\mu\}} dc_2 \quad (i = 0,1,2) \quad (12)$$

where $j \neq k$ and $\{\mu\}$ stands for a set of μ_0 and μ_1. If this is combined with eq. (9) for $q = 2$, we get

$$\left[M_2 - \sum_{k=0}^{1} M_k \left(\frac{\partial \mu_2}{\partial \mu_k}\right)_{T,\mu_j,c_2}\right] \omega_2 r = \left(\frac{\partial \mu_2}{\partial c_k}\right)_{T,\{\mu\}} \frac{dc_2}{dr} \quad (i = 1,2) \quad (13)$$

The solution density ρ is expressed by

$$\rho = (1/V) \sum_{k=0}^{2} M_k n_k \tag{14}$$

Therefore it follows that

$$\left(\frac{\partial \rho}{\partial c_2}\right)_{T,\{\mu\}} = \left[\frac{\partial(\sum_{k=0}^{2} n_k M_k)}{\partial n_2}\right]_{T,V,\{\mu\}}$$

$$= \sum_{k=0}^{1} M_k \left(\frac{\partial n_k}{\partial n_2}\right)_{T,V,\{\mu\},n_j} + M_2 \tag{15}$$

It can be shown (Appendix 1) that

$$\left(\frac{\partial n_k}{\partial n_2}\right)_{T,V,\{\mu\},n_j} = -\left(\frac{\partial \mu_2}{\partial \mu_k}\right)_{T,\mu_j,c_2} \tag{16}$$

Hence eq. (15) gives

$$\left(\frac{\partial \rho}{\partial c_2}\right)_{T,\{\mu\}} = M_2 - \sum_{k=0}^{1} M_k \left(\frac{\partial \mu_2}{\partial \mu_k}\right)_{T,\mu_j,c_2} \tag{17}$$

and eq. (13) can be simplified to

$$\left(\frac{\partial \rho}{\partial c_2}\right)_{T,\{\mu\}} \omega^2 r = \left(\frac{\partial \mu_2}{\partial c_2}\right)_{T,\{\mu\}} \frac{dc_2}{dr} \tag{18}$$

which is the famous relation derived by Casassa and Eisenberg (1964). It is important to recognize that the success in deriving such an elegant sedimentation equilibrium equation originates from having chosen T, V, μ_0, μ_1, and n_2 as independent variables. The elegance, however, does not always mean eq. (18) to be useful for practical purposes. For the actual use of this equation we have to evaluate the term $(\partial \rho/\partial c_2)_{T,\{\mu\}}$, which is usually called the density increment at constant chemical potential, but it is not a simple task, requiring density measurements on a series of solutions of different c_2 that are dialyzed against a mixture of components 0 and 1 maintained at given $\{\mu\}$. Discussing this practical problem is outside the scope of the present paper.

Compressible Solutions

For a compressible solution we have to replace the variable V for the incompressible case by pressure p (this is because V cannot be externally controlled) and to choose pressure-independent variables to specify the total amount of the system and its composition. Thus, when compressible, the best choice of variables for our ternary solution consists of T, p, n_0, m_1, and m_2, where m_i is the molality of component i. For this choice the chemical potential of each component becomes a function of the intensive variables T, p, m_1, and m_2. Hence, we have at constant temperature

$$\frac{d\mu_i}{dr} = \left(\frac{\partial \mu_i}{\partial p}\right)_{T,\{m\}} \frac{dp}{dr} + \sum_{k=1}^{2} \left(\frac{\partial \mu_i}{\partial m_k}\right)_{T,p,m_j} \frac{dm_k}{dr} \quad (i = 0, 1, 2) \tag{19}$$

where $\{m\}$ denotes a set of m_1 and m_2 and $j \neq k$. Substituting eq. (11) for dp/dr and using the relation

$$V_i = \left(\frac{\partial \mu_i}{\partial p}\right)_{T,\{m\}} \tag{20}$$

where V_i is the partial molar volume of component i, we obtain from eqs. (9) and (19)

$$M_i(1 - v_i\rho)\omega^2 r = \sum_{k=1}^{2} \left(\frac{\partial \mu_i}{\partial m_k}\right)_{T,p,m_j} \frac{dm_k}{dr} \quad (i = 1, 2) \qquad (21)$$

where v_i denotes the partial specific volume of component i.

Though exact, this sedimentation equilibrium equation is of little use for practical purposes, because the molality concentrations it contains are not experimentally accessible by optical methods. Thus, it would be desirable to have a corresponding equation in which the concentrations are expressed in terms of molarity. Differing from eq. (21), such an equation should reveal compressibility effects explicitly, because molarity changes with pressure through the pressure dependence of the solution volume. For simplicity we consider our solution containing no component 1. Replacing m_2 by c_2 and following the same logic as above, we obtain at constant temperature

$$\frac{d\mu_2}{dr} = \left(\frac{\partial \mu_2}{\partial p}\right)_{T,c_2} \frac{dp}{dr} + \left(\frac{\partial \mu_2}{\partial c_2}\right)_{T,p} \frac{dc_2}{dr} \qquad (22)$$

This is combined with eqs. (9) and (11) to give

$$M_2(1 - v_2^{\star}\rho)\omega^2 r = \left(\frac{\partial \mu_2}{\partial c_2}\right)_{T,p} \frac{dc_2}{dr} \qquad (23)$$

where $v_2^{\star}$ is defined by

$$v_2^{\star} = (1/M_2)(\partial \mu_2/\partial p)_{T,c_2} \qquad (24)$$

For compressible solutions the quantity $v_2^{\star}$ is not equal to the partial specific volume of component 2, v_2. In fact, as Appendix 2 shows, if terms of the second order in c_k are neglected in comparison with unity, we have

$$v_2^{\star} = v_2 - c_2\kappa \left(\partial \mu_2/\partial c_2\right)_{T,p} \qquad (25)$$

where κ denotes the compressibility of the pure solvent. With eq. (25) substituted into eq. (23), we find a sedimentation equilibrium equation that takes the compressibility effect into explicit account.

If the term containing κ is small enough as compared to v_2, eq. (23) reduces to

$$M_2(1 - v_2\rho)\omega^2 r = \left(\frac{\partial\mu_2}{\partial c_2}\right)_{T,p}\frac{dc_2}{dr} \tag{26}$$

which is the conventional basis for sedimentation equilibrium analysis for binary solutions. Very recently, Wills et al. (Wills, Comper, and Winzor, 1993) have shown that if the partial specific volume of each component is constant, eq. (26) can be transformed to

$$M_2(1 - v_2\rho_0)\omega^2 r = \left(\frac{\partial\mu_2}{\partial c_2}\right)_{T,\mu_0}\frac{dc_2}{dr} \tag{27}$$

where ρ_0 is the density of component 0 in the pure state. Interestingly, differing from eq. (26), not ρ but ρ_0 appears in the buoyancy factor in eq. (27). However, this does not mean that the view point that ρ should be used to calculate the buoyancy factor is incorrect. The difference simply concerns whether (T, p, c_2) or (T, μ_2, c_2) is used as the set of variables in the formulation. Equation (27) is included as a special case of the Casassa–Eisenberg equation, eq. (18), but the left-hand side of this equation for ternary solutions can no longer be written $M_2(1 - v_2\rho_0)\omega^2 r$ even when the partial specific volumes of all components are constant. Actually, it goes to $M_2[1 - v_2\rho_0 + (\partial c_1/\partial c_2)_{T,\mu_0,\mu_1}(1 - v_1\rho_0)]\omega^2 r$. In general, $(\partial c_1/\partial c_2)_{T,\mu_0,\mu_1}$, called the binding factor, does not vanish but plays a crucial role in determining sedimentation equilibrium behavior of ternary solutions.

APPENDIX 1

When T, V, μ_0, μ_1, and n_2 are chosen as independent variables, the characteristic thermodynamic function for ternary systems is $\mathbf{A}$ defined by (Kurata, 1982: p. 138)

$$\mathbf{A} = A - n_0\mu_0 - n_1\mu_1 \tag{1a}$$

where A is the Helmholtz free energy, i.e., $A = U - TS$. Since $dA = -SdT - pdV - \mu_0 dn_0 - \mu_1 dn_1 - \mu_2 dn_2$, the total derivative of $\mathbf{A}$ is given by

$$d\mathbf{A} = -SdT - pdV - n_0 d\mu_0 - n_1 d\mu_1 - n_2 d\mu_2 \tag{2a}$$

From this it follows that

$$n_0 = -\left(\frac{\partial \mathbf{A}}{\partial \mu_0}\right)_{T,V,\mu_1,n_2} \tag{3a}$$

$$n_1 = -\left(\frac{\partial \mathbf{A}}{\partial \mu_1}\right)_{T,V,\mu_0,n_2} \tag{4a}$$

$$\mu_2 = \left(\frac{\partial \mathbf{A}}{\partial n_2}\right)_{T,V,\mu_0,\mu_1} \tag{5a}$$

Therefore, we obtain

$$\left(\frac{\partial n_0}{\partial n_2}\right)_{T,V,\mu_0,\mu_1,n_1} = -\left(\frac{\partial^2 \mathbf{A}}{\partial n_2 \partial \mu_0}\right)_{T,V,n_1} = -\left(\frac{\partial \mu_2}{\partial \mu_0}\right)_{T,\mu_1,c_2} \tag{6a}$$

and

$$\left(\frac{\partial n_1}{\partial n_2}\right)_{T,V,\mu_0,\mu_1,n_0} = -\left(\frac{\partial^2 \mathbf{A}}{\partial n_2 \partial \mu_1}\right)_{T,V,n_0} = -\left(\frac{\partial \mu_2}{\partial \mu_1}\right)_{T,\mu_0,c_2} \tag{7a}$$

These can be summarized as eq. (16) in the text.

APPENDIX 2

Let z and u be functions of independent variables x and y, i.e., $z = z(x,y)$ and $u = u(x,y)$. Then, the following relation familiar in calculus holds:

$$\left(\frac{\partial z}{\partial x}\right)_u = \left(\frac{\partial z}{\partial x}\right)_y - \left(\frac{\partial z}{\partial y}\right)_x \frac{(\partial u/\partial x)_y}{(\partial u/\partial y)_x} \tag{8a}$$

Applying this with $z = \mu_2(p,c_2)$ and $u = m_2(p,c_2)$, we obtain

$$v_2^\star = v_2 + \left(\frac{\partial \mu_2}{\partial c_2}\right)_p \frac{(\partial m_2/\partial p)_{c_2}}{(\partial m_2/\partial c_2)_p} \tag{9a}$$

It can be shown that the factor multiplied by $(\partial \mu_2/\partial c_2)_p$ on the right-hand side is expressed as $-c_2[\kappa + O(c_2{}^2)]$, where κ denotes the compressibility of the pure solvent. Thus we obtain eq. (25) in the text.

GLOSSARY OF SYMBOLS

A = Helmholtz free energy.
$\mathbf{A}$ = characteristic thermodynamic function defined by eq. (1a).
c_i = molarity of component i.
i = running index for components.
k = running index for components.
M_i = molecular weight of component i.
m_i = molality of component i.
n_i = amount in moles of component i.
p = pressure.
p_b = pressure of air bubble.
p_e = external pressure.
q = number of thermodynamic components.
r = radial distance from the axis of rotation.
S = entropy.
T = absolute temperature.
T_b = temperature of air bubble.
T_e = external temperature.
T_0 = rotor temperature.
U = internal energy.
V = volume.
V_i = partial molar volume of component i.
v_i = partial specific volume of component i.
$v_2{}^*$ = specifc volume of component 2 defined by eq. (24).
α = running index for shells dividing the centrifuge cell.
δ = infinitesimal virtual displacement.
κ = solvent compressibility.
μ_i = chemical potential of component i.
ρ = solution density.
ρ_0 = solvent density.
ϕ = centrifugal potential of solution per unit mass.
ω = angular velocity.
$\{m\}$ = set of m_1 and m_2.
$\{\mu\}$ = set of μ_0 and μ_1.

REFERENCES

Casassa, E. T., and Eisenberg, H. (1964): *Adv. Protein Chem.*
19: 287.

Fujita, H. (1975): *Foundations of Ultracentrifugal Analysis*, New York: Wiley.

Goldberg, R. J. (1953): *J. Phys. Chem.* **57**: 194.

Kurata, M. (1982): *Thermodynamics of Polymer Solutions* (translated from the Japanese edition by H. Fujita), Chur, Switzerland: Harwood Academic.

Wills, P. R., Comper, W. D., and Winzor, D. J. (1993): *Arch. Biochem. Biophys.* **300**: 206.

ASSOCIATION OF REI IMMUNOGLOBULIN LIGHT CHAIN V_L DOMAINS: THE FUNCTIONAL LINEARITY OF PARAMETERS IN EQUILIBRIUM ANALYTICAL ULTRACENTRIFUGE MODELS FOR SELF-ASSOCIATING SYSTEMS

Ian Brooks, Ronald Wetzel, Winnie Chan, Grace Lee, Donald G. Watts, K. Karl Soneson and Preston Hensley

INTRODUCTION

The analytical ultracentrifuge has had wide application in the characterization of the structure, interaction and function of macromolecules in solution. This includes the determination of molecular weight, the characterization of shape, the determination of subunit stoichiometry, the quantification of ligand binding, the quantification of ligand-binding-promoted conformational changes and the characterization of macromolecular assembly processes (Schachman, 1992). In recent years, however, other technologies have been developed which can make some of the relevent measurements with equal or greater precision, often using less material. For instance, molecular weight can now be routinely determined by gene or protein sequencing and by mass spectroscopic methods, especially electrospray (ES) and matrix-assisted laser-desorption (MALD) MS (Carr, et al., 1991). Ligand binding can be quantified by classical spectroscopic and radiochemical methods as well as by microcalorimetry (Freire, et al., 1990). Ligand binding promoted conformational changes can be characterized by time-resolved fluorescence anisotropy (Lakowicz, 1983; Beechem, et al., 1986; Waxman, et al., 1994) although, the centrifuge still has a major impact in this area (Kirschner and Schachman, 1973a; Kirschner and Schachman, 1973b; Howlett and Schachman, 1977; Eisenstein, et al., 1990) The thermodynamic characterization of macromolecular assembly processes, however, is an area where the analytical ultracentrifuge has few rivals (Adams and Lewis, 1968; Roark and Yphantis, 1969; Hensley, et al., 1975b; Blackburn and Noltman, 1981; Minton and Lewis, 1981; Wilf and Minton, 1981; Correia, et al., 1985; Duong, et al., 1986; Hensley, et al., 1986; Lewis and Youle, 1986; Chatelier and Minton, 1987; Ross, et al., 1991; Rivas, et al., 1992).

To illustrate the power of the centrifuge in the analysis of assembly processes, we will present data from an analysis of the self-association of wild type and two sequence variants of the V_L domain (residues 1-107) of the immunoglobulin light chain, REI. The structure of the REI light chain dimer is shown in Figure 1.

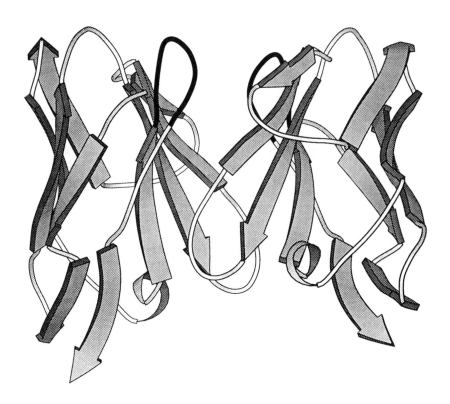

FIGURE 1. A ribbon drawing of the structure of the REI dimer as determined by X-ray crystallography (Epp, *et al.*, 1974). The CDR3 loop (residues 89 - 97) of the REI monomer is highlighted. Sequence variations of REI mutants will occur in this loop (see below).

Such V_L domains take part in a number of biologically important association processes. As monoclonal light chains present at high concentrations in the serum and urine of myoloma patients, Bence-Jones proteins vary in their association properties from monomers to non-covalent dimers (Stevens, *et al.*, 1980) and, in a few reports, non-covalent tetramers (Kosaka, *et al.*, 1989). As illustrated by X-ray crystallographic studies, the V_L dimer interface utilizes the same molecular surface as that involved in the V_L-V_H interface in an immunoglobulin molecule. In addition, V_L domains take part in less well-known self-associations in the formation of large insoluble aggregates (Wetzel, 1992; Wetzel, 1994). In light chain amyloidosis and light chain deposition disease, aggregates of monoclonal V_L domains cause lethal disruption of the function of the kidney, heart and other organs (Buxbaum, 1992). In bacterial expression of V_L domains, inclusion bodies composed of V_L can influence expression levels and require altered isolation procedures (Chan, *et al.*, 1993; Lee, *et al.*, 1993). In all of these association phenomena, primary sequence effects are very important. The ways in which these effects are mediated by tertiary structure, however, and the possible relationships between the various kinds of self-association possible, remain largely obscure (Wetzel, *et al.*, 1993).

Ultracentrifugation analysis promises to make important contributions to an understanding of these phenomena.

While the analysis of ultracentrifuge data in terms of models for interacting systems (assembly processes) is in principle straightforward, *ie* fitting the data to sums of exponentials, in fact, deciding between models is not always an easy task. These decisions are critical as the models define our understanding of the mechanisms of assembly. This is formally the area of the determination of the goodness of fit for nonlinear models. In deciding among models, an important task is the evaluation of the precision of the parameter estimates. Large errors in parameters, or confidence intervals which contain zero, are a signal that a particular aspect of a model, or the model itself, may not have physical significance. This task is computationally difficult because, in contrast to linear models, there are no closed-form expressions for the confidence intervals associated with the parameters and, therefore, empirical methods need to be employed. Here we discuss such an approach, profile analysis (Bates and Watts, 1988; Bates and Watts, 1991; Watts, 1991; Watts, 1994), which allows exact confidence intervals for parameters from nonlinear models to be rapidly determined. We apply this approach to the analysis of wt and mutant REI ultracentrifuge data. From these analyses, it can be seen that the models employed here, while formally nonlinear, are often functionally linear. When the models were nonlinear with respect to certain parameters, the profiling approach allowed us to rapidly determine and to visualize the extent of nonlinearity and to determine the effect this has on the determination of confidence intervals for those parameters. In one case discussed, certain parameters are moderately nonlinear.

DETERMINATION OF CONFIDENCE INTERVALS FOR LINEAR MODELS

Any linear model may be formally written,

$$y_n = \beta_0 + \beta_1 x_{1,n} + \beta_2 x_{2,n} + \dots + Z_n$$

or

$$y_n = f(x_{p,n}, \beta) + Z_n \quad \text{where } n = 1, \dots, N \text{ and } p = 1, \dots, P$$

where $x_{p,n}$ are the independent variables and β are the model-dependent parameters. This model can also be written in matrix notation,

$$Y = X\beta + Z \quad (1)$$

where Y is a N-dimensional column vector of response values, X is a N x P matrix of independent variables, β is a P-dimensional column matrix of parameter values and Z is an N-dimensional column vector accounting for the experimental noise at each value of n (Bevington, 1969; Draper and Smith, 1981; Press, *et al.*, 1986; Bates and Watts, 1988). Least squares estimates of the parameter values can be obtained by,

$$\hat{\beta} = (X^T X)^{-1} X^T Y$$

Here, the estimated standard errors and confidence intervals for a parameter, β_p, are given by,

$$se(\hat{\beta}_p) = s \text{ sqrt } [((X^T X)^{-1})_{pp}]$$

$$\hat{\beta}_p +/- t\,(\,N - P:\,\alpha/2)\,\text{se}\,(\hat{\beta}_p)$$

s is the square root of the variance estimate, $((X^T X)^{-1})_{pp}$ is the pth diagonal entry of the $(X^T X)^{-1}$ matrix and $t\,(\,N - P:\,\alpha/2)$ is the value that isolates an area $\alpha/2$ under the right tail of the Students t distribution with $N - P$ degrees of freedom. The $(1 - \alpha)$ joint parameter confidence region for all parameters is given by

$$(\beta - \hat{\beta})^T\,X^T\,X\,(\beta - \hat{\beta}) \leq P\,s^2\,F\,(P, N - P:\alpha)$$

where $F\,(\,P, N - P;\,\alpha)$ is the value which isolates an area $\alpha/2$ under the right tail of Fisher's F distribution with P and $N - P$ degrees of freedom.

Unfortunately, these closed form expressions for the standard errors and confidence intervals are valid only for linear models. Models are linear if the derivative of the expectation function with respect to any parameter, β_p, does not depend on any of the parameters. If this is not the case, the model is nonlinear. In the discussion below, the models (equations (4), (7) and (8)) satisfy this latter criteria and are nonlinear. In this discussion, nonlinear models are discussed in terms of parameters, Θ_p. For such nonlinear models, the sum of squares surface has no simple mathematically definable behavior and so empirical approaches, such as Monte Carlo methods (Press, *et al.*, 1986; Straume and Johnson, 1992) must be used to determine parameter error estimates. These methods are not generally available, or take too long, and as a result many modeling packages containing nonlinear least squares capabilities simply use linear approximation statistics to characterize the errors in parameter estimates and to determine the extent of numerical correlation among parameters. This can have undesired consequences occasionally producing qualitatively incorrect answers (Johnson and Faunt, 1992; Johnson and Staume, 1994)

PROFILE ANALYSIS OF NONLINEAR MODELS

With these concerns in mind, Bates and Watts described a method for *profiling* the sum of squares surface near the minimum. From this information, confidence intervals for the determined parameters can be determined directly. Profiling involves determining the conditional sum of squares surface as each parameter is varied around its estimated best fit value. The parameters are varied until enough information is obtained to define the 99% likelihood interval upper and lower endpoints. To obtain parameter inference regions, the parameter increment domain is normalized so that all parameters are plotted on the same relative axis (for easy comparison between models and data sets), *i.e.*, parameter values for nonlinear models, Θ_p, are converted to their Studentized values,

$$\delta(\Theta_p) = (\,\Theta_p - \hat{\Theta}_p\,)\,/\,\text{se}(\hat{\Theta}_p) \qquad (2)$$

where $\hat{\Theta}$ is the nonlinear least squares estimate of the nonlinear parameter value. The sum of squares response curve is then transformed to *profile t values* which are equal to the square root of the relative excess sum of squares, namely,

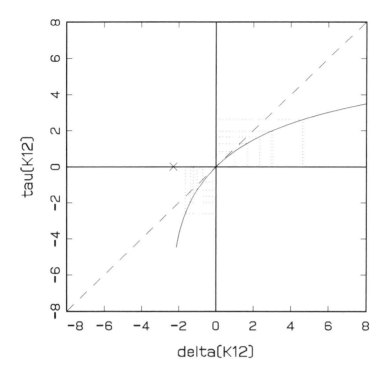

FIGURE 2. The profile t plot for the parameter, $K_{1,2}$, the dissociation equilibrium constant for the data set in Figure 9. This plot is taken from Figure 10. This figure is a plot of $\tau(K_{1,2})$ vs. $\delta(K_{1,2})$. These two terms are described in equations (2) and (3) (Brooks, et al., 1994b).

$$\tau(\Theta_p) = \text{sign} \,(\Theta_p - \hat{\Theta}_p) \,\text{sqrt} \,[\,(\tilde{S}(\Theta_p) - S(\hat{\Theta}))/ s^2 \,] \quad (3)$$

Here, $S(\hat{\Theta})$ is the sum of squares of the fit with parameters at the best fit values and $\tilde{S}(\Theta_p)$ is the sum of squares with all parameters, except Θ_p, at their best fit values, Θ_p having been incremented away from its best fit value. Then, $\tau(\Theta_p)$ is plotted versus $\delta(\Theta_p)$. Such a profile t plot is shown in Figure 2.

If the model is linear, the plot will be a straight line at a 45° angle, and the $(1 - \alpha)$ marginal confidence interval for $\delta(\Theta_p)$ is given by the point which corresponds to the points on the y-axis where

$$\tau(\Theta_p) = +/- \,t \,(N - P; \,\alpha/2)$$

In this case, the exact likelihood interval for $\delta(\Theta_p)$ is obtained by reflecting these values from the vertical scale through the 45° angle line onto the horizontal scale.

If the model is nonlinear, the profile t line will be curved and the extent to which the curve deviates from a straight line is a meaningful indicator of nonlinearity. In this case, the critical t-values are reflected from the vertical axis, through the profile t curve to the horizontal axis, yielding the exact likelihood interval for the nonlinear parameter without employing linear approximations. These values of $\delta(\Theta_p)$ are then un-normalized to express the likelihood interval endpoints in terms of the original parameter values.

Plots of the parameter joint likelihood regions can be obtained from pairwise plots of $\delta(\Theta_p)$ versus the profile parameter $S(\Theta_p)$ (Figure 3). The profile trace values (of 60, 80, 90, 95 and 99% confidence) from the profile trace plots correspond to positions on the respective line where the sum of squares contours have vertical or horizontal tangents. This information is useful in constructing the joint likelihood expectation contours (Bates and Watts, 1988). These plots are extremely useful and contain much information about the statistical behavior of the modeling effort. Each plot gives an empirically determined two-dimensional view of the minimum of the P-dimensional sum of squares surface (see Figures 6, 8 and 10). When the contours are elliptical, the model behaves linearly with respect to both parameters. When the contours are not elliptical, the model is nonlinear with respect to one or both parameters. Parameter correlation is related to the width of the ellipse. The narrower the ellipse the more highly correlated the parameters and conversely, the more open the ellipse (approaching circular), the less the parameters are correlated. In the discussion which follows, equilibrium analytical ultracentrifuge data are analyzed and confidence limits for parameters determined by the profiling approach.

PROFILE ANALYSIS OF ULTRACENTRIFUGE DATA FOR REI AND REI MUTANTS

For mono-disperse systems (single ideal species), the macromolecular distributions at sedimentation equilibrium can be described as follows (Cantor and Schimmel, 1980)

$$c_r = c_0 \exp [\, M \,(1 - \bar{v}\rho)\, \omega^2 \,(r^2 - r_0^2)\,/\,2RT\,] + base \quad (4)$$

Here, c_r and c_0 are the total concentrations of species at a radial position, r, and at the meniscus (a reference position). M and $\bar{v}$ are the molecular weight and partial specific volume of the macromolecule, ρ is the solvent density, ω is the angular velocity, r_0 is the reference radial position, R is the universal gas constant, T is the absolute temperature and base is a baseline term corresponding to optical density from a non-sedimenting material. This equation is often abbreviated

$$c_r = c_0 \exp [\, \sigma \,] + base \quad (5)$$

where

$$\sigma = M \,(1 - \bar{v}\rho)\, \omega^2 \,(r^2 - r_0^2)\,/\,2RT$$

The partial specific volume can be determined from the amino acid and/or carbohydrate composition (Zimyatnin, 1972; Laue, et al., 1992) or directly from density measurements (Kratky, et al., 1973). The other three fitted parameters, $c_{0,i}$, M and base, must be determined by non-linear least squares methods (Bevington, 1969; Draper and Smith, 1981; Press, et al., 1986; Johnson and Faunt, 1992). These equations are nonlinear in M and $c_{0,i}$, as the partial

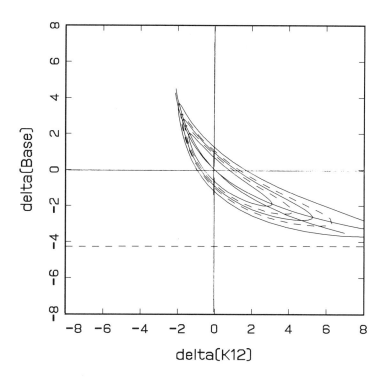

FIGURE 3. The profile trace plot for δ(base) vs. $\delta(K_{1,2})$. The concentric rings define the 60, 80, 90, 95 and 99% joint confidence regions for the base - $K_{1,2}$ pair (see Figure 10). Each of the two solid lines intersecting at the center of the plot are pairwise plots of $\delta(\Theta_p)$ versus the profile parameter, $\tilde{S}(\Theta_p)$ (Brooks, et al., 1994b).

derivative of equation (4) with respect to M or $c_{0,i}$ is a function of these parameters.

For interacting macromolecular systems, such as a dimerizing monomer, *i.e.,*

$$A + A \rightleftharpoons A_2$$

the distribution of protein at sedimentation equilibrium can be described as the sum of two exponentials, one for the monomer (i) and one for the dimer (j) (Hensley, *et al.*, 1975a; Hensley, *et al.*, 1975b) *i.e.,*

$$c_r = c_{0,i} \exp [\, \sigma_i \,] + c_{0,j} \exp [\, \sigma_j \,] + \text{base} \quad (6)$$

From the definition of the equilibrium dissociation constant, $K_{1,2} = [A] [A] / [A_2]$, the term $c_{0,j}$ may be recast as $c_{0,j} = c_{0,i}^2 / K_{1,2}$. With this identity, equation (6) may be recast as

$$c_r = c_{0,i} \exp [\, \sigma_i \,] + (\, c_{0,i}^2 / K_{1,2} \,) \exp [\, \sigma_j \,] + \text{base} \quad (7)$$

For a monomer-dimer-tetramer system, *i.e.*

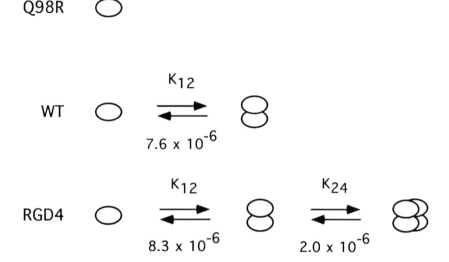

FIGURE 4. Summary of models for the self association of wt REI and two REI mutants (Brooks, *et al.*, 1994b).

$$4A \rightleftharpoons 2A_2 \rightleftharpoons A_4$$

equation (7) can be extended to equation (8),

$$c_r = c_{0,i} \exp [\sigma_i] + (c_{0,i}^2 / K_{1,2}) \exp [\sigma_j] +$$

$$((c_{0,i}^4 / K_{1,2}^2) / K_{2,4}) \exp [\sigma_k] + \text{base} \quad (8)$$

It is often the case that the molecular weights of the macromolecular species are known either from protein or DNA sequence and/or from mass spectroscopic analysis. Hence, the unknown parameters in, for instance, the monomer-dimer model (equation (7)) reduce to only $c_{0,i}$, $K_{1,2}$ and base. Such models are non-linear in $c_{0,i}$ and $K_{1,2}$ and are conditionally linear in base. While these models are formally nonlinear, it is important to determine empirically the extent to which they are nonlinear, the extent to which parameters are correlated and the effect that these considerations have on the determination of the best fit values of the parameters and their confidence intervals. Such determinations, as discussed above, are critical to defining the best model to describe the assembly process. We will do this by examining data from the REI system, where the self association is of increasing complexity. We will look at a system which is a pure monomer, then one which exhibits a monomer-dimer equilibrium and finally one which exhibits a monomer-dimer-tetramer equilibrium. The results of these analyses are summarized in Figure 4.

The first protein examined by this analysis was wt REI. This system was best described by a monomer-dimer model, equation (7). The data are shown in Figure 5 and the numerical results of a profile analysis are given in Table 1. From these results, we can make several observations. First, the parameters are all highly correlated. The correlation coefficients lie between 0.94 and 0.99. At first sight, these results suggest that unique values for the parameters may not be

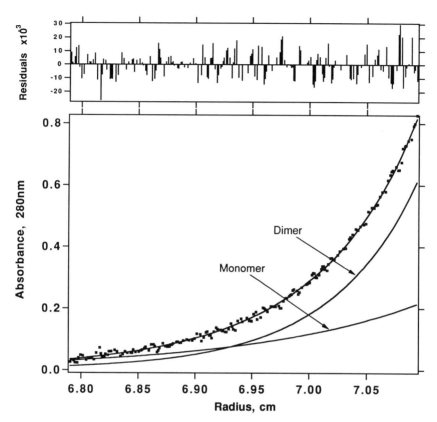

FIGURE 5. Equilibrium sedimentation data for wt REI. For all the proteins discussed here, the buffer was 20 mM Na phosphate, 150 mM NaCl, pH 7.4. The initial protein concentration was ~18 mM (in terms of monomer concentration, *i.e.*, $A_{280} \sim 0.3$). All data reported here were acquired at 25 °C and at 25,000 rpm in a Beckman Optima XL-A analytical ultracentrifuge. These data were fit best to a monomer- dimer model, equation (7). The value of $K_{1,2}$ was determined to be 7.6 x 10^{-6} M. The complete results of the analysis are summarized in Table 1 (Brooks, *et al.*, 1994b).

recoverable. However, experience with data such as this suggest that this is not the case. Data can be generated with parameters of known values and pseudo-random noise added. When these data are analyzed with the models discussed above, even with correlation coefficients of 0.99, the input parameter values are routinely recovered. However, when the parameters are this highly correlated, convergence proceeds more slowly and the analyses are less forgiving with respect to initial guesses for parameters and occassionally do not converge.

Profile t and profile trace plots for this analysis are given in Figure 6. In each case the profile t plots, the hatched plots on the diagonal, are nearly linear, demonstrating that this model for these data is functionally linear with respect to the three parameters. As discussed above, the model should be linear with respect to the base term. However, it is not clear that the model should be so linear with respect to c_0 and $K_{1,2}$. This was an unexpected finding. The linearity can also be seen by referring to Table 1. A comparison of the 60 - 99%

TABLE 1: Parameter Summary for Fit of REI Data to equation (7).

Sum of squares	d of f	variance	standard deviation
1.510255E-02	169	8.936421E-05	9.453264E-03

correlation matrix

c_0	1.0000		
$K_{1,2}$	0.9995	1.0000	
base	-0.9485	-0.9413	1.0000
	c_0	$K_{1,2}$	base

parameter	estimate	stderr
c_m[a]	1.228e-06	8.25e-08

	linear approximation x e06		exact likelihood x e06	
(1-α) %	lower	upper	lower	upper
60	1.158	1.298	1.158	1.298
80	1.122	1.334	1.122	1.334
90	1.092	1.364	1.092	1.364
95	1.065	1.391	1.065	1.391
99	1.013	1.443	1.013	1.443

parameter	estimate	stderr
$K_{1,2}$[a]	7.601e-06	1.19e-06

	linear approximation x e06		exact likelihood x e06	
(1-α) %	lower	upper	lower	upper
60	6.597	8.605	6.640	8.649
80	6.071	9.132	6.168	9.235
90	5.634	9.569	5.794	9.741
95	5.253	9.950	5.479	10.20
99	4.502	10.70	4.891	11.14

parameter	estimate	stderr
base[b]	-0.009477	0.003197

	llinear approximation		exact likelihood	
(1-α) %	lower	upper	lower	upper
60	-0.01217	-0.00678	-0.01217	-0.00678
80	-0.01359	-0.005364	-0.01359	-0.005364
90	-0.01476	-0.00419	-0.01476	-0.00419
95	-0.01579	-0.003166	-0.01579	-0.003166
99	-0.01781	-0.001149	-0.01781	-0.001149

a. Units are moles / L. The molar extinction coefficient used was $16,800 \text{ M}^{-1} \text{ cm}^{-1}$, estimated from amino acid composition (Creighton, 1984). b. Units are absorbance at 280 nm (Brooks, *et al.*, 1994b).

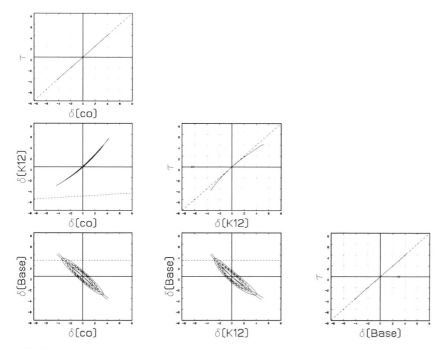

FIGURE 6. Profile t and trace plots for the parameters determined from analysis of wt REI data in Figure 5. The plots on the diagonal are the profile t plots and the off-diagonal plots are the profile trace plots. The profile t plots are plots of $\tau(\Theta)$ vs. $\delta(\Theta)$ for the parameters c_m, $K_{1,2}$ and base, respectively. In these plots, the dashed line is a linear reference line and the superimposed solid line is the computed profile t curve. In all three cases, the two curves almost superimpose, demonstrating that all three parameters behave essentially linearly. The profile trace plots, $\delta(\Theta_1)$ vs. $\delta(\Theta_2)$, show the 60, 80, 90, 95 and 99% joint likelihood regions for the three pairwise combinations of parameters. In all cases, the regions are essentially elliptical and symmetric, consistent with the functional linearity of the parameters. For the plot of $\delta(K_{1,2})$ vs. $\delta(c_0)$, the ellipses are very narrow, consistent with the high parameter correlation, as can be seen in Table 1 (Brooks, et al., 1994b).

confidence intervals determined using linearity assumptions or by exact likelihood (profile) methods gives essentially the same numerical result. There are small deviations for $K_{1,2}$ in the 60 to 99% confidence interval range, but these are second order. The effect of high parameter correlation can be clearly seen in the profile trace plot for $K_{1,2}$ vs c_0. In this plot the ellipses are very narrow.

The next data set to be examined was from a mutant of REI, Q89R, where a glutamine has been substituted for an arginine at position 89. This mutation occurs at the beginning of the CDR3 loop in an area which is near the REI dimer interface and which, therefore, might be expected to effect dimerization, see Figure 1. This is, in fact, what was found, i.e. the molecule was completely monomeric. Data from this analysis is given in Figure 7 and the results of the profile analysis of this data are given in Table 2.

The model described by equation (5) is qualitatively different from that described by equation (7), a monomer-dimer model. In the former, we are

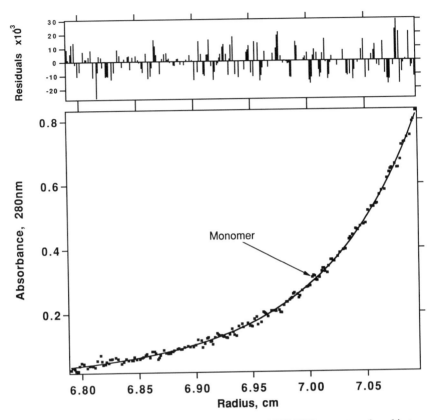

FIGURE 7.   Equilibrium sedimentation data for the Q89R REI mutant analyzed in terms of equation (5), *i.e.* fitting just for the monomer term.   The glutamine for arginine substitution occurs just before the beginning of the CDR3 loop in the REI monomer (see Figure 1). The determined monomer molecular weight was 11,940 +/- 101 Da compared to the predicted value of 11,884 Da (Brooks, *et al.*, 1994b).

fitting for the monomer molecular weight and this term is in the exponential. In the latter, we are fitting a monomer-dimer equilibrium and are only fitting for pre-exponential terms.   One might expect the profile analysis of the data in Figure 7 to be qualitatively different from that in Figure 5.   However this is not the case.   Figure 8 shows the profile t and profile trace plots for the data in Figure 7.   These data are slightly more linear than those for the REI wt protein (see Figures 5 and 6 and Table 1.).

The final mutant in this series was made by replacing the wt residues in positions 90 - 97 of the CDR3 loop with the RGD containing sequence, GKISRIP**RGD**MPDDRS.   This mutant is called RGD4.   Aside from changing the sequence, eight additional amino acids have been added.   If a single amino acid substitution at position 89 (the Q89R mutant discussed above) dissociated the dimer to a monomer, what might this more drastic alteration do to the assembly properties of the REI monomer?

Surprisingly, these data fit best to a monomer-dimer-tetramer, equation (8). The primary data from this experiment are shown in Figure 9. The results of the

TABLE 2: Parameter Summary for Fit to Q89R Data to equation (4).

sum of squares	d of f	variance	standard deviation
2.986981E-03	183	1.632230E-05	4.040087E-03

correlation matrix

c_m	1.0000		
M	-0.9958	1.0000	
base	-0.9811	0.9637	1.0000
	c_m	M	base

parameter	estimate	stderr
$c_m{}^a$	3.393e-06	7.211e-08

	linear approximation x e06		exact likelihood x e06	
(1-α) %	lower	upper	lower	upper
60	3.332	3.453	3.332	3.454
80	3.300	3.485	3.301	3.487
90	3.273	3.512	3.275	3.514
95	3.250	3.535	3.253	3.538
99	3.200	3.580	3.210	3.586

parameter	estimate	stderr
M	1.194e+04	101.2

	linear approximation x e-4		exact likelihood x e-4	
(1-α) %	lower	upper	lower	upper
60	1.186	1.203	1.186	1.203
80	1.181	1.207	1.181	1.207
90	1.178	1.211	1.178	1.211
95	1.174	1.214	1.174	1.214
99	1.168	1.221	1.168	1.221

parameter	estimate	stderr
base	0.04325	0.002326

	linear approximation		exact likelihood	
(1-α) %	lower	upper	lower	upper
60	0.04129	0.04521	0.04127	0.04520
80	0.04026	0.04624	0.04022	0.04622
90	0.03940	0.04710	0.03935	0.04706
95	0.03866	0.04784	0.03858	0.04778
99	0.03719	0.04930	0.03706	0.04920

a. Units are moles / L. The molar extinction coefficient used was 16,800 M^{-1} cm^{-1}, estimated from amino acid composition (Creighton, 1984). b. Units are absorbance at 280 nm (Brooks, *et al.*, 1994b).

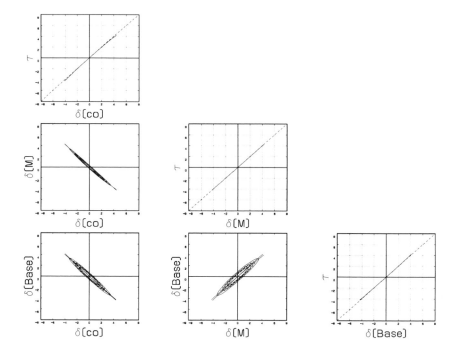

FIGURE 8. Profile t and trace plots for the parameters determined from analysis of the Q89R REI data in Figure 7 (Brooks, *et al.*, 1994b).

profile analysis of the data are given in Table 3 and the profile t and profile trace plots are shown in Figure 10. From the profile t plots in Figure 10, it is clear that $K_{1,2}$ and to some extent, $K_{2,4}$ are nonlinear. This has the effect of distorting nearly all the profile trace plots from their normal elliptical shapes. As these plots are two-dimensional views of the minimum of the five-dimensional error space associated with this fit, the consequence of these non-elliptical contours is asymmetric joint confidence intervals for the parameters. This is more dramatic for some parameters than others. Comparing the linear approximation to the exact likelihood confidence intervals for $K_{1,2}$ in Table 3, for instance, the quantitative consequences of the nonlinearity are made evident. For the upper and lower 99% confidence intervals the linear assumption statistics give 1.102 and 17.61 x 10^{-6} M, while the exact likelihood method gives 2.285 and 24.86 x 10^{-6} M., which is a significant difference, *i.e.*, 24.86 x 10^{-6} M is 4.6 standard errors above the mean whereas 17.61 x 10^{-6} M is only 2.6.

SUMMARY

In the above discussion, we have seen that equilibrium analytical ultracentrifugation can be a powerful tool in the quantitative analysis of the relationship between structure and assembly in the light chain Bence-Jones V_L domain dimer. A single amino acid substitution, Q89R, at the beginning of the CDR3 loop and near the dimer interface of the protein, destabilized the

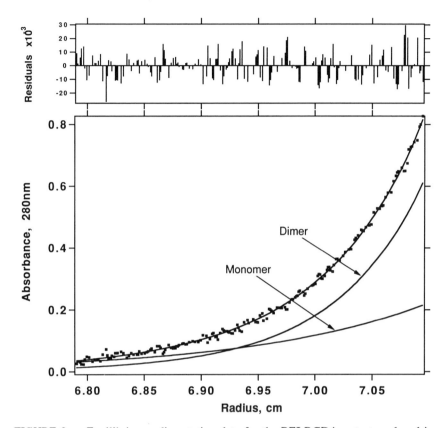

FIGURE 9.   Equilibrium sedimentation data for the REI RGD4 mutant analyzed in terms of a monomer-dimer-tetramer model (equation (8)). $K_{1,2} = 8.25 \times 10^{-6}$ and $K_{2,4} = 1.98 \times 10^{-6}$ M (Brooks, *et al.*, 1994b))

dimer resulting in an apparently stable monomer. In contrast, the extension of the CDR3 loop by eight residues and the inclusion of eight new amino acids (a sixteen amino acid substitution) apparently had the effect of stabilizing the dimer and promoting a monomer-dimer-tetramer equilibrium. It remains to be determined, however, whether the dimer formed by RGD4 is related to the native dimer formed by the wt (Figure 1.) An intriguing possibility is that the tetramer formed at equilibrium by RGD4 under these conditions may be a key intermediate in the formation of inclusion bodies when this mutant is produced in *E. coli* (Chan, *et al.*, 1993). RGD4 also resembles in some of its properties an REI mutant associated with formation of inclusion bodies in human tissue in light chain deposition disease (Helms and Wetzel, 1994).

This series of wt and mutant proteins also gave us the opportunity to implement the profiling approach for the determination of parameter confidence intervals developed by Bates and Watts on a set of increasingly numerically complex systems. We found that for the most part the parameters in the models used to analyze this equilibrium analytical ultracentrifuge data were functionally linear. However, in some cases, the deviation from functional linearity was

TABLE 3: Parameter summary for fit to REI RGD4 data.

sum of squares	d of f	variance	standard deviation
9.551946E-03	192	4.974972E-05	7.053348E-03

correlation matrix

c_m	1.0000			
$K_{1,2}$	0.9980	1.0000		
$K_{2,4}$	0.9789	-0.9898	1.0000	
base	-0.9609	-0.9462	0.8994	1.0000
	c_m	$K_{1,2}$	$K2,4$	base

parameter	estimate	stderr
c_m[a]	8.64e-07	1.284e-07

	linear approximation x e07		exact likelihood x e07	
(1-α) %	lower	upper	lower	upper
60	7.557	9.724	7.557	9.724
80	6.989	10.29	6.989	10.29
90	6.517	10.76	6.517	10.76
95	6.107	11.17	6.107	11.17
99	5.299	11.98	5.299	11.98

parameter	estimate	stderr
$K_{1,2}$[a]	8.256e-06	3.597e-06

	linear approximation x e06		exact likelihood x e06	
(1-α) %	lower	upper	lower	upper
60	5.222	11.29	5.656	11.84
80	3.630	12.88	4.586	14.25
90	2.311	14.20	3.824	16.62
95	1.161	15.35	3.241	19.01
99	1.102	17.61	2.285	24.86

parameter	estimate	stderr
$K_{2,4}$[a]	1.978e-06	5.783e-07

	linear approximation x e06		exact likelihood x e06	
(1-α) %	lower	upper	lower	upper
60	1.491	2.466	1.523	2.499
80	1.235	2.722	1.309	2.799
90	1.022	2.934	1.145	3.062
95	0.837	3.119	1.012	3.301
99	0.473	3.483	0.774	3.803

parameter	estimate	stderr
Base[b]	0.01298	0.003056

	linear approximation		exact likelihood	
(1-α) %	lower	upper	lower	upper
60	0.010400	0.01556	0.010400	0.01556
80	0.009050	0.01691	0.009050	0.01691
90	0.007929	0.01803	0.007929	0.01803
95	0.006952	0.01901	0.006952	0.01901
99	0.005030	0.02093	0.005030	0.02093

a. Units are moles / L. The molar extinction coefficient used was 16,800 M^{-1} cm^{-1}, estimated from amino acid composition (Creighton, 1984). b. Units are absorbance at 280 nm (Brooks, *et al.*, 1994b).

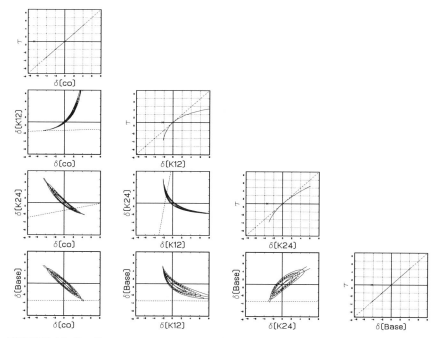

FIGURE 10. Profile t and trace plots for the parameters determined from analysis of the RGD4 REI data in Figure 9 (Brooks, *et al.*, 1994b).

significant. An example was found in the monomer - dimer - tetramer model. The parameter in question, $K_{2,4}$ is found in the pre-exponential term for the concentration of tetramer in equation (8). This term is a complicated function of three parameters, *i.e.* $(c_{o,i}{}^4 / K_{1,2}{}^2) / K_{2,4}$. This may be the reason for the nonlinearity. However, with simpler data sets, at least two of these parameters are functionally linear (Brooks and Hensley, unpublished), so this nonlinearity may not be a general result.

While we have seen functional linearity previously (Brooks, *et al.*, 1994a), we emphasize that these results represent only a small set of very well behaved systems. While we did not expect to find them to be as linear as they were, we caution that this should not be taken as a general result and that, for even moderately more complex systems, empirical approaches to the determination of parameter confidence interval such as the one employed here, Monte Carlo (Straume and Johnson, 1992) or bootstrapping methods should be routinely employed. Using such approaches one can always be confident that errors in parameters are rigorouly determined and, hence, models based on interpretation of the parameters are less likely to be in error.

Finally, one aspect of these analyses which has been evident throughout this discussion is the high correlation of parameter values, especially c_m and $K_{i,j}$ (see Tables 1,2 and 3). This problem confounds any analysis of equilibrium sedimentation data. We have shown that while simultaneous analysis of data from multiple rotor speeds helps to ameliorate this problem (Brooks, *et al.*, 1994a), it does not eliminate it. In this volume, Minton (Minton, 1994) discusses a novel conservation of signal approach which greatly reduces or, in some cases, eliminates the problem.

ACKNOWLEDGMENTS

The authors gratefully acknowledge Larry Helms and Byung-Ha Oh for constructing Figure 1.

GLOSSARY OF SYMBOLS

α	The probability that an event will occur.
β_p	A linear parameter, p.
$\hat{\beta}_p$	The best fit value for a linear parameter, p.
base	The baseline term in equibrium sedimentation.
Bence-Jones	A class of IgG chains found free in the serum
CDR3	Complimentarity determining region 3 in an immunoglubulin.
$c_{0,i}$	The concentration at the meniscus of species, i.
$\delta(\Theta_p)$	Trace function of a nonlinear parameter, p.
$F(P, N - P)$	F-statistic for P parameters and N data points.
$K_{1,2}$	The monomer-dimer equilibrium constant.
$K_{2,4}$	The dimer-tetramer equilibrium constant.
M_i	Molecular weight of species i.
P	Probability
Θ_p	A nonlinear parameter, p.
$\hat{\Theta}_p$	The best fit value for the nonlinear parameter, p.
r	Radial position in centimeters.
R	The universal gas constant.
REI	A class of Bence Jones light chains.
RGD4	An REI variable domain mutant where the sequence RGD has been inserted in the CDR3 loop.
r_0	Radial position of the meniscus in cm.
$\tilde{S}(\Theta_p)$	Sum of squares with all nonlinear parameters except p at their best fit values.
$S(\hat{\Theta})$	Sum of squares with all nonlinear parameters at their best fit values.
s^2	Variance.
σ_i	Exponential term in the equilibrium sedimentation equation, $ie\ M_i(1-\bar{v}_i\rho)w^2(r^2-r_0^2)/2RT$.
T	Temperature.
$\tau(\Theta_p)$	Trace function for the nonlinear parameter, Θ_p.
$\bar{v}_i$	Partial specific volume for species, i.
V_L	A light chain variable domain.
$V_L\text{-}V_H$	A light chain-heavy chain interaction.
ω	Angular velocity.
$x_{p,n}$	An independent variable with dimensions, p,n.
X	A matrix of independent variables with elments $x_{p,n}$.

$(X^T X)^{-1}$	Inverse of the $X^T X$ matrix.
$Y = X \beta + Z$	Matrix formulation of a typical linear equantion. The Y matrix elements are observables, the X matrix elements are independent variables, and the b matrix are linear parameters and the Z matrix elements are error terms.
y_n	An observable with index n.

REFERENCES

Adams, E. T. J. and Lewis, M. S. (1968): Sedimentation equilibrium in reacting systems. VI. Some applications to indefinite self-associations. Studies with beta-lactoglobulin A, *Biochemistry* **7**, 1044-53.

Bates, D. M. and Watts, D. G. (1988): Nonlinear Regression Analysis and Its Applications, New York, Wiley Interscience.

Bates, D. M. and Watts, D. G. (1991): Model building in chemistry using profile *t* and trace plots., *Chemometrics and Intelligent Laboratory Systems* **10**, 107-116.

Beechem, J. M., Knutson, J. R., and Brand, L. (1986): Global analysis of multiple dye fluorescence anisotropy experiments on proteins, *Biochem Soc Trans* **14**, 832-5.

Bevington, P. R. (1969): Data Reduction and Error Analysis for the Physical Sciences, New York, McGraw Hill Book Company.

Blackburn, M. N. and Noltman, E. A. (1981): Evidence for an intermediate in the denaturation and assembly of phosphoglucose isomerase., *Archives of Biochemistry and Biophysics* **212**, 162-169.

Brooks, I., Watts, D. G., Soneson, K. K., and Hensley, P. (1994a): Determining the Confidence Intervals of Parameters from Analysis of Equilibrium Analytical Ultracentrifuge Data. **In:** *Methods in Enzymology*, L. Brand and M. L. Johnson, ed.s, New York, Academic Press, *(in press)*.

Brooks, I., Wetzel, R., Chan, W., Lee, G., Watts, D. G., Soneson, K. K., and Hensley, P. (1994b): A Mutational Analysis of the Relation Between Domain-Domain Interactions in Solution, Inclusion Body and Fibril Formation in vitro for REI, an Immunoglobulin VL(Bence-Jones) Domain Expressed in *E. coli*, *(submitted)*.

Buxbaum, J. (1992): Mechanisms of disease: Monoclonal immunoglobulin deposition, *Hemat./Oncol. Clin. North America* **6**, 323-346.

Cantor, C. R. and Schimmel, P. R. (1980): Biophysical chemistry, pt. 2: Techniques for the study of biological structure and function., San Francisco, W.H. Freeman and Co.

Carr, S. A., Hemling, M. E., Bean, M. F., and Roberts, G. D. (1991): Integration of mass spectrometry in analytical biotechnology, *Analytical Chemistry* **63**, 2802-2824.

Chan, W., Hensley, P., Lee, G., and Wetzel, R. (1993): Secretion into the *Escherichia coli* periplasm of the immunoglobulin V_L domain REI: inclusion body formation, purification, and dimerization of a series of point mutants, *Ms. in preparation*.

Chatelier, R. C. and Minton, A. P. (1987): Sedimentation equilibrium in macromolecular solutions of arbitrary concentration. I. Self-associating proteins, *Biopolymers* **26**, 507-24.

Correia, J. J., Shire, S., Yphantis, D. A., and Schuster, T. M. (1985): Sedimentation equilibrium measurements of the intermediate-size tobacco mosaic virus protein polymers, *Biochemistry* **24**, 3292-7.

Creighton, T. E. (1984): Proteins: Structures and Molecular Properties, New York, W. H. Freeman.

Draper, N. and Smith, H. (1981): Applied Regression Analysis, Second Edition, New York, Wiley-Interscience.

Duong, L. T., Eisenstein, E., Green, S. M., Ornberg, R. L., and Hensley, P. (1986): The quaternary structure of ornithine transcarbamoylase and arginase from Saccharomyces cerevisiae, *J Biol Chem* **261**, 12807-13.

Eisenstein, E., Markby, D. W., and Schachman, H. K. (1990): Heterotropic effectors promote a global conformational change in aspartate transcarbamoylase, *Biochemistry* **29**, 3724-31.

Epp, O., Colman, P., Fehlhammer, H., Bode, W., Schiffer, M., Huber, R., and Palm, W. (1974): Crystal and molecular structure of a dimer composed of the variable portions of the Bence-Jones protein REI, *Eur. J. Biochem.* **45**, 513-524.

Freire, E., Mayorga, O. L., and Straume, M. (1990): Isothermal Titration Calorimetry, *Analytical Chemistry* **62**, 950A-959A.

Helms, L. R. and Wetzel, R. (1994): Dramatic reduction in domain folding stability by a point mutation, Asp82 to Val82, associated with light chain deposition disease., *(submitted)*.

Hensley, P., Edelstein, S. J., Wharton, D. C., and Gibson, Q. H. (1975b): Conformation and spin state in methemoglobin, *J Biol Chem* **250**, 952-60.

Hensley, P., Moffat, K., and Edelstein, S. J. (1975a): Influence of inositol hexaphosphate binding on subunit dissociation in methemoglobin, *J Biol Chem* **250**, 9391-6.

Hensley, P., OKeefe, M. C., Spangler, C. J., Osborne, J. C. J., and Vogel, C. W. (1986): The effects of metal ions and temperature on the interaction of cobra venom factor and human complement factor B, *J Biol Chem* **261**, 11038-44.

Howlett, G. J. and Schachman, H. K. (1977): Allosteric regulation of aspartate transcarbamoylase. Changes in the sedimentation coefficient promoted by the bisubstrate analogue N-(phosphonacetyl)-L-aspartate, *Biochemistry* **16**, 5077-83.

Johnson, M. L. and Faunt, L. M. (1992): Parameter estimation by least-squares methods. **In:** *Methods Enzymol*, L. Brand and M. L. Johnson, ed.s, New York, Academic Press, 1-37.

Johnson, M. L. and Staume, M. (1994): Comments on the Analysis of Sedimentation Equilbrium Experiments. **In:** *Modern Analytical Ultracentrifugation: Acquisition and Interpretation of Data for Biological and Synthetic Polymer Systems*, T. M. Schuster and T. M. Laue, ed.s, Boston, MA, Birkhauser Boston, Inc., (this volume).

Kirschner, M. W. and Schachman, H. K. (1973a): Local and gross conformational changes in aspartate transcarbamylase, *Biochemistry* **12**, 2997-3004.

Kirschner, M. W. and Schachman, H. K. (1973b): Conformational studies on the nitrated catalytic subunit of aspartate transcarbamylase, *Biochemistry* **12**, 2987-97.

Kosaka, M., Iishi, Y., Okagawa, K., Saito, S., Sugihara, J., and Muto, Y. (1989): Tetrameric Bence-Jones protein in the immunoproliferative diseases, *Am. J. Clin. Path.* **91**, 639-646.

Kratky, O., Leopold, H., and Stabinger, H. (1973): Determination of the partial specific volume of proteins by the mechanical oscillator tecnhique. **In:** *Methods in Enzymology. Enzyme Structure, part D*, C. H. W. Hirs and S. N. Timasheff, ed.s, New York, Academic Press, 98-110.

Lakowicz, J. R. (1983): Principles of Fluorescence Spectroscopy, New York, Plenum.

Laue, T. M., Shah, B. D., Ridgeway, T. M., and Pelletier, S. L. (1992): Computer-Aided Interpretation of Analytical Sedimentation Data for Proteins. **In:** *Analytical Ultracentrifugation in Biochemistry and Polymer Science*, S. E. Harding, A. J. Rowe and J. C. Horton, ed.s, Cambridge, Great Britain, The Royal Society of Chemistry, 90-125.

Lee, G., Chan, W., Hurle, M. R., DesJarlais, R. L., Watson, F., Sathe, G. M., and Wetzel, R. (1993): Strong inhibition of fibrinogen binding to platelet receptor αIIbβIII by RGD sequences installed into a presentation scaffold, *Protein Engineering* **6**, 745-754.

Lewis, M. S. and Youle, R. J. (1986): Ricin subunit association. Thermodynamics and the role of the disulfide bond in toxicity, *J Biol Chem* **261**, 11571-7.

Minton, A. P. (1994): Conservation of signal: a new algorithm for the elimination of reference concentration as an independent parameter in the analysis of sedimentation equilbrium data. **In:** *Modern Analytical Ultracentrifugation: Acquisition and Interpretation of Data for Biological and Synthetic Polymer Systems*, T. M. Shuster and T. M. Laue, ed.s, Boston, MA, Birkhauser Boston, Inc., (This volume).

Minton, A. P. and Lewis, M. S. (1981): Self-association in highly concentrated solutions of myoglobin: a novel analysis of sedimentation equilibrium of highly nonideal solutions, *Biophys Chem* **14**, 317-24.

Press, W. H., Flannery, B. P., Teukolsky, S. A., and Vetterling, W. T. (1986): Numerical Recipies - The Art of Scientific Computing, New York, Cambridge University Press.

Rivas, G., Ingham, K. C., and Minton, A. P. (1992): Calcium-linked self-association of human complement C1s, *Biochemistry* **31**, 11707-12.

Roark, D. E. and Yphantis, D. A. (1969): Studies of self-associating systems by equilibrium ultracentrifugation, *Ann N Y Acad Sci* **164**, 245-78.

Ross, P. D., Howard, F. B., and Lewis, M. S. (1991): Thermodynamics of antiparallel hairpin-double helix equilibria in DNA oligonucleotides from equilibrium ultracentrifugation, *Biochemistry* **30**, 6269-75.

Schachman, H. K. (1992): Is There a Future for the Ultracentrifuge? **In:** *Analytical Ultracentrifugation in Biochemistry and Polymer Science*, S. E. Harding, A. J. Rowe and J. C. Horton, ed.s, Cambridge, Great Britain, The Royal Society of Chemistry, 3-15.

Stevens, F. J., Westholm, F. A., Solomon, A., and Schiffer, M. (1980): Self-association of human immunoglobulin κI light chains: Role of the third hypervariable region, *Proc. Natl. Acad. Sci. USA* **77**, 1144-1148.

Straume, M. and Johnson, M. L. (1992): Monte Carlo method for determining complete confidence probability distributions of estimated model parameters. **In:** *Methods Enzymol*, L. Brand and M. L. Johnson, ed.s, New York, Academic Press, 117-29.

Watts, D. G. (1991): Model building in chemistry using profile t and trace plots., *Chemometrics and intelligent laboratory systems* **10**, 107-116.

Watts, D. G. (1994): How good are parameter estimates from nonlinear models? **In:** *Methods in Enzymology*, L. Brand and M. L. Johnson, ed.s, New York, NY, Academic Press, *(in press)*.

Waxman, E., Laws, W. R., Laue, T. M., and Ross, J. B. A. (1994): Refining Hydrodynamic Shapes of Proteins: The Combination of Data from Analytical Ultracentrifugation and Time-Resolved Fluorescence Anisotropy Decay. **In:** *Modern Analytical Ultracentrifugation: Acquisition and Interpretation of Data for Biological and Synthetic Polymer Systems*, T. M. Schuster and T. M. Laue, ed.s, Boston, MA, Birkhouser Boston, Inc, (This Volume).

Wetzel, R. (1992): Protein aggregation *in vivo*: Bacterial inclusion bodies and
   mammalian amyloid. **In:** *Stability of Protein Pharmaceuticals: In Vivo
   Pathways of Degradation and Strategies for Protein Stabilization*, T. J.
   Ahern and M. C. Manning, ed.s, New York, Plenum Press, 43-88.
Wetzel, R. (1994): Aggregation - The Dark Side of Protein Folding, *Trends in
   Biotechnology* **12**, *(in press)*.
Wetzel, R., Hurle, M. R., Li, L., Helms, L., and Chan, W. (1993): Molecular
   basis of sequence effects in light chain amyloidosis and light chain deposition
   disease, *Submitted*.
Wilf, J. and Minton, A. P. (1981): Evidence for protein self-association induced
   by excluded volume. Myoglobin in the presence of globular proteins,
   Biochim Biophys Acta **670**, 316-22.
Zimyatnin, A. Á. (1972): Protein volume in solution., *Prog. Biophys. Mol. Biol.*
   24, 109-123.

COMMENTS ON THE ANALYSIS OF SEDIMENTATION EQUILIBRIUM EXPERIMENTS

Michael L. Johnson and Martin Straume

INTRODUCTION

The normal method for the analysis of data from sedimentation equilibrium experiments is to "fit" the experimental data to a functional form, described below. The functional form is in terms of parameters, such as molecular weights and equilibrium constants, which describe the chemistry of the solution. There are three goals of this fitting operation. The first goal is to evaluate the parameter values with the highest probability (maximum likelihood) of being correct. The second objective is to provide a statistically based measure of the precision to which the maximum likelihood parameter values were determined. The third goal is to provide criteria to answer the question, "Do the functional form and the maximum likelihood parameters provide a good description of the data?" In other words, test for goodness-of-fit.

The desired parameter values are those values that have the highest probability of being correct. With certain assumptions these maximum likelihood parameter values will correspond to the parameter values at the minimum variance-of-fit. If these assumptions are met then the proper method of analysis is weighted nonlinear least-squares (WNLLS). If these assumptions are not justified, the parameter values with the highest probability of being correct will not correspond to a minimum variance-of-fit and thus it is not statistically valid to use WNLLS to analyze the data.

The desired maximum likelihood parameter values have comparatively little meaning without a valid measure of their precision. For example, if we estimated the molecular weight of a protein to be 85,000 what is actually known? For a molecular weight of 85,000 $\pm$

80,000 we know relatively little. We could guess at the molecular weight with equal precision without performing an experiment. If, however, it was 85,000 ± 1,000 we know a lot about the protein. These measures of the precision of estimated parameter values enable us to test hypotheses about mechanisms, etc.

Goodness-of-fit criteria are important to evaluate if we want to compare molecular mechanisms. If a particular molecular mechanism does not provide a statistically valid fit of the data then either the mechanism is wrong or one, or more, of the assumptions made about the data is wrong.

The purpose of this chapter is to discuss basic assumptions and methods mentioned above. This discussion includes methods of constraining parameter values, finding roots of transcendental equations, the simultaneous analysis of multiple sets of data, data smoothing, goodness-of-fit criteria, confidence interval evaluation and some pitfalls that may be encountered in this type of analysis.

THE FORM OF THE FITTING FUNCTION

In the analytical ultracentrifuge solute molecules sediment in the direction perpendicular to the axis of rotation of the rotor. Once sedimentation occurs, the solute molecules are redistributed with increasing concentrations at the higher radii of rotation. Further sedimentation is slowed by diffusion of the solute in the direction of lower concentration, $i.e.$, toward lower radii. Sedimentation equilibrium is obtained when the forces of sedimentation, due to the rotation of the rotor, are exactly balanced by the back diffusion, due to the established concentration gradient. At equilibrium the concentration distribution as a function of radius, r, of the ith component of an ideal system, $c_{r,i}$, is given by (Johnson et al, 1981)

$$\frac{\partial \ln c_{r,i}}{\partial (r^2/2)} = \sigma_i = \frac{M_i (1 - \bar{v} \rho) \omega^2}{RT} \tag{1}$$

where M_i is the molecular weight of the ith component, R is the gas constant, T is the absolute temperature, v is the partial specific volume, ρ is the density and ω is the angular velocity. The concentration distribution of the ith species is obtained by integration of equation 1:

$$c_{r,i} = c_{0,i} \exp[\sigma_i (r^2/2 - r_0^2/2)] \tag{2}$$

where $c_{0,i}$ is the concentration of the ith component at a radius r_0, and *exp* is the exponential function. The exact choice of r_0 is arbitrary. r_0 could be zero, but that would cause difficulties when performing calculation on computers with finite precision and dynamic range. r_0 is usually taken to be the radius of the meniscus of the sample solution.

For an associating system, the total concentration at any radius can be expressed in terms of the monomer concentration and association constants by the normal mass action equations. For example, for a monomer n-mer associating system the total concentration at any radius is given by

$$c_{r,t} = c_{r,1} + K_n(c_{r,1})^n \tag{3}$$

where K_n is the n-mer association constant. Two implicit assumptions are made in equation 3. It was assumed that no volume change occurs on association (partial specific volume is a constant) and that the activity coefficients of the monomer, γ_1, and n-mer, γ_n, are related by equation 4:

$$n \ln \gamma_1 = \ln \gamma_n \tag{4}$$

The form of the concentration distribution as a function of the square of the radius is simply the sum of exponential terms. There is a large literature about the analysis of experimental data with this functional form. Least-squares parameter estimation with this functional form is the classic example of an ill posed mathematical problem.

For nonideal systems describable by communal virial coefficients the equivalent of equation 1 is

$$\frac{\partial \ln c_i}{\partial (r^2/2)} = \frac{\sigma_i}{1 + \dfrac{\partial \ln \gamma_i}{\partial \ln \sum c_i}} \tag{5}$$

$$\frac{\partial \ln c_i}{\partial (r^2/2)} = \frac{\sigma_i}{1 + 2B_1 \sum c_i + \cdots} \tag{6}$$

where $\sum c_i$ is the total solute concentration, γ_i is the activity coefficient of the ith species, B_1 is the first virial coefficient, and the $\cdots$ refers to the higher virial terms.

The nonideal form of equation 2 can be obtained from integration of equations 5 and 6 as

$$c_{r,i} = c_{0,i} \exp[\sigma_i (r^2/2 - r_0^2/2) - 2B_1 \sum c_{r,i} - \cdots] \tag{7}$$

The form of equation 3 is also valid for associating nonideal systems.

Equation 7 is a transcendental equation. This simply means that the concentration appears on both sides of the equal sign. To calculate the concentration of any, or all, of the species directly from equation 7 all of the concentrations must already be known. Since all maximum likelihood analysis methods require the calculation of the concentration as a function of the radius a numerical method for finding the root of this transcendental equation is required. A method is discussed later in this chapter.

For heterogeneous systems the total concentration is describable as a linear combination of terms of the form of equations 2 and/or 7. Heterogeneous systems are usually not described in a form analogous to equation 3.

One additional parameter, δc, must be included for the analysis of actual experimental data, as in equation 8:

$$c_{obs} = c_{true} + \delta c \qquad (8)$$

where the "obs" and "true" subscripts refer to the observed and true concentrations. For scanner optics the δc term describes a zero offset of the spectrophotometer. For Rayleigh interferometer data the δc is required because the interferometer measures relative instead of absolute concentrations. The δc term must be included for all experiments, even for the "meniscus depletion" method where it will be close to, but generally not equal to, zero. In this latter case the non-zero value of δc would arise from our inability to evaluate its value to infinite precision from the experimental data.

The maximum likelihood parameter values for sedimentation equilibrium experiments are estimated by "fitting" equations of the form of equations 3, 8, and 2 or 7 to the experimental data. This fit is usually done by a WNLLS algorithm because certain assumptions about the experimental data are acceptable. It is, however, important that the investigator be aware of those assumptions. The investigator can then attempt to design the experimental protocol such that these assumptions are valid.

MAXIMUM LIKELIHOOD AND LEAST-SQUARES METHODS

Assumptions

The investigator wants the numerical values of a series of fitting parameters, *e.g.*, K_n σ_1 δc etc., which have the highest probability, *i.e.*, maximum likelihood, of being correct. If a few assumptions about the nature of the experimental data are met then WNLLS methods will provide these maximum likelihood parameter values. Those assumptions are (Johnson and Frasier, 1985; Johnson and Faunt, 1992; Johnson, 1983):

1) All experimental uncertainties are in the dependent variables.
2) These uncertainties follow a Gaussian distribution, mean = zero.
3) Independent observations.
4) Large number of data points.
5) The fitting equation is correct.

It is important to note that the first three of these assumptions pertain to the experimental uncertainties in the observations.

Assumption one literally means that the X-axis (*i.e.*, the independent variables, in the present case the radius) is known to infinite precision. For practical purposes, the radius need only be known with significantly greater precision than the observed concentration. One consequence of this assumption is that any convenient transformations of the independent variables are allowed. For example, the fitting can be done in terms of $(r^2/2-r_0^2/2)$ instead of r. Another pertinent example is that if σ_1 is known and the virial coefficients are zero then a transformation of the independent variable can be done such that:

$$X_r = \exp[\sigma_1(r^2/2-r_0^2/2)] \qquad (9)$$

Then the observed data can be fit to an equation of the form:

$$c_{r,t} = c_{0,1}X_r + K_n(c_{0,1}X_r)^n + \delta c \qquad (10)$$

This is a much simpler equation to evaluate and thus requires decidedly less computer time to perform the WNLLS operation. Any desired transformation of the independent variable, *i.e.*, the radius in the current example, is allowed by the assumptions of the least-squares method.

Assumption two is that all of the experimental measurement errors, or uncertainties, can be described by a Gaussian, or bell-shaped, probability distribution with a mean of zero. This, however, does not require that the uncertainties be of the same magnitude for all of the data points. This is a reasonable assumption for sedimentation equilibrium data.

One consequence of this assumption is a limitation on the types of transformation that can be applied to the Y-axis (*i.e.*, the dependent variable, concentration in the current example) of the experimental observations. For example, if the original concentration data included Gaussian distributed random noise then it would not be proper to analyze the data by a logarithmic plot. The logarithmic transform of a Gaussian distribution is not a Gaussian distribution. Generally it is not good to make nonlinear transformations on the dependent variables. The only exceptions are those transformations that will transform a non-Gaussian uncertainty distribution into a Gaussian uncertainty distribution (Abbott and Gutgesell, 1994). In the section on goodness-of-fit criteria we discuss some methods to test for non-Gaussian error distributions.

Another consequence of this assumption is that the experimental data cannot contain any systematic errors. For example, normally δc is evaluated by the fitting procedure simultaneously with all of the other parameters. However, the analysis could be done in a two-step process. First, evaluate δc from a "meniscus depletion" experiment. Second, fit the data while using the value of δc estimated in the first step. Since δc cannot be evaluated to infinite precision it will have an associated experimental error. In this case any error in the evaluation of δc will appear as a systematic error in the second step. By including the δc term in the fitting equation and evaluating δc simultaneously with the other parameters this systematic error in the data is eliminated. There will still be an uncertainty in the evaluation of δc, but the evaluation of the other parameters, and their associated uncertainty estimates, will correctly include the consequences of the uncertainties of the evaluation of δc. All the parameters should be simultaneously evaluated from the original, $i.e.$, untransformed, experimental data.

The next assumption is that the data points are independent observations. Another way of stating this assumption is that the random experimental uncertainties that are superimposed upon the data are not correlated. This is a standard statistical assumption that is made for virtually every method of data analysis.

If the optical components in either the scanner and/or Rayleigh interferometer of the analytical ultracentrifuge are dirty it will appear as a correlated uncertainty in the experimental data points. If the scanner is operated too fast, as compared to the response time of its RC filters, the data will have correlated uncertainties. Consequently, the operator must be careful to operate the ultracentrifuge correctly. In the section on goodness-of-fit criteria we discuss a method to test for correlation in the uncertainties of the data points.

The fourth assumption is that there are a large number of data points as compared to the number of parameters being estimated. The theoretical minimum number of data points is equal to the number of parameters being estimated. However, since each data point has some experimental uncertainty, it is best to have substantially more data points than parameters being estimated. With a large number of independent data points the fitting process will, to a first approximation, average out the contributions of the noise. Unfortunately, there is no theory to tell us how many data points are needed in excess of the number of fitting parameters. For most sedimentation equilibrium experiments this assumption is not a problem since many data points can be recorded automatically.

The last assumption is that the fitting function is correct. The previous section described the theoretical form of the concentration distribution in the analytical ultracentrifuge. However, there are an infinite number of other equations that can be used to fit the data with equal precision. For example, the data could be fit to a twenty degree polynomial with quite good precision. It would be extremely difficult, if not impossible, to interpret the coefficients of this polynomial in terms of molecular weights and association constants. Another example would be to fit the experimental data to a monomer-dimer-tetramer association when the actual molecular mechanism is a monomer-trimer association. For this second example the molecular weights and equilibrium constants estimated for the 1-2-4 model have little meaning in terms of the 1-3 model. The parameter values estimated by any analysis procedure are dependent on the assumed model of the underlying chemistry that is taking place. Consequently, the investigator must have a reason for the choice of a particular functional form. Arbitrary choices should not be made.

Why are these assumptions required in order for the WNLLS method to provide parameter values with the highest probability of being correct? With assumptions 1 (uncertainties in X-axis only), 2 (Gaussian distribution) and 5 (correct fitting equation) the probability of observing a particular data point, $P_i(\alpha)$, can be written as

$$\ln P_i(\alpha) = -\ln[\sigma_i\sqrt{2\pi}] - \frac{1}{2}[(Y_i - g_i(\alpha))/\sigma_i]^2 \qquad (11)$$

where Y_i is the observed value of the ith data point, σ_i is the uncertainty (standard deviation) of the ith data point (*i.e.*, the weighting factor), $g_i(\alpha)$ is the fitting function evaluated for the ith data point at a set of parameter values α for the fitting equation. For the present example of sedimentation equilibrium data the form of the fitting function is described in the previous section and the vector α would consist of the

$$c_{r,t} = c_{r,1} + \exp[Z_n](c_{r,1})^n \tag{31}$$

and fit to Z_n instead of K_n. The value of K_n is thus constrained to always be positive. In this example, Z_n can have positive or negative value, but the exponential of Z_n, *i.e.*, K_n, is always positive. By performing this transformation of the fitting function we have defined a system where only positive values of the equilibrium constant can exist.

In a similar fashion, to constrain a variable S to have a value between A and B all that is needed is to substitute a variable, T, of the form

$$S = B + (A-B)\frac{\exp[T]}{1 + \exp[T]} \tag{32}$$

and fit to T instead of S. Even though T can have any real value, the resulting value of S must be between A and B.

A sedimentation equilibrium experiment will typically consist of multiple channels of data. These multiple channels of data can originate from loading different sample concentrations in different cells, or different channels in a multiple channel cell, or from collecting data from different temperatures or rotor speeds. The least-squares procedure, as outlined above, was designed for the analysis of only a single set of data. It is, of course, possible to analyze each set individually and this should always be done first. This individual analysis will provide many individual estimates of the fitting parameters, such as molecular weight and association constants. There is no obvious method to combine these multiple estimates of fitting parameters into the values that have the highest probability of describing all of the data simultaneously. However, all these channels of data can be simultaneously analyzed to provide the most probable values based on all of the data. This simultaneous analysis requires that the fitting parameters be divided into two classes, global and non-global. Global fitting parameters are those that are expected to have the same value for all the data channels, for example association constants, virial coefficients and molecular weights. Non-global fitting parameters are those that are unique to individual channels of data, for example the concentration of monomer at the meniscus of the sample solution. For the previously outlined least-squares method we assumed that the data had only a single independent variable, the radius. However, a single independent variable is not a requirement of the method. To simultaneously analyze multiple sets of data each data point must consist of four quantities; an observed concentration Y_i, its corresponding precision σ_i, its radius r, and an index set of values for the fitting parameters, *e.g.*, δc K_n M_i $c_{0,1}$ etc. With assumption 3 (independent observations) the complete probability of observing all of the data points, $P(\alpha)$, is given by

$$\ln P(\alpha) - \sum_{i-1}^{N} \ln P_i(\alpha) \tag{12}$$

$$\ln P(\alpha) - \ln Q - \frac{1}{2} \sum_{i-1}^{N} [(Y_i - g_i(\alpha))/\sigma_i]^2 \tag{13}$$

where Q is a constant given by

$$\ln Q - -\sum_{i-1}^{N} \ln [\sigma_i \sqrt{2\pi}] \tag{14}$$

The basic objective is to find a set of values of α that will maximize the complete probability, $P(\alpha)$. This is accomplished when the negative summation on the right of equation 13 in minimized. The summation on the right of equation 13 is simply the weighted sum of squared residuals, *WSSR*. If we denote the weighted residuals as R_i

$$WSSR(\alpha) - \sum_{i-1}^{N} [(Y_i - g_i(\alpha))/\sigma_i]^2 - \sum_{i-1}^{N} R_i^2 - \chi^2 \tag{15}$$

WSSR is also referred to as χ^2. The sampled variance-of-fit of the weighted residuals, s^2, is given by

$$s^2 - \frac{1}{N-M} \sum_{i-1}^{N} R_i^2 \tag{16}$$

Therefore, given a few assumptions about the nature of the fitting equation and the experimental data a WNLLS method provides the parameter values with the highest probability of being correct. However, if these assumptions cannot be met then WNLLS should not be used for the analysis of data. Consequently, it is not always the case that the best answers will correspond to the answers at the minimum variance.

Algorithms

There are many algorithms for performing a weighted nonlinear least-squares (WNLLS) analysis (Bevington, 1969; Bard, 1974; Johnson and Frasier, 1985; Bates and Watts, 1988; Straume *et al*, 1991; Johnson and Faunt, 1992). When fitting to an arbitrary equation, $g_i(\alpha)$, there is no closed form solution. For arbitrary equations these algorithms are all successive approximation methods. Specifically the algorithm is provided with an initial estimate of the values of the fitting parameters, the vector α, and the algorithm returns a better estimate of the values of the fitting parameters, α'. These better estimates of the fitting parameters are then used as a subsequent "initial" estimate, the algorithm is repeated, and an even better estimate is obtained. This process is repeated until the values, α', do not change within some specified tolerance. Some algorithms will have computational advantages for some types of problems. However, when correctly applied all of the methods should provide equivalent answers. Only our preferred algorithm, a damped Gauss-Newton iteration, will be described here.

The basic Gauss-Newton algorithm is based on a first order series expansion of the fitting equation. For the ith data point the experimental observation, Y_i, is approximated by the fitting function evaluated at an improved set of answers, α'.

$$Y_i = g_i(\alpha') + noise_i \tag{17}$$

where $noise_i$ is the random experimental uncertainty. The fitting function evaluated at an improved set of answers, α', is then expressed as a series expansion about the previous values of the parameters, α.

$$\frac{1}{\sigma_i} g_i(\alpha') = \frac{1}{\sigma_i} [g_i(\alpha) + \sum_{k-1}^{M} \frac{\partial g_i(\alpha)}{\partial \alpha_k} (\alpha'_k - \alpha_k) + \cdots] \tag{18}$$

σ_i is the appropriate weighting factor for the ith data point, the k subscript refers to a particular fitting parameter, M is the total number of fitting parameters, and the $\cdots$ represents the higher order derivatives. For linear least-squares methods these higher order derivatives are all zero. For nonlinear least-squares methods these higher order derivatives are not equal to zero but they are assumed to be small and are neglected. As a consequence, least-squares is an iterative technique when the second and higher order derivatives are not zero. Combining equations 17 and 18 and neglecting the higher order derivatives yields

$$\frac{1}{\sigma_i} [Y_i - g_i(\alpha)] = \frac{1}{\sigma_i} \sum_{k-1}^{M} \frac{\partial g_i(\alpha)}{\partial \alpha_k} (\alpha'_k - \alpha_k) \tag{19}$$

It is important to realize that equation 19 is a system of N equations, one for each data point (the i subscript). This system contains M unknowns, one for the better approximation of each parameter being estimated (the k subscript). This system of equations can, in principal, be solved for the better estimate of α' if N is greater than or equal to M (the fourth assumption) and the data points are independent observations (the third assumption). The solution of equation 19 can best be shown in a matrix notation, as in

$$Y^* = P\epsilon \tag{20}$$

$$Y_i^* = \frac{1}{\sigma_i}[Y_i - g_i(\alpha)] \tag{21}$$

$$P_{i,k} = \frac{1}{\sigma_i}\frac{\partial g_i(\alpha)}{\partial \alpha_k} \tag{22}$$

$$\epsilon_k = (\alpha'_k - \alpha_k) \tag{23}$$

The P matrix is not a square matrix so it cannot be directly inverted to solve for ϵ. However, if both sides of equation 20 are multiplied by the transpose of P the resulting $P^T P$ matrix can be inverted.

$$P^T Y^* = (P^T P)\epsilon \tag{24}$$

$P^T P$ is a symmetric positive definite matrix that is usually almost singular. Therefore, the choice of the matrix inversion method is critical to avoid truncation and roundoff errors. We recommend the square root method (Faddeeva, 1959).

The solution of equations 19-24 is

$$\epsilon = (P^T P)^{-1}(P^T Y^*) \tag{25}$$

The value of α is subsequently updated as

$$\alpha \leftarrow \alpha + \epsilon \tag{26}$$

This process is repeated until α does not change (*i.e.*, ϵ becomes zero).

The standard Gauss-Newton algorithm, equations 20-26, is fast and easy to program. It only suffers from one minor problem. It does not always converge! The source of this is the higher order derivatives in equation 18 that were assumed to be negligible in equation 19. For nonlinear fitting functions these higher order derivatives can sometimes be large and cannot be neglected. When these are large the algorithm will diverge instead of converge. The Marquardt-Levenberg algorithm and the damped Gauss-Newton algorithm are both modifications of the basic Gauss-Newton algorithm that eliminate this failure to converge. Our preference is the damped Gauss-Newton algorithm.

When the higher order derivatives in equation 18 are large then the Gauss-Newton algorithm will still predict a direction, ϵ, in which to alter α that will lower the *WSSR*. The problem is that the predicted distance, the magnitude of ϵ, may be too large. The damped Gauss-Newton algorithm solves the divergence problem by searching in the direction of ϵ for a point where the *WSSR* has decreased. This point is subsequently used as the starting value for the next iteration. This is accomplished by replacing equation 26 with the following steps

1) $\lambda \leftarrow 1.$
2) $Z \leftarrow WSSR(\alpha).$
3) $\alpha \leftarrow \alpha + \lambda\epsilon.$
4)   If $WSSR(\alpha) \geq Z$ then
            $\lambda \leftarrow \lambda/2$
            $\alpha \leftarrow \alpha - \lambda\epsilon$
            If convergence criteria are satisfied
                  stop the convergence.
            If convergence criteria are not satisfied
                  return to step 4.
5)   If $WSSR(\alpha) < Z$ use the current value of α to start the next iteration (*i.e.*, back to equation 25).

Why is this a least-squares technique? In the derivation of the Gauss-Newton algorithm no reference is made to maximizing probabilities or minimizing the *WSSR*. When convergence is reached ϵ in equations 20-26 is equal to zero and $P^T Y^*$ must also be equal to zero since $P^T P$ will, in general, be finite. The elements of $P^T Y^*$ are proportional to the derivatives of the *WSSR* with respect to the fitting parameters:

$$(P^T Y^*)_k \propto \frac{\partial\, WSSR\,(\alpha)}{\partial\alpha_k} \tag{27}$$

When $P^T Y^*$ is equal to zero the derivatives of the WSSR with respect to the fitting parameters are also zero. Therefore, this procedure will provide parameter values that minimize the weighted sum of squared residuals. Thus, the Gauss-Newton procedure is a least-squares method.

A word of caution is in order about the general properties of weighted nonlinear least-squares fitting procedures. They will all fail occasionally.

Possibly the most common cause for failure is to use a fitting function with parameters that are perfectly correlated with each other. For example, when two fitting parameters appear in the fitting equation only as a product they are totally correlated with each other and cannot be estimated simultaneously. For any value of one parameter a corresponding value for the other parameter can be found that yields the same WSSR. If we were to fit the optical density distribution, OD, in the ultracentrifuge to an equation that includes a molar concentration, $c_{0,i}$, and an extinction coefficient, A, as in

$$OD_{r,i} = A\,c_{0,i} \exp\left[\sigma_i\,(r^2/2 - r_0^2/2)\right] \tag{28}$$

it would be impossible to determine both A and $c_{0,i}$ simultaneously. These two parameters are completely correlated with each other.

For a nonideal solute the fitting function is transcendental (i.e., depends on itself) as in equation 7. The concentration distribution is evaluated from such transcendental equations by finding a root of the equation. The simplest way to find the root of an equation is by a Newton's iteration. If we define a new function G as the difference between the calculated concentration and the function evaluated at the calculated concentration, as in

$$G(c_r) = c_r - c_0 \exp\left[\sigma\,(r^2/2 - r_0^2/2) - 2B_1 c_r\right] \tag{29}$$

Then the desired c_r is the value where G is equal to zero. This value can be found by a Newton's iteration as

$$c_r^{q+1} = c_r^q - \frac{G(c_r^q)}{G'(c_r^q)} \tag{30}$$

where the q superscript refers to the iteration number and the prime is the derivative with respect to c_r. This iteration is continued until the value of c_r does not change within some specified limit. Thus the process of performing a WNLLS fit to a functional form that includes nonideality is a pair of nested iterations. One iteration to evaluate the concentration distribution at each point nested inside another iteration to perform the least-squares operation.

It is often useful to constrain a parameter to have only a positive value. For example, it is desirable to constrain association constants, such as K_n in equation 3, to be positive. Negative association constants have no physical meaning. The simplest method to constrain parameters to be positive is to redefine the coordinate system. If we define K_n as the exponential of Z_n, we can rewrite equation 3 as to specify from which channel of data the point is chosen. The global parameters are treated exactly as outlined above. Every non-global parameter will be an individual fitting parameter for each channel of data. The index specifies which of those parameters to use for the evaluation of the function. For example, if three channels of data are being analyzed for a monomer-dimer association the procedure would fit to a single monomer molecular weight, a single association constant and three monomer concentrations at the meniscus (one for each channel).

Furthermore, there is no requirement that every data point will use a fitting function of the same mathematical form. It is quite possible, and perfectly valid, to combine multiple sets of data from totally different type of experiments. The method would be as outlined for the analysis of multiple channels of data, except that a different functional form would be used for the different types of experiments.

A comment on convergence criteria is in order. It is best to use multiple criteria. For example, we usually continue the *WNLLS* iteration until the estimated parameter values do not change by more that one part in ten to the fourth and the calculated *WSSR* does not change by more that one part in ten to the fourth. We accept the convergence when these criteria have been satisfied for two, or more, successive iterations. With less stringent convergence criteria there is a good chance that the algorithms may prematurely converge before the minimum variance is located.

THE PRECISION OF ESTIMATED PARAMETERS

It is critically important that whenever a parameter value is estimated that a measure of the precision of that parameter value is also estimated. These are sometimes expressed as confidence intervals for the determined parameter. For example, if a molecular weight is estimated to be 85,000 and the molecular weight of a polymer of this monomer is measured to be 350,000 we might assume that the associated species is a tetramer ($n=4$). If our molecular weights are ± 3000 this is a valid assumption, but if the molecular weights are estimated to $\pm 40,000$ precision the conclusion cannot be made that the polymer is a tetramer. Either a 3-mer or a 5-mer would be almost equally likely.

The most commonly used, and by far the least accurate, measures of the precision of least-squares fitted parameters are the classical asymptotic standard errors (*ASE*) of the parameters. The asymptotic standard errors are square roots of the corresponding diagonal elements of the covariance matrix (*CM*):

$$ASE_i = \sqrt{CM_{i,i}} \qquad (33)$$

This covariance matrix is calculated from the observed variance-of-fit, s^2, and the P matrix evaluated at the least-squares minimum values as:

$$CM = s^2 (P^T P)^{-1} \qquad (34)$$

The reason for the popularity of this method of estimating confidence intervals is that both s^2 and $(P^T P)^{-1}$ have already been evaluated by the preceding least-squares procedure.   Consequently, the asymptotic standard errors are very easy to evaluate.   Unfortunately, they almost always significantly underestimate the true confidence intervals of the determined parameters (Box, 1960; Johnson and Frasier, 1985; Bates and Watts, 1988; Bates and Watts, 1991; Johnson, 1992; Johnson and Faunt, 1992).   If the confidence intervals are underestimated then the researcher will wrongly conclude that the estimated values are known more precisely than is justified by the experimental data.   This can lead to wrong conclusions about significance or results and/or molecular mechanisms.

The source of the problems with using ASE as measures of the precision of estimated parameters is the implicit assumptions of the ASE. The most serious problem is that the ASEs assume that the fitting parameters are orthogonal, $i.e.$, they ignore the covariance terms in the CM. ASEs assume that there are a large number of data points, so that s^2 is a good estimate of the true experimental uncertainties of the experimental data.   This last assumption is equivalent to assumptions 4 (large number of points) and 5 (correct fitting equation) that were listed in the earlier discussion of least-squares parameter estimation methods. The ASEs also assume that the fitting equation is a linear equation, $i.e.$, that the second and higher derivatives of the fitting function with respect to the fitting parameters are zero.   This last assumption implies that the variance for any values of the fitting parameters, α, is related to the variance at the maximum likelihood parameter values, α', by

$$\frac{s^2(\alpha)}{s^2(\alpha')} = 1 + \frac{M}{N-M} F(M, N-M, 1-PROB) \qquad (35)$$

where F is the F-statistic and $PROB$ is the probability that α and α' are equivalent.   Thus, equation 35 can be used to evaluate the probability that any two sets of estimated parameter values are equivalent.

The covariance between two estimated parameters is given by the off diagonal elements of the CM and is equivalent to

$$COVARIANCE_{k,l} = s^2 \sum_{i-1}^{N} \frac{\partial g_i(\alpha)}{\partial \alpha_k} \frac{\partial g_i(\alpha)}{\partial \alpha_l} \qquad (36)$$

where the summation is over all of the data points, and α is the maximum likelihood set of determined parameter values. For a few types of fitting equations like orthogonal polynomials (Acton, 1959) and Fourier series (Faunt and Johnson, 1992) the summation in equation 36 is equal to zero and the parameters are orthogonal. However, for the vast majority of fitting equations the summation in equation 36 is not equal to zero and the fitting parameters are not orthogonal. The assumption that the fitting parameters are orthogonal can introduce errors in the evaluation of the precision of the fitting parameters as large as a factor of three or more. Clearly, the *ASE*s cannot be used as a reliable measure of the precision of the determined parameters.

A schematic example of the non-orthogonal nature of fitting parameters is shown in Figure 1. This figure presents a series of contours of constant variance, s^2, as a function of the fitting parameters. It should be noted that, in general, these contours will be in more than two dimensions and that only the simplest of fitting equations will generate such a simple two dimensional contour. The probability that corresponds to any of the contours can be calculated from equation 35. The probability refers to the likelihood of the actual values of the parameters being within the given contour. If the contour is evaluated at a probability of 0.68 the contour would correspond to the $\pm$ one standard deviation confidence interval. Any set of parameters within that contour would be acceptable at 68%, or $\pm$ one standard deviation, confidence. If the parameters are orthogonal then the major and minor axes of these "elliptically" shaped regions will correspond to the axes of the parameter coordinate system. When these two sets of axes do not align the parameters are correlated. Neglecting the covariance terms, as with the use of the *ASE*s, is equivalent to finding that portion of this contour that is aligned with the axes and ignoring the remaining regions.

For non-orthogonal parameters a small change, or uncertainty, in the estimated value of one of the estimated parameters can be partially compensated for by a change in the other parameters. This compensation is shown in Figure 1 where changes of the parameters along the major diagonal of the elliptical regions will result in only a small change in the overall variance-of-fit. For orthogonal cases, these ellipses are aligned with the parameter axes and thus there is no compensation for one parameter by the other parameters. Consequently, for orthogonal cases each of the parameters can be evaluated independently. A mathematical definition of orthogonal parameters is that the off-diagonal elements of CM in equation 34 are zero. Another is that the $COVARIANCE_{k,l}$ in equation 36 is zero for $k \neq l$.

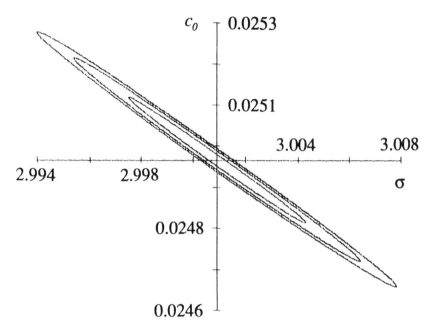

FIGURE 1. Contours of constant variance, s^2, as a function of reduced molecular weight, σ, and concentration at the meniscus, c_0.

If the fitting function is a linear equation then these multi-dimensional contours in Figure 1 are exactly elliptically shaped and symmetrical about the maximum likelihood values. For linear systems this multi-dimensional ellipsoid, α^*, is given by (Box, 1960)

$$(\alpha^* - \alpha')^T P^T P (\alpha^* - \alpha') \leq M s^2 F(M, N-M, 1-PROB) \qquad (37)$$

where α' corresponds to the maximum likelihood parameter values, and the other terms are as defined previously. If, however, the fitting function is not linear then these contour regions will be neither elliptical nor symmetrical. In many nonlinear cases equation 37 is still a reasonable approximation to the actual shape of the confidence contours and is the basis of a second method of estimating the uncertainties of fitted parameters known as "joint confidence intervals." It must be noted that for arbitrary nonlinear fitting equations there are no closed form simple equations, or theories, that define the confidence region for the determined parameters. The most commonly used, and least accurate, method is to assume that the system is approximately linear and then use the results estimated by linear methods.

Some authors (Bates and Watts, 1988; Bates and Watts, 1991; Johnson *et al*, 1981; Johnson and Frasier, 1985; Johnson and Faunt, 1992; Johnson, 1992; Straume *et al*, 1991) have developed search techniques that correct for the asymmetry and non-elliptical nature of the confidence regions that are observed with nonlinear fitting equations. In both methods the variance surface is searched to define the acceptable ranges of parameter values within some probability as given by equation 35. The method that we developed (Johnson *et al*, 1981; Johnson and Frasier, 1985; Johnson and Faunt, 1992; Johnson, 1992; Straume *et al*, 1991) searches the axes of the ellipse defined by equation 35 looking for significantly different values of the parameters. The method developed by Bates and Watts (1988; 1991) searches each of the estimated parameters independently. For each of the Bates and Watts searches a series of values for a particular parameter are specified while the remaining parameters and apparent variance-of-fit, s^2, are evaluated by repeating the least-squares fit while holding the one parameter constant. The probability of the particular value of the searched parameter is then evaluated. Our method requires substantially less computer time and provides slightly less accurate values for the confidence regions of the determined parameters. For most applications the results are approximately equivalent, but for highly correlated systems (*i.e.*, extremely non-orthogonal) or very nonlinear systems the increased computer time required by the Bates and Watts methods may be justified. It should be noted that both search methods use equation 35 to calculate the probability of a particular set of parameters being different from the maximum likelihood values. The derivation of equation 35 is dependent on the assumption that the fitting equation is linear (*i.e.*, the second and higher order derivatives are zero).

Monte Carlo methods for confidence interval estimation (Straume and Johnson, 1992a) do not require the use of equation 35 and thus have no implicit assumption that the fitting equation is linear. However, they require substantially more computer time. For most problems the increased precision of the Monte Carlo method is probably not justified. We will describe the Monte Carlo method since it is simple to implement, accurate, interesting, instructive and provides a method for the evaluation of the complete probability distribution for each of the determined parameters.

The first step of the Monte Carlo procedure for evaluating confidence intervals is to calculate a perfect set of synthetic data based on the least-squares, or maximum likelihood, estimates of the parameters and the values of the independent variables from a previous fit of an actual data set. The second step is to add a realistic amount of pseudo random experimental uncertainties (Knuth, 1969) to this set of synthetic data and repeat the least-squares fit. This second step is repeated 500, or more, times. These 500, or more, fits of the synthetic data provide multiple estimates for each of the determined parameters that are subsequently organized as individual probability histograms. These histograms are the complete probability distributions for the determined parameters. Clearly, this procedure will require 500, or more, times the amount of computer time than that required by the original least-squares fit!

This section has presented several different algorithms for the evaluation of confidence intervals for determined parameters. The order of presentation is in order of increasing precision of the estimate of the confidence intervals. The order of presentation is also in order of increasing amounts of computer time required to evaluate the confidence intervals of the determined parameters. Clearly, for most applications the 500, or more, fold increase in computer time required for the Monte Carlo is clearly not justified by the modest increase in precision of the confidence intervals that it provides. The simplest, fastest, and least accurate method is the asymptotic standard error. However, these are so inaccurate that they are of little use. We recommend the joint confidence interval method for linear fitting equations. For routine applications with nonlinear fitting equations both the Bates and Watts search and our search methods are recommended. The Bates and Watts method requires significantly more, but not an unreasonable amount of, computer time and provides only a small increase in precision of the confidence intervals for most problems.

GOODNESS-OF-FIT CRITERIA

The use of goodness-of-fit criteria can be characterized by an attempt to answer either, or both, of the following two questions: Does a particular fitting function provide a good description of the experimental data? Does fitting to function A provide a better description (*i.e.*, fit) of the data than does fitting to function B? Clearly, if either, or both, fitting functions fail on the first question then the answer to the second question is either trivial or meaningless. Therefore, we will start with the first question.

Does the fitting function adequately describe the data?

The most common approach for answering this question is to examine the randomness of the residuals. The residuals are the differences between experimental data and the fitting function evaluated at the maximum likelihood values of the estimated parameters. The assumptions in the section on the theory of least-squares methods state that all of the uncertainties are in the Y-axis, that the uncertainties followed a Gaussian distribution with a mean of zero, that the data points are independent observations, and that the fitting function corresponds to the correct mechanism. If these assumptions are met then the residuals should follow a Gaussian distribution with a mean of zero and have no autocorrelation. Therefore, the methods for answering the question about adequacy of fit will test to see if the residuals are random, do they follow a Gaussian distribution and do they have any autocorrelation. If the residuals fail one, or more, of these tests, then one, or more, of the assumptions were violated. Since the first three assumptions are independent of the form of the fitting function a failure of the residuals to pass one, or more, of these tests is most likely a failure of the assumption about the correct fitting function.

The simplest test of randomness of residuals is simply to plot a graph of the weighted residuals as a function of the independent variable, the radius squared, and/or the dependent variable, the concentration. Look at the graph. The human brain is very good at locating trends in a series of random numbers. Does such a graph look random? If it does not look random then it is clearly not random and the question is answered. A simulated example is shown in Figure 2. It is clear from the residual plots presented in this figure that panel A is not a good fit of the data and panel B is a much better fit of the data.

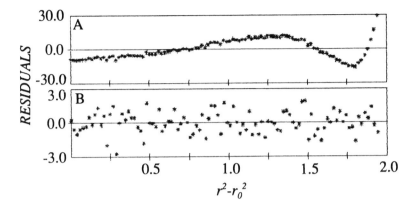

FIGURE 2. Residual plots for a simulated sedimentation equilibrium experiment. The data for this example was generated for a monomer-dimer association at sedimentation equilibrium with a 3 mm solution column, a monomer reduced molecular weight, σ, of 3, and a $\delta c=0$. Panel A was fit to a monomer only model with an apparent $\sigma=3.98\pm0.03$, an apparent $\delta c=0.0105\pm0.0016$, and a weighted variance-of-fit $s^2=71.8$.
Panel B was for a monomer-dimer fit with an apparent $\sigma=2.992\pm0.010$, $\delta c=(-0.2\pm2.0)\times10^{-3}$, and a weighted $s^2=0.874$.

If a more rigorous statistical test is needed we suggest the runs test (Bard, 1974). This test is designed to answer the question: Are the residuals random? A run is one, or more, residuals in a row with the same sign. For the example in Figure 2, panel A had 6 runs and panel B had 48 runs. The expected number of runs for both panels was approximately 50 ± 5. The expected number of runs, R, and its standard deviation, σ_R, are calculated by

$$R = \frac{2N_p N_n}{N_p + N_n} + 1 \qquad (38)$$

$$\sigma_R = \sqrt{\frac{2N_p N_n (2N_p N_n - N_p - N_n)}{(N_p + N_n)^2 (N_p + N_n - 1)}} \qquad (39)$$

where N_p is the number of positive runs and N_n is the number of negative runs. The observed number of runs can be related to a standard normal deviate, Z, and thus to a probability by

$$Z = |\frac{N_p + N_n - R + 0.5}{\sigma_R}| \qquad\qquad (40)$$

where the vertical bars denote the absolute value function. For this case, panel A had a $Z=8.935$ ($p < < 0.0001$) and panel B had a $Z=0.468$ ($p \approx 0.82$). Clearly, panel A is not random and panel B does not fail this test of randomness.

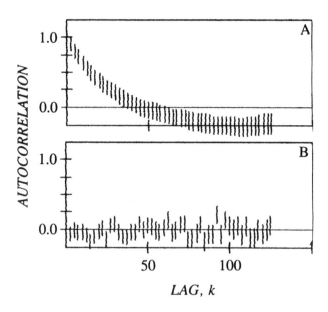

FIGURE 3. Autocorrelation of the residuals presented in Figure 2.

To address the question of autocorrelation of the residuals the autocorrelation function is calculated. Figure 3 presents such an autocorrelation plotted as a function of lag, k. Such a plot is a series of autocorrelation tests with different lags. Autocorrelation with a lag of one refers to autocorrelation between successive residuals. A lag of two refers to autocorrelation between every other residual, etc. If no autocorrelation exists within the residuals then the autocorrelation at each lag will be zero, within its standard error. Clearly, panel A exhibits a large amount of autocorrelation and panel B does not exhibit significant autocorrelations.

The autocorrelation for a lag of k is calculated as (Draper and Smith, 1981)

$$\hat{\rho}(k) = \frac{\dfrac{1}{N}\displaystyle\sum_{i=1}^{N-k}(R_i - \hat{\mu})(R_{i+k} - \hat{\mu})}{\dfrac{1}{N}\displaystyle\sum_{i=1}^{N}(R_i - \hat{\mu})^2} \tag{41}$$

where N is the number of residuals, R_i is the ith residual, and

$$\hat{\mu} = \frac{1}{N}\sum_{i=1}^{N} R_i \tag{42}$$

The variance of the autocorrelation at a lag of k is

$$\sigma^2\{\hat{\rho}(k)\} = \frac{N-k}{N(N-k)} \tag{43}$$

The simple tests outlined above are sufficient for most purposes. However, more elaborate tests of the randomness of residuals are described elsewhere. Therefore, the reader is referred elsewhere for a discussion of other test such as; cumulative frequency plots, chi-square statistics, outliers, influential observations, Durbin-Watson tests, Kolmogorov-Smirnov tests, etc. (Bard, 1974; Bates and Watts, 1988; Bevington, 1969; Draper and Smith, 1981; Straume *et al*, 1991; Straume and Johnson, 1992b).

Does one function describe the data better than a second function?

For the sedimentation equilibrium data this question suggests a more general class of questions. Examples of this class of questions are: Does a monomer-dimer association model describe the data better than a non-association model? This is the example presented in Figures 2 and 3. Does a monomer-dimer-tetramer association model describe the data better than a monomer-dimer model? Does a monomer-dimer describe the data better than a monomer-trimer association model? These questions only have meaning if both functions fit the data sufficiently well that the tests of randomness of the residuals outlined in the previous section demonstrate that the residuals cannot be shown to be non-random. If either function, or both functions, fail the previous tests for random residuals the answer to this question is either trivial, or meaningless. Under these conditions the following tests need not be performed.

There are two possible F-statistic tests to compare the quality of two fits of the same data set. Both employ an F-statistic to compare two weighted sums of squared residuals, χ_1^2 and χ_2^2 with ν_1 and ν_2 degrees of freedom, respectively. If the data points are truly independent observations then the number of degrees of freedom for a fit, ν, is the number of data points minus the number of fitted parameters, N-M.

One of these tests, the "test for an additional term" (Bevington, 1969) is based on the idea that the weighted sum of squared residuals can be divided into two parts. One part is the difference between them, $\Delta\chi^2 = |\chi_1^2 - \chi_2^2|$, with $\Delta\nu = |\nu_1 - \nu_2|$ degrees of freedom. The other part corresponds to the fit with the most terms, χ^2, and its number of degrees of freedom, ν. The weighted sums of squared residuals are thus divided into terms corresponding to the improvement in going from the less complete model to the more complete model and that due to the more compete model. In this case we are testing to see if the improvement in the variance is statistically significant. These can be compared by an F-test where the probability is given by

$$\frac{\Delta\chi^2/\Delta\nu}{\chi^2/\nu} = F(\Delta\nu, \nu, 1 - PROB) \tag{44}$$

This test has three distinct problems. First, the ability to separate the variance into two parts, as in this test, is based on the assumption that a linear fitting equation is being used. This is rarely the case for sedimentation equilibrium experiments. The second problem is that sometimes the two fitting equations being compared cannot easily be classified as one being the same as the other plus an additional term. For example, try to compare the same association model with and without nonideality. The third problem is that the researcher will sometimes wish to consider a pair of models with the same number of fitting parameters, in which case $\Delta\nu = 0$.

The other test is to directly compare the two weighted sums of squared residuals by an F-test. In this case we are testing to see if the two variances are different. If they are different then the two fits are not equivalent. This comparison is given by

$$\frac{\chi_1^2/\nu_1}{\chi_2^2/\nu_2} = F(\nu_1, \nu_2, 1 - PROB) \tag{45}$$

This test does not suffer from any of the problems of the previous test and is consequently the recommended test.

This section presented several statistical tests for the goodness-of-fit and for the comparison of multiple fits. These are not all of the available tests for these types of problems. The reader is referred to more complete works on the subject (Bard, 1974; Bates and Watts, 1988; Bevington, 1969; Draper and Smith, 1981; Straume *et al*, 1991; Straume and Johnson, 1992b).

PITFALLS AND PROBLEMS

There are many problems that an investigator can encounter in performing least-squares, maximum likelihood, analyses of sedimentation equilibrium data. However, the problems to be discussed are general problems of data analysis and are not limited to the analysis of sedimentation equilibrium data.

The most significant problem is that of using an invalid assumption. This problem has, however, been discussed throughout this chapter and will not be discussed further in this section.

A significant problem that we have not yet discussed is the occurrence of relative (*i.e.*, local) minima in the variance space. For some sets of data and nonlinear fitting functions the investigator is going to find that there are multiple sets of the fitting parameters, α, that correspond to different but approximately equivalent minima in the variance. A question then arises about which of these minima is the desired one and which are extraneous. It should be understood that there is no method of parameter estimation that is guaranteed to find the desired minimum and exclude all other possible relative minima. The only way to be sure that relative minima do not exist is to start the least-squares procedure at many different initial points, different values of α, that span the whole range of physically meaningful values of the parameters. If the least-squares procedure always converges to the same set of values for the parameters then the investigator can be reasonably certain that relative minima do not exist.

If multiple minima are encountered how are they distinguished? The investigator should always examine the parameter values that result from any type of data analysis to see if they are physically meaningful. Quite often the unwanted relative minima will have unrealistic parameter values. The previously discussed method of constraining the parameters to physically meaningful values is one method of eliminating multiple relative minima. Sometimes the unwanted relative minima will have a significantly higher variance-of-fit and thus can be eliminated. However, cases do occur where there are equivalent relative minima that have physically reasonable values for the parameters. These multiple minima cannot be distinguished on solely the basis of the particular data being analyzed and all should be reported by the investigator. An interesting example of such relative minima has been shown while investigating models of cooperativity in human hemoglobin (Johnson *et al*, 1984).

We have already discussed the problem of the nonlinear least-squares procedure prematurely deciding that convergence has been reached. All nonlinear least-squares algorithms will occasionally encounter situations where their test for convergence will fail. That was why we stressed the importance of a very good matrix inversion method for the calculation of the inverse of the $P^T P$ matrix. It is also important that some calculation be done with double precision arithmetic. For example, each element of the $P^T P$ matrix is the sum of many products of derivatives. Most matrix inversion methods also use sums of many products. Calculations of this type are particularly prone to truncation and roundoff errors. Truncation and roundoff errors can cause least-squares algorithms to prematurely converge. Clearly all calculations of this type should be done in double precision. Even if these precautions are taken all algorithms still have the possibility of premature convergence. Therefore, it is important that the data analysis algorithm have a check for premature convergence. Such a check is implicit in the Bates and Watts (1988; 1991) and our (Johnson *at al*, 1981; Johnson and Frasier, 1985; Johnson and Faunt, 1992; Johnson, 1992; Straume *et al*, 1991) search methods for finding the confidence intervals of the estimated parameters. These methods search the variance space for a significant increase in the variance in an attempt to define the precision of the determined parameters. A premature convergence will be obvious since the search will encounter variances that are lower than the apparent minimum variance. Consequently, the use of these methods for confidence interval estimation can provide an excellent check for premature convergence of the fitting algorithm.

Data smoothing is included in this section because it is literally a pitfall that traps many investigators. The process of smoothing experimental data will remove most of the information within the data set that pertains to the magnitude of the uncertainties inherent in the experimental data. Thus, smoothing will preclude the evaluation of realistic estimates of the precision of the determined parameters. In addition, there is no data smoothing algorithm that can smooth the experimental data without simultaneously introducing some systematic error in the data that was not present in the original experimental data. The better smoothing algorithms will introduce less systematic error, but they all introduce some systematic error. Consequently, the smoothed experimental data contains less information than the original data. There is no statistically valid reason to smooth experimental data before it is analyzed. Sometimes an investigator may wish to smooth the data for presentation purposes, but even that is deceptive.

CONCLUSION

Computers are not oracles. Computer programmers, and programs, will sometimes make invalid assumptions and outright errors. Consequently, the investigator must always question if the results provided by a computer program make sense. The investigator should also devise realistic tests of any computer program to be sure that the program is working properly for the particular application. It is important that the investigator be aware of any assumptions of the data analysis method, and the particular computer program being used, to be sure that these assumptions are realistic for the particular data being analyzed.

ACKNOWLEDGEMENTS

This work was supported in part by the University of Virginia Diabetes Endocrinology Research Center Grant NIH DK-38942, the National Science Foundation Science and Technology Center for Biological Timing at the University of Virginia, the Center for Fluorescence Spectroscopy at the University of Maryland NIH RR-08119, and NIH grant GM-35154.

GLOSSARY OF SYMBOLS

α	Vector of parameter values being estimated
α'	Vector of parameter values that correspond to the minimum variance-of-fit
ASE	Asymptotic Standard Errors
B_i	First virial coefficient
CM	Covariance matrix
c_{obs}	Observed concentration
c_{true}	Actual concentration
$c_{r,i}$	Concentration of the ith macromolecular component at a radius of r
$c_{r,t}$	Total concentration at a radius of r
χ^2	Weighted Sum of Squared Residuals
δc	Zero offset of the optical system

ϵ	Vector of changes in the values of the α vector for the current iteration
F	F-statistic
$g_i\,(\alpha)$	Fitting function evaluated at the ith data point and the values of α
γ_i	Activity coefficient of the ith component
k	Autocorrelation lag
K_n	Association constant to form an n-mer from monomers
M_i	Molecular weight of the ith component
P	Matrix of partial derivatives
$P_i\,(\alpha)$	Probability of observing the ith data point given the values of α
$P(\alpha)$	Probability of observing all of the data points given the values of α
r	Radius of rotation in the centrifuge
r_0	Reference radius
R	Gas constant
R	Also, the expected number of runs of the residuals
R_i	ith residual
s^2	Sample variance
ρ	Density of the solution
$\hat{\rho}(k)$	Autocorrelation function
σ_i	Reduced Molecular Weight
σ_R	Standard deviation of the expected number of runs of the residuals
$\sigma^2\{\hat{\rho}(k)\}$	Variance of the autocorrelation function
ω	Angular velocity of the rotor
ν	Partial Specific Volume
ν	Number of degrees of freedom for use with F-statistic
WNLLS	Weighted Nonlinear Least-Squares
WSSR	Weighted Sum of Squared Residuals
Y^*	Vector of current weighted residuals

REFERENCES

Abbott RD and Gutgesell HP (1994): The effects of heteroscedasticity and skewness on prediction in regression: An example from modeling growth of the human heart. *Methods in Enzymology*, in press.

Acton FS (1959): Analysis of Straight Line Data. New York: John Wiley and Sons.

Bard Y (1974): Nonlinear Parameter Estimation. New York: Academic

Press.

Bates DM and Watts DG (1988): Nonlinear Regression Analysis and Its Applications. New York: John Wiley and Sons.

Bates DM and Watts DG (1991): Model building in chemistry using *t* and trace plots. *Chemometrics & Intelligent Lab. Systems* 10: 107.

Box GEP (1960): Fitting Empirical Data. *Ann. N.Y. Acad. Sci.* 86:792.

Bevington PR (1969): Data Reduction and Error Analysis for the Physical Sciences. New York: McGraw-Hill Book Company.

Draper NR and Smith H (1981): Applied Regression Analysis. New York: Academic Press.

Faddeeva VN (1959): Computational Methods of Linear Algebra. New York: Dover Publications.

Faunt LM and Johnson ML (1992): Analysis of discrete, time-sampled data using Fourier series methods. *Methods in Enzymology* 210: 340.

Johnson ML (1992): Why, when and how biochemists should use least squares. *Analytical Biochemistry* 206: 215.

Johnson ML, Correia JJ, Yphantis DA and Halvorson HR (1981): Analysis of data from the analytical ultracentrifuge by nonlinear least-squares techniques. *Biophysical Journal* 36: 575.

Johnson ML and Faunt LM (1992): Parameter estimation by least-squares methods. *Methods in Enzymology* 210: 1.

Johnson ML and Frasier SG (1985): Nonlinear least-squares analysis. *Methods in Enzymology* 117: 301.

Johnson ML, Turner BW and Ackers GK (1984): A quantitative model for the cooperative mechanism of human hemoglobin. *Proc. Natl Acad. Sci. USA* 81: 1093.

Knuth DE (1969): The art of computer programming, volume 2, seminumerical algorithms. Reading, Massachusetts: Addison-Wesley Publishing.

Straume M, Frasier-Cadoret SG and Johnson ML (1991): Least-squares analysis of fluorescence data. In: *Topics in Fluorescence Spectroscopy*, Lakowicz JR ed. New York: Plenum Press.

Straume M and Johnson ML (1992a): Monte Carlo method for determining complete probability distributions of estimated model parameters. *Methods in Enzymology* 210: 117.

Straume M and Johnson ML (1992b): Analysis of residuals: criteria for determining goodness-of-fit. *Methods in Enzymology* 210: 87.

THE OMEGA ANALYSIS AND THE CHARACTERIZATION OF SOLUTE SELF-ASSOCIATION BY SEDIMENTATION EQUILIBRIUM

Donald J. Winzor and Peter R. Wills

INTRODUCTION

For many years the analytical ultracentrifuge was considered to be the major source of information on the heterogeneity and molecular weight of macromolecules, a role that has now been taken over largely by gel electrophoretic and gel chromatographic techniques. The main use of analytical ultracentrifugation in protein chemistry is now the determination of equilibrium constants for macromolecular interactions. The traditional approach to ultracentrifugal analysis undoubtedly influenced initial attempts to employ the ultracentrifuge for characterizing solute self-association. Those analyses were based on the concentration-dependence of weight-average molecular weights derived from sedimentation equilibrium distributions (Adams and Williams, 1964; Roark and Yphantis, 1969; Hoagland and Teller, 1969). Although such practices still prevail (Bucci, 1986; Chatelier and Minton, 1987; Minton, 1990), more direct procedures have also been developed for analyzing the sedimentation equilibrium distribution, which is an experimental record of total solute concentration, $c(r)$, as a function of radial distance, r.

Two approaches are available for the characterization of solute self-association by direct analysis of the equilibrium concentration distribution. Whereas the omega analysis (Milthorpe *et al.*, 1975) employs the form of the experimental distribution to deduce the magnitude(s) of the self-association constant(s), the other approach (Johnson *et al.*, 1981) entails comparison of the experimental pattern with those generated by simulation for a series of possible models of the self-association phenomenon. Comparable characterization of an experimental system can be achieved by either method (Jacobsen and Winzor, 1992) because both procedures are based on direct application of the integrated sedimentation equilibrium expression. However, analysis based on the omega function is more convenient in that a simple polynomial fitting procedure allows all the complexities of thermodynamic nonideality to be taken into account in the evaluation of equilibrium constants for solute self-association.

THE OMEGA FUNCTION

The omega function, $\Omega(r)$, is an experimental parameter defined by the relationship (Milthorpe et al., 1975)

$$\Omega(r) = [c(r)/c(r_F)]\exp[\phi M_1(r_F^2 - r^2)] \tag{1a}$$

$$\phi = (1 - \bar{v}\rho_s)\omega^2/(2RT) \tag{1b}$$

in which $c(r)$ and $c(r_F)$ denote the respective total concentrations of macromolecular solute (g/liter) at radial distance r and reference radial position r_F. The exponent in Eq. (1a) is calculated on the basis of a molecular weight M_1 for self-associating monomeric component and a partial specific volume $\bar{v}$, which is considered to be the same for monomer and all oligomeric states. R is the universal gas constant and T the temperature in a sedimentation equilibrium experiment conducted at angular velocity ω: ρ_s is the solvent density (Wills and Winzor, 1992; Wills et al., 1993) for multi-component as well as single-solute systems. From the condition of sedimentation equilibrium for the distribution of the thermodynamic activity of monomer, $z_1(r)$, namely,

$$z_1(r) = z_1(r_F)\exp[\phi M_1(r^2 - r_F^2)] \tag{2}$$

it is evident that the exponential term in Eq. (1a) describes $z_1(r_F)/z_1(r)$, the ratio of thermodynamic activities of monomer at the respective radial positions. We may therefore rewrite Eq. (1a) as

$$\Omega(r) = [c(r)z_1(r_F)]/[c(r_F)z_1(r)] \tag{3}$$

For situations where the self-associating solute is the sole source of nonideality, $c(r) \rightarrow z_1(r)$ as $c(r) \rightarrow 0$, whereupon it follows that

$$\lim_{c(r) \rightarrow 0} \Omega(r) = \Omega_0 = z_1(r_F)/c(r_F) \tag{4}$$

The thermodynamic activity of monomer at the reference radial position, $z_1(r_F)$, is thus given by the ordinate intercept, Ω_o, of a plot of $\Omega(r)$ *versus* $c(r)$. Consequently, the thermodynamic activity of monomer, $z_1(r)$, throughout the concentration distribution may be calculated from Eq. (2).

From the viewpoint of elucidating and characterizing the self-association pattern, the ability of the omega analysis to provide the magnitude of the thermodynamic activity of monomer, $z_1(r)$, as a function of total solute concentration, $c(r)$, is a feature that places the method at a decided advantage over other analyses of sedimentation equilibrium distributions. Evaluation of this dependence of $z_1(r)$ upon $c(r)$ without any assumptions about the stoichiometry of the association phenomenon is a unique feature of the omega analysis: the magnitude of $\Omega(r)$ is dependent only upon knowledge of the product ϕM_1, which may be determined from the limiting slope, as $c(r) \to 0$, of a plot of $\ln c(r)$ *versus* r^2 derived from a sedimentation equilibrium experiment of meniscus-depletion design (Yphantis, 1964).

Clearly, the aim of the next step in the characterization of solute self-association is to make optimal use of this information that is afforded by the omega analysis — an endeavor that is still evolving despite the passage of almost two decades since the omega function was formulated. We now consider the progression of procedures that have been proposed to take advantage of the experimentally available dependence of $z_1(r)$ upon $c(r)$.

OMEGA ANALYSIS OF IDEAL SELF-ASSOCIATION

To illustrate developments in the use of the omega function for characterizing solute self-association, we begin with the simple situation in which solute self-association occurs under conditions approaching thermodynamic ideality. Self-association to a single polymeric state is considered first.

Ideal Two-State Self-Association

The standard procedure for evaluating $z_1(r)$ as a function of $c(r)$ follows the line of reasoning developed in the preceding section in that $\Omega(r)$ is extrapolated to zero solute concentration to obtain Ω_o and hence $z_1(r_F)$ via Eq. (4). This procedure is illustrated in Figure 1a, which summarizes results from five sedimentation equilibrium experiments on α-chymotrypsin in acetate-chloride buffer, pH 3.9, I 0.2 (Tellam *et al.*, 1979). From the indicated ordinate intercept (Ω_o) of 0.32 for a reference concentration, $c(r_F)$, of 1.87 g/liter, $z_1(r_F)$ is calculated to be 0.60 g/liter. Combination of this value with the reference radial

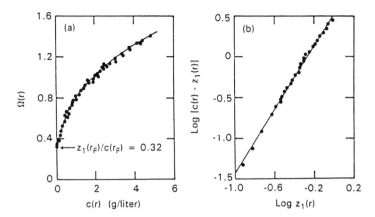

FIGURE 1. Analysis of sedimentation equilibrium experiments on α-chymotrypsin (pH 3.9, I 0.2) by means of the omega function. (a) Evaluation of the monomer activity, $z_1(r)$, in solutions with a reference concentration, $c(r_F)$ of 1.87 g/liter. (b) Test of the results for conformity with an ideal monomer-dimer equilibrium. Adapted from Tellam R, de Jersey J and Winzor DJ (1979) *Biochemistry* 18: 5316-5321 with permission of the American Chemical Society, Washington DC.

position (r_F) associated with $c(r_F)$ in each sedimentation equilibrium distribution then allows definition of the relationship between $z_1(r)$ and $c(r)$. Provided that thermodynamic ideality is assumed, $z_1(r)$ may be equated with the corresponding weight-concentration of monomer, $c_1(r)$, an approximation that allows the results to be tested for conformity with two-state self-association. From a plot of the results in accordance with the logarithmic form of the law of mass action for such a system, namely,

$$\log c_n = \log [c(r) - c_1(r)] = \log X_n^{app} - n \log c_1(r) \qquad (5)$$

the self-association of α-chymotrypsin under these conditions is described adequately by reversible dimerization ($n = 2$) and an association constant, $X_2^{app} = c_2(r)/c_1(r)^2$, of 3.5 liter/g (Figure 1b).

A drawback of such application of the omega analysis is the extent of reliance placed upon the accuracy of the extrapolation to obtain Ω_0 as the ordinate intercept in Figure 1a (Morris and Ralston, 1985). However, because the extrapolation is an obvious potential source of error, the usual practice has been to evaluate dependences of $z_1(r)$ upon $c(r)$ for a range of plausible values of $z_1(r_F) = \Omega_0 c(r_F)$, and to identify the best estimate as the value which leads to the smallest standard error in X_n^{app} (Shearwin and Winzor, 1988). A comparable procedure has been devised by Morris and Ralston (1985, 1989), who recommend nonlinear regression analysis of the entire $\Omega(r)$ *versus* $c(r)$ plot to guide the extrapolation and hence improve definition of the ordinate intercept.

A subsequent development has been the recent suggestion (Jacobsen and Winzor, 1992) that Eqs. (2) - (4) be combined to give

$$z_1(r) = \Omega_o c(r_F) \exp[\phi M_1(r^2 - r_F^2)] \; ; \tag{6}$$

and that the total solute concentration, $c(r)$, then be expanded in terms of individual species activities, $z_i(r)$, as

$$c(r) = z_1(r)/\gamma_1(r) + X_n\{z_1(r)\}^n/\gamma_n(r) \tag{7a}$$

where the thermodynamic activity of the polymeric state is expressed as the product of the association constant, X_n, and the monomeric activity raised to the appropriate power. Because $\gamma_1(r)$ and $\gamma_n(r)$, the respective activity coefficients of monomeric and polymeric species, are unity in the present context, the expression may be written as

$$c(r) = c_1(r) + X_n^{app}\{c_1(r)\}^n \tag{7b}$$

Combination of Eqs. (6) and (7) then gives

$$\Omega(r) = \Omega_o + X_n^{app}\Omega_o^n[c(r_F)\exp\{\phi M_1(r^2 - r_F^2)\}]^{n-1} \tag{8}$$

For the α-chymotrypsin system (in which $n = 2$) a plot of $\Omega(r)$ *versus* $c(r_F)\exp[\phi M_1(r^2 - r_F^2)]$ is thus linear (Figure 2a), thereby allowing Ω_o to be obtained by linear least-squares calculations. A value of 2.6 ($\pm$ 0.8) liter/g for X_2^{app} is then obtained by combining Ω_o and the slope $(X_2\Omega_o^2)$.

Although this procedure provides a more realistic estimate of the uncertainty in the self-association constant inasmuch as X_n and Ω_o are evaluated concurrently, this benefit is gained by forsaking the model-independence of the standard analysis. The appropriate value for n must therefore be determined either as the value for which the plot of results according to Eq. (8) is best described by a linear relationship (Jacobsen and Winzor, 1992), or as the value indicated by preliminary analysis according to the standard, model-independent omega method.

Ideal Multiple-State Self-Association

In circumstances where polymeric species larger than dimer are found, there is a distinct possibility that the self-association is not simply a two-state equilibrium between monomer and a single higher polymeric species. Under those conditions the omega analysis

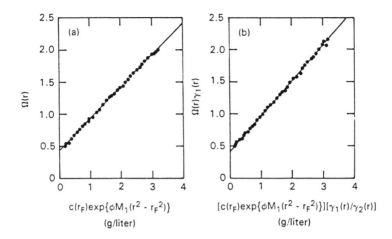

FIGURE 2. Determination of the dimerization constant for α-chymotrypsin (pH 3.9, I 0.2) by a modified form of omega analysis. (a) Analysis in accordance with Eq. (8) on the basis of ideal behavior. (b) Revised analysis after incorporating allowance for effects of thermodynamic nonideality via Eqs. (13) and (15). Adapted from Jacobsen MP and Winzor DJ (1992) *Biophys. Chem.* 45: 119-132 with permission of Elsevier Science Publishers, Amsterdam.

continues to provide the thermodynamic activity of monomer, $z_1(r)$, as a function of total concentration, $c(r)$, of solute. However, Eq. (7) ceases to describe the dependence of $z_1(r)$ upon $c(r)$, which now becomes a polynomial in $z_1(r)$, or $c_1(r)$ in an ideal system [Eq. (9)].

$$c(r) = z_1(r)/\gamma_1(r) + \sum_2^n X_i\{z_1(r)\}^i/\gamma_i \qquad (9a)$$

$$= c_1(r) + \sum_2^n X_i^{app}\{c(r)\}^i \qquad (9b)$$

$X_i = z_i(r)/\{z_1(r)\}^i$ (liter^{i-1}g^{1-i}) is the stoichiometric association constant for the formation of i-mer from i monomers. When all activity coefficients are effectively unity under conditions approaching thermodynamic ideality [Eq. (9b)], $X_i^{app} \simeq X_i$. For the alternative form of omega analysis considered above, the counterpart of Eq. (8) becomes (Jacobsen and Winzor, 1992)

$$\Omega(r) = \Omega_0 + X_2^{app}\Omega_0^2\{c(r_F)\exp[\phi M_1(r^2 - r_F^2)]\}$$

$$+ X_3^{app}\Omega_0^3\{c(r_F)\exp[\phi M_1(r^2 - r_F^2)]\}^2$$

$$+ X_4^{app}\Omega_0^4\{c(r_F)\exp[\phi M_1(r^2 - r_F^2)]\}^3 + \dots \qquad (10)$$

Although it is feasible, in principle, to extend the polynomial series in Eqs. (9) and (10) indefinitely, experience indicates a practical limit of two or three terms, beyond which the uncertainty in the value for X_i^{app} is likely to exceed its magnitude. This failure of the polynomial approach to define all but a few of the stoichiometric constants (X_i^{app}) with an acceptable degree of precision has prompted the introduction of models which place no restriction on the ultimate extent of self-association (upper limit of summation, n), but which necessarily impose restrictions on the relative magnitudes of equilibrium constants describing successive self-association steps (Van Holde and Rossetti, 1967; Adams et al., 1975; Garland and Christian, 1975; Tang et al., 1977; Beckerdite et al., 1980). In isodesmic models all successive self-association steps are governed by the same molar equilibrium constant, whereas in attenuated models the successive equilibrium constants decrease to allow for varying entropy changes at each step. For the simplest isodesmic model (Van Holde and Rossetti, 1967), involving indefinite self-association of monomer, the expression relating monomer concentration, $c_1(r)$, to the corresponding total solute concentration, $c(r)$, is

$$c(r) = c_1(r)/[1 - k_i^{app}c_1(r)] \; ; \; k_i^{app}c_1(r) < 1 \qquad (11)$$

where k_i^{app} is the apparent molar isodesmic association constant divided by M_1 to allow expression of the relationship in terms of weight-concentration. Combination of Eq. (11) with Eqs. (2) and (3) leads to the expression (Jacobsen and Winzor, 1992)

$$1/\Omega(r) = (1/\Omega_o) - k_i^{app}c(r_F)\exp[\phi M_1(r^2 - r_F^2)] \, , \qquad (12)$$

whereupon conformity with this simplest isodesmic model may be recognized experimentally by a linear dependence of $1/\Omega(r)$ upon $c(r_F)\exp[\phi M_1(r^2 - r_F^2)]$. Comparable analyses for other isodesmic models have also been suggested (Jacobsen and Winzor, 1992).

Models based on indefinite self-association are clearly open to criticism on the grounds that the major justification for their application is simply the availablity of a quantitative description of the $c_1(r)$ versus $c(r)$ dependence in closed form [e.g., Eq. (11)]. However, retention of all possible polymeric species can only be accommodated by imposing restrictions on the relative magnitudes of successive self-association constants. Consequently, resort to some such simplification of the relationship(s) between the magnitudes of successive equilibrium constants is mandatory if a reasonably definitive description of solute self-association involving a large number of polymeric states is to be obtained.

ALLOWANCE FOR THERMODYNAMIC NONIDEALITY

If it is unjustified to assume thermodynamic ideality, Eq. (9a) must be used for total solute concentration, which is no longer the sum of the thermodynamic activities of all species. From the consequent counterpart of Eq. (10), namely (Jacobsen and Winzor, 1992)

$$\Omega(r)\gamma_1(r) = \Omega_o + \sum_{2}^{n} X_i \Omega_o{}^i \{c(r_F)\exp[\phi M_1(r^2 - r_F{}^2)]\}^{i-1} \{\gamma_1(r)/\gamma_i(r)\}$$

(13)

it is evident that estimates of the activity coefficients of all states of solute are required for unequivocal estimates of the X_i to be obtained. Three approaches to this problem have been suggested.

Adams-Fujita Approximation

By far the most common method of allowing for thermodynamic nonideality in solute self-association uses the assumption (Adams and Fujita, 1963) that the composition-dependence of activity coefficients may be expressed as

$$\gamma_i(r) = \exp[iBM_1 c(r)]$$

(14)

where B is an empirical constant. This description of the composition-dependence of activity coefficients makes the apparent equilibrium constants equal to the true thermodynamic constants based on thermodynamic activities. For self-association of globular proteins, however, the thermodynamic and apparent equilibrium constants should differ in magnitude, whereupon their enforced identity becomes manifested as a dependence of B upon total solute concentration (Ogston and Winzor, 1975). Such a dependence clearly defeats the object of the Adams-Fujita approximation, the application of which relies upon constancy of B so that it may be determined as a curve-fitting parameter of fixed magnitude. We therefore prefer to calculate the various γ_i on a rigorous statistical-mechanical basis.

Statistical Mechanical Evaluation of Activity Coefficients

To illustrate the statistical-mechanical approach we consider, for simplicity, a solute for which self-association is restricted to reversible dimer formation and for which nonideality is described adequately by confining consideration to nearest-neighbor interactions. On the basis of spherical geometry for all species the expressions for the activity coefficients become (Wills and Winzor, 1992)

$$\gamma_1(r) = \exp\{\alpha_{11}c_1(r)/M_1 + \alpha_{12}[c(r) - c_1(r)]/M_2 + ...\} \quad (15a)$$

$$\gamma_2(r) = \exp\{\alpha_{22}[c(r) - c_1(r)]/M_2 + \alpha_{12}c_1(r)/M_1 + ...\} \quad (15b)$$

$$\alpha_{ij} \simeq 4\pi N(R_i + R_j)/3$$

$$+ Z_iZ_j(1 + \kappa R_i + \kappa R_j)/[2I(1 + \kappa R_i)(1 + \kappa R_j)] \quad (15c)$$

where $c_1(r)$ is the weight concentration of monomer associated with total solute concentration $c(r)$, and the charge-charge term in α_{ij} ($i,j \in \{1,2\}$) is defined in standard Debye-Hückel nomenclature: Z_1, Z_2 and R_1, R_2 are the respective net charges (valences) and radii of monomer and dimer with molecular weights M_1 and M_2; and the inverse screening length (κ) at 20°C may be calculated in units of cm^{-1} from the molar ionic strength (I) by the expression $\kappa = 3.27 \times 10^7 I^{1/2}$.

Because the evaluation of activity coefficients requires knowledge of $c_1(r)$, the concentration of monomer at radial distance r, the approach that has been adopted (Jacobsen and Winzor, 1992) is to assume thermodynamic ideality [$\gamma_1(r) = \gamma_2(r) = 1$] initially to allow first estimates of X_2 and hence $c_1(r)$ to be obtained for substitution into Eq. (15) to obtain $\gamma_1(r)$ and $\gamma_2(r)$ for each $c(r)$. Availability of these values then allows reassessment of the results in terms of Eq. (11) with $n = 2$ and hence iterative refinement of the values of Ω_0 and X_2. This procedure is illustrated in Figure 2b, where the experimental points have been obtained by such iterative application of Eqs. (11) and (15) to the results for α-chymotrypsin (Jacobsen and Winzor, 1992). An alternative means of incorporating composition-dependent activity coefficients into the characterization of solute self-association is to employ the original version of the omega analysis to obtain the $z_1(r)$ versus $c(r)$ relationship, from which the association equilibrium constant(s) may be determined by iterative, nonlinear curve-fitting in terms of Eqs. (10) and (15) (Wills et al., 1980).

An interesting outcome of the omega analysis arises in the event that nonideality of a reversibly dimerizing solute is dominated by the presence of a dialyzable solute such as sucrose (Shearwin and Winzor, 1988). Under conditions where the molar concentration of the small inert solute, c_M/M_M, greatly exceeds that of either state of self-associating solute, the expressions for the activity coefficients of monomer and dimer become

$$\gamma_1(r) \simeq \exp[\alpha_{1M}c_M(r)/M_M] \qquad (16a)$$

$$\gamma_2(r) \simeq \exp[\alpha_{2M}c_M(r)/M_M] \qquad (16b)$$

Because $c_M(r)$ is essentially invariant across the entire macromolecular distribution at sedimentation equilibrium, $\gamma_1(r) \simeq \gamma_1(r_F)$ and hence the counterpart of Eq. (4) becomes

$$\lim_{c(r)\to 0} \Omega(r) = c_1(r)/c(r_F) \qquad (17)$$

In other words, Ω_0 is now defining the concentration of monomer (rather than its thermodynamic activity) at the reference total concentration, $c(r_F)$. Consequently, the equilibrium constant evaluated by omega analysis under these circumstances is an apparent value that exhibits a dependence on small solute concentration governed by the relationship

$$X_2{}^{app} = X_2\exp[(2\alpha_{1M} - \alpha_{2M})c_M(r)/M_M + ...] \qquad (18)$$

Such dependence of the measured dimerization constant for α-chymotrypsin upon the concentration of sucrose included in the dialysis medium (pH 3.9, I 0.2) is illustrated in Figure 3, which also includes the theoretical dependence predicted by Eq. (18) on the basis of radii of 2.44 and 3.07 nm for monomeric and dimeric enzyme respectively (Shearwin and Winzor, 1988), the radius of sucrose, r_M, having been taken as 0.31 nm (Wills *et al.*, 1993).

Hill-Chen Approach

Whereas resort to the above statistical-mechanical approach leads to the expression of a species activity coefficient as a series expansion that requires specification of the concentrations of all solute states present, Hill and Chen (1973) have regarded solute self-association as merely another form of nonideality. Adoption of such a viewpoint allows the thermodynamic activity of monomer, $z_1(r)$, to be expressed as a series expansion in terms of the total concentration of the single component via Eq. (19).

$$z_1(r) = c(r)\exp\{2B_2c(r)/M_1 + (3/2)B_3[c(r)/M_1]^2 + ..\} \qquad (19)$$

In this expression the coefficients B_i are defined from the virial expansion for the osmotic pressure, Π, in terms of the base-molar concentration of macromolecular solute, namely,

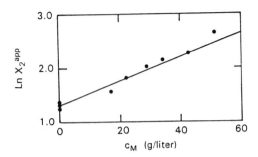

FIGURE 3. Effect of sucrose concentration on the dimerization constant evaluated for α-chymotrypsin (pH 3.9, I 0.2) by omega analysis: the line denotes the relationship predicted on the basis of Eqs. (15) and (18). Adapted from Shearwin KE and Winzor DJ (1988) *Biophys. Chem.* 31: 287-294 with permission of Elsevier Science Publishers, Amsterdam.

$$\Pi/RT = c(r)/M_1 + B_2[c(r)/M_1]^2 + B_3[c(r)/M_1]^3 + .. \qquad (20)$$

Although the use of Eq. (19) to describe nonideality in terms of total solute concentration is reminiscent of the Adams and Fujita (1963) approach [Eq. (14)], it differs therefrom in that the coefficients in Eq. (19) are now defined on rigorous statistical-mechanical grounds. Specifically,

$$B_2 = (\alpha_{11} - X_2M_1)/2 \qquad (21a)$$

$$B_3 = (2/3)\alpha_{111} - X_2M_1(2\alpha_{11} - \alpha_{12}) + (X_2M_1)^2 - (2/3)X_3M_1^2 \qquad (21b)$$

where α_{11} and α_{12} remain defined by Eq. (15) and α_{111} is the virial coefficient for the corresponding second-order term (Wills *et al.*, 1980).

The omega analysis is ideally suited to take advantage of the Hill and Chen approach because of its ability to provide the required relationship between between $z_1(r)$ and $c(r)$. Incorporation of Eq. (19) into the omega analysis is achieved by writing the equivalent expression for $z_1(r_F)$, the monomeric activity corresponding to the reference total concentration, to give

$$\Omega(r) = \exp\{2B_2[c(r_F) - c(r)]/M_1$$

$$+ (3/2)B_3[c(r_F)^2 - c(r)^2]/M_1^2 + ..\} \qquad (22)$$

In principle the magnitudes of the various B_i could be obtained by nonlinear regression analysis of $[\Omega(r), c(r)]$ results in accordance with Eq. (22) or its logarithmic counterpart.

Although the use of Eq. (19) has been illustrated by characterizing the self-association of lysozyme (Wills and Winzor, 1992), that analysis was made without full appreciation of the limitations implicit in its application to experimental data. This may be understood by considering the case of an ideal monomer-dimer system, for which all excluded volume terms are zero and Eq. (19) simplifies to

$$\ln [z_1(r)/c(r)] = -X_2 c(r) + (3/2)[X_2 c(r)]^2 - (10/3)[X_2 c(r)]^3 + ..$$

(23)

Evaluation of X_2 by direct polynomial fitting of $\ln [z_1(r)/c(r)]$ as a function of $c(r)$ would rely upon identity of the estimates of X_2 from each polynomial coefficient of Eq. (23). However, numerical examples have shown that fitting to a cubic, or even a quartic, polynomial is likely to be required in order to obtain a reliable estimate of the linear coefficient and hence X_2 for this simple self-associating system.

This problem can be overcome completely by using the equation which expresses the inverse relationship between $z_1(r)$ and $c(r)$, namely

$$c(r) = z_1(r)\{1 + 2b_2 z_1(r)/M_1 + 3b_3[z_1(r)/M_1]^2 + ..\} \qquad (24)$$

where the coefficients b_2, b_3, etc., come from the relationships [Eqs. 6 and 8 of Hill (1959)]

$$\Pi/RT = (z_1/M_1) + b_2(z_1/M_1)^2 + b_3(z_1/M_1)^3 + ... \qquad (25)$$

and

$$c = z_1 \partial(\Pi/RT)/\partial(z_1/M_1) \qquad (26)$$

These coefficients (b) in Eq. (24) are related to the more familiar osmotic (B) and nonideality (α) virial coefficients through the expressions

$$b_2 = -B_2 = (X_2/M_1 - \alpha_{11})/2 \qquad (27a)$$

$$b_3 = -B_3/2 + 2B_2^2$$

$$= -(1/3)\alpha_{111} + (1/2)(\alpha_{11}^2 - X_2 M_1 \alpha_{12}) + (1/3)X_3 M_1^2 \qquad (27b)$$

For omega analysis Eqs. (3),(4) and (24) are combined to obtain

$$\Omega(r) = \Omega_o + 2b_2\Omega_o^2[c(r)/\Omega(r)] + 3b_3\Omega_o^3[c(r)/\Omega(r)]^2 + \ldots \quad (28)$$

On noting that Eq. (1) provides an alternative experimental expression for the ratio $[c(r)/\Omega(r)]$, values of Ω_o, b_2, b_3, etc., are, in principle, obtainable by regarding $\Omega(r)$ as a polynomial expansion in $c(r_F)\exp[\phi M_1(r^2 - r_F^2)]$.

The suggested dependence of $\Omega(r)$ upon $c(r_F)\exp[\phi M_1(r^2 - r_F^2)]$ has already been recommended (Jacobsen and Winzor, 1992) as a means of obtaining the initial estimate of X_2^{app} for a nonideal monomer-dimer system (Fig. 2a) required in a subsequent iterative allowance for composition-dependence of activity coefficients to obtain X_2 (Fig. 2b). We now see that a far more direct procedure is to regard $\Omega(r)$ as a polynomial (rather than approximately linear) function of $c(r_F)\exp[\phi M_1(r^2 - r_F^2)]$. Resort to this method yields values of 2.9 ($\pm$ 0.1) and 3.5 ($\pm$ 0.9) liter/g for X_2 from the linear ($2b_2$) and quadratic ($3b_3$) coefficients of Eq. (23). Although preliminary, these findings are sufficiently encouraging to suggest a bright future for the quantitative characterization of solute self-association by this combination of the omega analysis and the Hill-Chen approach.

CONCLUDING REMARKS

In this review we have drawn attention to the potential of the omega function (Milthorpe et al., 1975) as a means of characterizing solute self-association. Its major virtue is undoubtedly the access afforded to the magnitude of the thermodynamic activity of monomer, $z_1(r)$ in a solution with known total concentration, $c(r)$. To our knowledge no other current method of analyzing sedimentation equilibrium distributions has this ability to define $z_1(r)$ as a function of $c(r)$. The sparcity of investigations to have taken advantage of this information (unique to the omega analysis) can, we feel, be traced to a general reluctance of researchers to grapple realistically with the problem of allowance for effects of thermodynamic nonideality.

The second endeavor of this review has been to illustrate ways in which activity coefficients based on the statistical mechanical interpretation of virial coefficients may be incorporated into the omega analysis to allow evaluation of the thermodynamic constant(s) describing solute self-association. The most thoroughly explored statistical-mechanical approach (Wills et al., 1980; Jacobsen and Winzor, 1992) requires the specification of composition-dependent activity coefficients [Eq. (15)]. However, as we have now shown, this need not entail an iterative analysis that is undeniably tedious. The alternative method (Hill and Chen, 1973) that seems to have languished unheeded for two decades can be adapted to allow the estimation of thermodynamic association constants by fitting a simple polynomial to the dependence of $\Omega(r)$ upon $c(r_F)\exp[\phi M_1(r^2 - r_F^2]$ and interpreting the coefficients in terms of Eq. (28).

Finally, we hope that this review will convince a wider audience of the unique power of the omega function for the characterization of solute self-association by sedimentation equilibrium; and thereby end the situation wherein the omega analysis is regarded as an oddity of the southern Pacific region (Minton, 1990).

ACKNOWLEDGEMENT

Financial support of this investigation by the Australian Research Council is gratefully acknowledged.

GLOSSARY OF SYMBOLS

B_i classical osmotic virial coefficient
b_i alternative osmotic virial coefficient (Hill)
$c(r)$ total weight-concentration of solute at radial distance r
$c_i(r)$ weight-concentration of species i at radial distance r
I ionic strength
k_i^{app} apparent isodesmic association constant
M_i molecular weight of species i
n stoichiometry of reversible solute self-association
R gas constant
R_i effective thermodynamic radius of species i
r radial distance
r_F reference radial distance in omega analysis
T temperature
$\bar{v}$ partial specific volume
X_i association constant (liter^{i-1}g^{1-i}) for formation of species i
Z_i net charge (valence) of species i
$z_i(r)$ thermodynamic activity of species i at radial distance r
α_{ij} nonideality virial coefficient for species i and j
$\gamma_i(r)$ activity coefficient of species i at radial distance r
κ inverse screening length
Π osmotic pressure
ρ density of solution
ρ_s density of solvent
$\Omega(r)$ omega function as defined by Eq. (1)
Ω_o limiting value of $\Omega(r)$ as $c(r) \to 0$
ω angular velocity

REFERENCES

Adams ET and Fujita H (1963): In *Ultracentrifugal Analysis in Theory and Experiment*, Williams JW, ed., pp. 119-128. New York: Academic Press.

Adams ET and Williams JW (1964): *J Am Chem Soc* 3454-3461.

Adams ET, Ferguson WE, Wan PJ, Sarquis JL and Escott BM (1975): *Sep Sci* 10: 175-244.

Beckerdite JM, Wan CC and Adams ET (1980) *Biophys Chem* 12: 199-214.

Bucci E (1986): *Biophys Chem* 24: 47-52.

Chatelier RC and Minton AP (1987): *Biopolymers* 26: 507-524.

Garland F and Christian SD (1975): *J Phys Chem* 79: 1247-1252.

Hill TL (1959): *J Chem Phys* 30: 93-97.

Hill TL and Chen YD (1973): *Biopolymers* 12: 1285-1312.

Hoagland VD and Teller DC (1969): *Biochemistry* 8: 594-602.

Jacobsen MP and Winzor DJ (1992): *Biophys Chem* 45: 119-132.

Johnson ML, Correia JJ, Yphantis DA and Halvorson HR (1981): *Biophys J* 36: 575-588.

Milthorpe BJ, Jeffrey PD and Nichol LW (1975): *Biophys Chem* 3: 169-176.

Minton AP (1990): *Anal Biochem* 190: 1-6.

Morris M and Ralston GB (1985): *Biophys Chem* 23: 49-61

Morris M and Ralston GB (1989): *Biochemistry* 28: 8561-8567.

Ogston AG and Winzor DJ (1975): *J Phys Chem* 79: 2496-2500.

Ralston GB and Morris MB (1992): In *Analytical Ultracentrifugation in Biochemistry and Polymer Science*, Harding SE, Rowe AJ and Horton JC, eds., pp. 253-274. Cambridge: R. Soc. Chem.

Roark DE and Yphantis DA (1969): *Ann NY Acad Sci* 164: 245-278.

Shearwin KE and Winzor DJ (1988): *Biophys Chem* 31: 287-294.

Tang LH, Powell DR, Escott BM and Adams ET (1977): *Biophys Chem* 7: 121-139.

Tellam R, de Jersey J and Winzor DJ (1979): *Biochemistry* 24: 5316-5321.

Van Holde KE and Rossetti GP (1967): *Biochemistry* 6: 2189-2194.

Wills PR and Winzor DJ (1992): In *Analytical Ultracentrifugation in Biochemistry and Polymer Science*, Harding SE, Rowe AJ and Horton JC, eds., pp. 311-330. Cambridge: R. Soc. Chem.

Wills PR, Nichol LW and Siezen RJ (1980): *Biophys Chem* 11: 71-82.

Wills PR, Comper WD and Winzor DJ (1993): *Arch Biochem Biophys* 300: 206-212.

Yphantis DA (1964): *Biochemistry* 3: 297-317.

CONSERVATION OF SIGNAL: A NEW ALGORITHM FOR THE ELIMINATION OF THE REFERENCE CONCENTRATION AS AN INDEPENDENTLY VARIABLE PARAMETER IN THE ANALYSIS OF SEDIMENTATION EQUILIBRIUM

Allen P. Minton

INTRODUCTION

The equation describing the radial distribution of a single solute species at sedimentation equilibrium in an ideal solution (Cantor & Schimmel, 1980) may be written in integrated form as

$$w(r) = w(r_{ref}) \times \exp\left[\frac{M^*\omega^2}{2RT}(r^2 - r_{ref}^2)\right] \qquad (1)$$

where w(r) denotes the weight/volume concentration of the solute at radial position r, ω the angular velocity of the rotor, R the molar gas constant, T the absolute temperature, and r_{ref} an arbitrarily selected reference position. M^* denotes the buoyant molecular weight of solute, defined as

$$M^* \equiv M(1-\overline{v}\rho)$$

where M and $\overline{v}$ respectively denote the molecular weight and partial specific volume of solute, and ρ denotes the density of solvent. The results of a sedimentation equilibrium experiment are ordinarily obtained as an experimental dependence of w (or, more properly, some measurable quantity that is proportional to w) upon r. Such data are customarily analyzed by fitting equation (1) to the data by the method of nonlinear least-squares (Johnson & Faunt, 1992) in order to obtain best-fit values of M^* and $w(r_{ref})$. Note that only the best-fit value of M^* is sought by the investigator; the presence of a second undetermined variable (the reference concentration) may, under certain circumstances, significantly reduce the precision with which the value of M^* can be determined by least-squares fitting of the data. It has been pointed out on several occasions (Nichol & Ogston, 1965; Lewis, 1991; Hsu & Minton, 1991) that if $w(r)$ is known over the entire length of the solution column, then the condition of conservation of mass may be utilized, together with the known loading concentration of solute, to eliminate the reference concentration $w(r_{ref})$ as an independently variable parameter. However, one may not always be able to obtain reliable data for $w(r)$ over the entire length of the solution column, and in some experiments (particularly those involving unstable macromolecules) conservation of mass may not obtain over the duration of the experiment.

The purpose of the present communication is to demonstrate another technique for eliminating the reference concentration as an independently variable parameter. This technique does not depend upon either conservation of mass or the acquisition of data spanning the entire solution column. I call this technique "conservation of signal" to distinguish it from conservation of mass. It may be used in ideal solutions of associating as well as non-associating solutes, providing that the partial specific volume of an associating solute is independent of the state of association (i.e., equilibrium association constants are independent of position within the centrifuge cell). This seems to be a realistic assumption for most sedimentation equilibrium studies of protein associations carried out at low to moderate angular velocity (Hsu & Minton, 1991).

The remainder of this communication will be divided into three sections. In the first, the conservation of signal algorithm is applied to the simple case, referred to above, of a single non-associating solute. Next the algorithm is generalized to treat a single self-associating solute, and in the last of the three sections the algorithm is further generalized to treat the case of two solute components that may either self or hetero-associate.

SINGLE NONASSOCIATING SOLUTE

Let the solution property that is measured as a function of radial position at sedimentation equilibrium be called the *signal*, and the magnitude of the signal be denoted by $S(r)$. The signal may be absorbance, activity of radiolabel, or any other quantity such that $S(r)$

varies linearly with the mass concentration of solute molecules at radial position r. Allowance is also made for the presence of experimental artifacts that lead to a background level of signal, denoted by δ, in the absence of solute. Hence

$$S(r) = \alpha w(r) + \delta \qquad (2)$$

The values of α and δ may be independently determined via appropriate calibration of technique and instrumentation. Let us suppose that data on the dependence of S upon r has been obtained for all radial positions between two arbitrarily selected radial positions r_a and r_b, such that $r_a < r_b$, $r_a \geq r_{meniscus}$, and $r_b \leq r_{base}$. For a sector-shaped cell, the condition of conservation of signal is expressed as

$$I_g \equiv \int_{r_a}^{r_b} S(r) \ r \ dr = \alpha \int_{r_a}^{r_b} w(r) \ r \ dr + \delta \int_{r_a}^{r_b} r \ dr \qquad (3)$$

I_S is evaluated from the experimental data by numerical quadrature of the left-hand side of equation (3). For convenience, r_{ref} is selected to be equal to r_a. Then combination of equations (1) and (3) yields

$$J \ w(r_a) + \delta (r_b^2 - r_a^2)/2 - I_S = 0 \qquad (4)$$

where

$$J \equiv \alpha \left\{ \frac{\exp[\phi(r_b^2 - r_a^2)] - 1}{2\phi} \right\}$$

and

$$\phi \equiv \frac{M^* \omega^2}{2RT}$$

The experimenter knows, or can determine independently, the values of all variables appearing in equation (4) except M^* and $w(r_a)$. Hence for any trial value of M^* employed in the least-squares procedure for fitting equation (1) to the data, equation (4) may be used to calculate the unique value of $w(r_a)$ that satisfies the condition of signal conservation; $w(r_a)$ is no longer an independently variable fitting parameter.

SINGLE SELF-ASSOCIATING SOLUTE COMPONENT

Consider an ideal solution of a solute *component A* that may exist as a monomeric *species* a_1, and self-associate to form the oligomeric *species* denoted by a_2, a_3, and so forth. It is assumed, as stated above, that the partial specific volume of the component A is independent of the state of association. It follows that when sedimentation and chemical equilibrium are simultaneously satisfied at all points within the centrifuge cell,

$$w_i(r) = w_i(r_{ref}) \times \exp\left[\frac{i M_1^* \omega^2}{2RT}(r^2 - r_{ref}^2)\right] \qquad (5)$$

for all i, and

$$w_i(r) = K_i \, w_1(r)^i \qquad (6)$$

where w_i is the w/v concentration of a_i, and K_i is the equilibrium association constant for the formation of a_i from monomeric A (which, by virtue of the above assumption, is independent of radial position). It is assumed that the magnitude of the measured signal associated with each oligomeric species is proportional to the w/v concentration of the species, i.e.,

$$S_i(r) = \alpha \, w_i(r) \qquad (7)$$

If the signal is optical absorbance, then the above assumption is equivalent to assuming the absence of significant hyper- or hypochromicity upon selfassociation. The validity of this assumption would, of course, have to be checked by the experimenter prior to application of the present analysis. By analogy to equation (2), the total signal is then

$$S(r) = \sum_i S_i(r) + \delta = \alpha \sum_i w_i(r) + \delta$$

$$= \alpha \sum_i \left\{ K_i \, w_1(r_{ref})^i \times \exp\left[\frac{i \, M_1^{*} \omega^2}{2RT} (r^2 - r_{ref}^2) \right] \right\}$$

$$+ \delta \qquad (8)$$

Consider the set of all data acquired between the radial positions r_a and r_b. For a sector-shaped cell, the condition of signal conservation is expressed as

$$I_S \equiv \int_{r_a}^{r_b} S(r) \, r \, dr = \alpha \sum_i \int_{r_a}^{r_b} w_i(r) \, r \, dr + \delta \int_{r_a}^{r_b} r \, dr \qquad (9)$$

Combining equations (5), (6), and (9), one obtains

$$\sum_i J_i \, w_1(r_a)^i + \delta (r_b^2 - r_a^2)/2 - I_S = 0 \qquad (10)$$

where

$$J_i \equiv \alpha \left\{ \frac{\exp\left[i\phi_1\left(r_b{}^2 - r_a{}^2\right)\right] - 1}{2\phi_1} \right\} \times K_i$$

and

$$\phi_1 \equiv \frac{M_1{}^* \omega^2}{2RT}$$

The experimenter knows, or can determine independently, the values of all variables appearing in equation (10) except for $w_1(r_a)$, $M_1{}^*$, and the K_i. The objective of the analysis is to obtain best-fit values of $M_1{}^*$ and the K_i for a given association scheme by fitting equation (8) to the data. For any given set of trial values of $M_1{}^*$ and the K_i, equation (10) may be solved for the unique value of $w_1(r_a)$ that satisfies the condition of signal conservation. Hence the reference concentration has been eliminated as an independently variable fitting parameter.

TWO SELF- AND/OR HETERO-ASSOCIATING SOLUTE COMPONENTS

Consider an ideal solution containing two solute *components*, A and B, that may self- and/or hetero-associate to form molecular *species* denoted by a, a_2, a_3, ..., b, b_2, b_3, ..., ab, a_2b, ab_2, a_2b_2, and so forth. Each species is denoted by the general form a_ib_j, where at least one of the two subscripts must be nonzero, and a property of a given species is denoted by subscripting the property variable with the corresponding values of i and j. For example, w_{10}, w_{01}, and w_{11} respectively denote the w/v concentrations of the species a, b, and ab. It is assumed, as before, that the partial specific volumes of components A and B are independent of the state of association of each component. It follows that when the conditions of sedimentation and chemical equilibrium are simultaneously satisfied at all radial positions in the centrifuge cell,

$$w_{ij}(r) = w_{ij}(r_{ref}) \times \exp\left[\phi_{ij}\left(r^2 - r_{ref}^2\right)\right] \tag{11}$$

where

$$\phi_{ij} \equiv \frac{(i M_{10}{}^* + j M_{01}{}^*)\,\omega^2}{2RT}$$

$$M_{10}^* = M_{10}\,(1 - \overline{v_A}\,\rho)$$

$$M_{01}^* = M_{01}\,(1 - \overline{v_B}\,\rho)$$

and

$$w_{ij}(r) = K_{ij}\,w_{10}(r)^i\,w_{01}(r)^j \qquad (12)$$

where K_{ij} is the (position independent) equilibrium association constant for the formation of $a_i b_j$ from monomeric A and B. By definition, $K_{00} = 0$, and $K_{10} = K_{01} = 1$.

Let us postulate the existence of two distinct and independent experimental signals, denoted by S^I and S^{II}, that vary linearly with the total concentrations of one or both components:

$$S^I(r) = \alpha_A^I w_A(r) + \alpha_B^I w_B(r) + \delta^I \qquad (13a)$$

$$S^{II}(r) = \alpha_A^{II} w_A(r) + \alpha_B^{II} w_B(r) + \delta^{II} \qquad (13b)$$

where $w_A(r)$ and $w_B(r)$ respectively denote the total w/v concentrations of components A and B at radial position r, related to the concentrations of individual species by:

$$w_A(r) = \sum_i \sum_j f_{A,ij} \, w_{ij}(r) \qquad (14a)$$

$$w_B(r) = \sum_i \sum_j f_{B,ij} \, w_{ij}(r) \qquad (14b)$$

and $f_{A,ij}$ and $f_{B,ij}$ denote the mass fractions of components A and B respectively in species $a_i b_j$:

$$f_{A,ij} \equiv \frac{i \, M_{10}}{i \, M_{10} + j \, M_{01}}$$

$$f_{B,ij} \equiv \frac{j \, M_{01}}{i \, M_{10} + j \, M_{01}}$$

The qualification that S^I and S^{II} be "distinct and independent" is equivalent to the requirement that

$$\begin{vmatrix} \alpha_A^I & \alpha_B^I \\ \alpha_A^{II} & \alpha_B^{II} \end{vmatrix} \neq 0$$

Examples of such signals include the following: (i) S^I and S^{II} could be absorbance at two different wavelengths of light. In this case α_A^I would correspond to the product of the molar extinction coefficient of component A at wavelength I and the length of the light path. The assumption that hyper- and hypochromicity are absent is implicit in the formulation of equations (13ab). (ii) Components A and B could each be labeled with different radioisotopes having distinct energy emission spectra. Then S^I and S^{II} may be defined as the total radioactivity of sample in each of two different energy windows on a scintillation counter (Attri & Minton, 1987), and α_A^I (for example) would be proportional to the molar radioactivity of the isotope labeling A in energy window I. As before, δ^I and δ^{II} represent baseline levels of signals I and II. (iii) A combination of two different types of signals might be employed. For example, A might be radiolabeled with an isotope, and S^I would be radioactivity (in which case $\alpha_B{}^I = 0$), and S^{II} would be absorbance, with contributions from both A and B. (iv) Other techniques utilizing more than two signals (Hofrichter & Henry, 1992; Lewis et al, this volume) may be used to determine $w_A(r)$ and $w_B(r)$ more precisely than would be the case with only two measurable signals. In this event, S^I and S^{II} can be defined equal to $w_A(r)$ and $w_B(r)$ respectively. The formalism to follow still holds, with $\alpha_A{}^I = \alpha_B{}^{II} = 1$, and $\alpha_A{}^{II} = \alpha_B{}^I = \delta^I = \delta^{II} = 0$.

For a sector-shaped cell, the condition of conservation of signals is then expressed as the dual relations

$$I_{S^I} \equiv \int_{r_a}^{r_b} S^I(r)\, r\, dr = \alpha_A^I \int_{r_a}^{r_b} w_A(r)\, r\, dr$$

$$+\ \alpha_B^I \int_{r_a}^{r_b} w_B(r)\, r\, dr + \delta^I \int_{r_a}^{r_b} r\, dr \qquad (15a)$$

$$I_{S^{II}} \equiv \int_{r_a}^{r_b} S^{II}(r)\, r\, dr = \alpha_A^{II} \int_{r_a}^{r_b} w_A(r)\, r\, dr$$

$$+\ \alpha_B^{II} \int_{r_a}^{r_b} w_B(r)\, r\, dr + \delta^{II} \int_{r_a}^{r_b} r\, dr \qquad (15b)$$

As before, the values of I_{S^I} and $I_{S^{II}}$ may be evaluated from the experimental data, via the expressions on the left-hand sides of equations(15a) and (15b) respectively. Then combination of equations (11-15) yields

$$\sum_i \sum_j J_{ij}^I \, w_{10} (r_a)^i \, w_{01} (r_a)^j + \delta^I (r_b^2 - r_a^2)/2$$

$$- I_{S^I} = 0 \qquad\qquad (16a)$$

$$\sum_i \sum_j J_{ij}^{II} \, w_{10} (r_a)^i \, w_{01} (r_a)^j + \delta^{II} (r_b^2 - r_a^2)/2$$

$$- I_{S^{II}} = 0 \qquad\qquad (16b)$$

where

$$J_{ij}^X \equiv (i\alpha_A^X + j\alpha_B^X) \left\{ \frac{\exp [\phi_{ij} (r_b^2 - r_a^2)] - 1}{2\phi_{ij}} \right\} K_{ij}$$

(X = I or II). The experimenter knows, or can determine independently, the values of all of the parameters appearing in equations (11)-(16), except M_{10}, M_{01}, $w_{10}(r_a)$, $w_{01}(r_a)$, and the K_{ij}. The objective of the analysis is to obtain best-fit values of M_{10}, M_{01}, and the K_{ij} for a given association scheme by simultaneously fitting equations (13a) and (13b) to the data. For any given trial values of M_{10}, M_{01}, and the K_{ij}, equations (16a) and (16b) may be solved for the unique values of $w_{10}(r_a)$ and $w_{01}(r_a)$ satisfying the condition of conservation of signals. Hence $w_{10}(r_a)$ and $w_{01}(r_a)$ have been eliminated as independently variable fitting parameters.

DISCUSSION

One of the factors leading to uncertainty in the values of parameters obtained by least-squares fitting of model functions to data is correlation between the values of independently variable parameters. It is well known that, depending upon the structure of a model and the associated data set (including experimental error), it may be possible for the dependence of the sum of squared residuals or χ^2 upon the value of one parameter, P_i, to be compensated to a greater or lesser extent by changes in the value of one or more other parameters, P_j, P_k, ... (Hsu & Minton, 1991; Hensley et al, this volume). When this is the case, the dependence of χ^2 upon the values of each of these parameters is lessened, and the investigator's ability to define independent best-fit values of each of the parameters via least-squares fitting is compromised. Hence reduction in the number of independently variable fitting parameters can significantly reduce uncertainty in the best-fit values of remaining parameters.

As an example, consider the simple associating system A + B ⇌ AB. Let us assume that the buoyant molecular weights of the individual components, M_{10}^* and M_{01}^*, have been previously determined in sedimentation equilibrium experiments on solutions of the individual components. In the absence of constraints such as conservation of mass or conservation of signal, analysis of sedimentation equilibrium experiments on solution mixtures of A and B utilizing equations (11)-(14) would require the existence of three independently variable parameters, $w_{10}(r_{ref})$, $w_{01}(r_{ref})$, and K_{11}. While it may or may not be possible to utilize conservation of mass as a constraint, conservation of signal may almost always be utilized, particularly when non-optical signals are measured. If only a single signal is measured as a function of radial distance, then equation (16a) may be used to eliminate one of the two reference concentrations as an independently variable parameter, and if two distinct and independent signals are measured as a function of radial distance, then equations (16a) and (16b) may be solved simultaneously to yield both $w_{10}(r_{ref})$ and $w_{01}(r_{ref})$ as functions of K_{11}, leaving only a single parameter to be determined by least-squares fitting.

Elimination of the reference concentration of solute as an independent fitting parameter via numeric solution of equation (4) has been implemented in the data analysis program XLAEQ, and elimination of the reference concentration of monomeric solute as an independent fitting parameter via numeric solution of equation (10) has been implemented in the data analysis program EQASSOC. These programs are distributed by Beckman Instruments to purchasers of the XLA analytical ultracentrifuge.

GLOSSARY OF SYMBOLS

w	weight/volume concentration
r	radial position
r_{ref}	reference radial position
M^*	buoyant molecular weight
ω	angular velocity of rotor
R	molar gas constant
T	absolute temperature
M	molecular weight
$\bar{v}$	partial specific volume
ρ	solvent density
α	constant of proportionality between concentration and magnitude of signal
δ	baseline level of signal
I_S	integrated signal
K	equilibrium association constant
S	magnitude of measured signal
f	mass fraction

REFERENCES

Attri AK and Minton AP (1987): Simultaneous determination of the individual concentration gradients of two solute species in a centrifuged mixture. *Anal. Biochem.* 162: 409-419.

Cantor CR and Schimmel PR (1980): *Biophysical Chemistry*. San Francisco, Academic Press. Sec. 11-3.

Henry ER and Hofrichter J (1992): Singular value decomposition: application to analysis of experimental data. *Meth. Enzymol.* 210: 129-192.

Hsu CS and Minton AP (1991): A strategy for efficient characterization of macromolecular heteroassociations via measurement of sedimentation equilibrium. *J. Mol. Recognition* 4: 93-104.

Johnson ML and Faunt LM (1992): Parameter estimation by least squares methods. *Meth. Enzymol.* 210: 1-37.

Lewis MS (1991): Ultracentrifugal analysis of mixed associations. *Biochemistry* 30: 11716-11719.

Nichol LW and Ogston AG (1965): Sedimentation equilibrium in reacting systems of the type mA + nB ⇌ C. *J. Phys. Chem.* 69: 4365-4367.

ANALYSIS OF PROTEIN–NUCLEIC ACID AND PROTEIN–PROTEIN INTERACTIONS USING MULTI-WAVELENGTH SCANS FROM THE XL–A ANALYTICAL ULTRACENTRIFUGE.

Marc S. Lewis

Richard I. Shrager

Soon-Jong Kim

INTRODUCTION

Protein–nucleic acid and protein–protein interactions have been studied by a variety of biochemical and physicochemical techniques, each of which has distinctive advantages and disadvantages. In particular, analytical ultracentrifugation has the advantages that it is rigorously based upon reversible thermodynamics, and the reactants and product or products of an interaction each have uniquely defined concentration gradients that can be resolved by appropriate mathematical analysis to give the desired values of the natural logarithms of the equilibrium constants with a minimum of assumptions.

Prior to the development of the Beckman XL–A analytical ultracentrifuge, the ultracentrifugal analysis of protein–protein and protein–nucleic acid interactions was limited by the inherent deficiencies of the absorption optical system of most Beckman Model E analytical ultracentrifuges. The need for wide monochromator slit openings to pass adequate light for the photomultiplier tube gave mediocre monochromaticity. Limitation to wavelengths above 260 nm created a potential problem with hyper–or hypochromism of the protein–nucleic acid complex that could limit the accuracy of the results of the analysis unless studies were undertaken to evaluate the extent of change in the absorbance of the complex. The practical limit of about one absorbance unit as a maximum for solute absorbances created difficulties in the analysis of some systems. Optimal analysis of the data from either the absorption optical system or the refractometric optical system required a complex analytical approach such as the method of implicit constrains, utilizing conservation of mass within the ultracentrifuge cell to limit values possible for parameters and requiring verification that this condition was met (Lewis, *et al.*, 1991, 1992).

The new Beckman XL–A analytical ultracentrifuge has significantly enhanced operational wavelength range (200–800 nm), monochromaticity ($\pm$ 2 nm bandpass), and absorbance range (0–3) as specified by the manufacturer. The analytical method to be described here utilizes these properties, particularly those of wavelength range and monochromaticity, to obtain results with enhanced accuracy without the experimentally stringent requirement of conservation of mass in the ultracentrifuge cell. A new method for the utilization of these attributes has been developed for protein–nucleic acid interactions (Lewis, *et al.*, 1994). This chapter enlarges upon this method and extends it to the study of protein–protein interactions as well.

THEORY

The method of analysis to be described here requires that the two reactants have significantly different absorption spectra over an experimentally appropriate range of wavelengths. The practical aspects of selecting this wavelength range and the appropriate concentrations of reactants will be discussed later. Let us begin by considering the molar concentrations of reactants and product at a single radial position, r, within the solution column in the ultracentrifuge cell, and for this discussion, denote these reactants as α and β. The total molar concentration of α will then be the sum of the concentrations of the uncomplexed α and of $\alpha\beta$, the complex; the total molar concentration of β will likewise be the sum of the concentrations of the uncomplexed β and of the complex; the concentration of the $\alpha\beta$ complex is thus the product of the concentrations of α and β and the equilibrium constant for the formation of the complex:

(1) $C_{\alpha, T, r} = C_{\alpha, r} + C_{\alpha\beta, r} = C_{\alpha, r} + K_{\alpha\beta} C_{\alpha, r} C_{\beta, r}$

(2) $C_{\beta, T, r} = C_{\beta, r} + C_{\alpha\beta, r} = C_{\beta, r} + K_{\alpha\beta} C_{\alpha, r} C_{\beta, r}$

If the Beer–Lambert law is obeyed and there are no contributions from light scattering or fluorescence at a particular wavelength and there are no changes in scattering and fluorescence, then the absorbance of a solution at that particular wavelength is the product of the molar concentration and the extinction coefficient of the solute at that wavelength. Thus, the absorbances, A (in AU), of α, β, and the $\alpha\beta$ complex at the wavelength λ are given by:

(3) $A_{\alpha, r, \lambda} = C_{\alpha, r} E_{\alpha\lambda}$

(4) $A_{\beta, r, \lambda} = C_{\beta, r} E_{\beta, \lambda}$

(5) $A_{\alpha\beta, r, \lambda} = C_{\alpha\beta, r} E_{\alpha\beta, \lambda} = C_{\alpha\beta, r} (E_{\alpha, \lambda} + E_{\beta, \lambda})$

Equation 5 explicitly assumes that hyper- or hypochromism is not significant at the wavelengths being observed.

The optical system of the ultracentrifuge only permits the observation of the total absorbance as a function of radial position at a particular wavelength, which is given by:

(6) $A_{T, r, \lambda} = A_{\alpha, r, \lambda} + A_{\beta, r, \lambda} + A_{\alpha\beta, r, \lambda}$

and substituting equations 3, 4, and 5 in equation 6 and rearranging then gives:

(7) $A_{T, r, \lambda} = E_{\alpha, \lambda} C_{\alpha, T, r} + E_{\beta, \lambda} C_{\beta, T, r}$

Consider now the case of measuring the absorbance at several wavelengths, but still at only one radial position. We now have A at wavelengths λ_i, $i = 1$ to m, and equation 7 becomes a set of m equations in the two unknowns $C_{\alpha, T, r}$

and $C_{\beta, T, r}$. The m A's form a row vector A, the E's form a 2 x m matrix of coefficients, **E**, and the C's form a 2 element row vector, **C**, such that $A = CE$. The least–squares solution to this set of equations is $C = AE^+$, where E^+ is the Moore–Penrose pseudoinverse of **E** (Strang, 1986).

Extending this now to n radial positions, let A be an n x m matrix of absorbances, such that $A_{i,j}$ is the absorbance for radius r_i and wavelength λ_j. Thus, each column of A is the radial distribution of absorbance at a fixed value of λ; each row of A is the absorption spectrum of the solution at a given value of radius over the wavelength range of the values of λ. In similar fashion, **E** is a 2 x m matrix of molar extinction coefficients, where row 1 is for α and row 2 is for β. In this context, **C** becomes an n x 2 matrix computed by $C = AE^+$, where the two columns of **C** contain total molar concentrations, uncomplexed and complexed, of α and β, respectively. and each column of **C** is a function of r. We now concatenate **C**, the dependent variables, with a vector of the radial positions, the independent variable, to form a 3–column data matrix suitable for analysis by non–linear, least–squares curve–fitting. We use as mathematical models the equations:

$$(8) \quad C_{r, \alpha, T} = C_{b, \alpha} \exp(A_\alpha M_\alpha (r^2 - r_b^2))$$

$$+ C_{b, \alpha} C_{b, \beta} \exp(ln\, K + (A_\alpha M_\alpha + A_\beta M_\beta)(r^2 - r_b^2)) + e_\alpha$$

$$(9) \quad C_{r, \beta, T} = C_{b, \beta} \exp(A_\beta M_\beta (r^2 - r_b^2))$$

$$+ C_{b, \alpha} C_{b, \beta} \exp(ln\, K + (A_\alpha M_\alpha + A_\beta M_\beta)(r^2 - r_b^2)) + e_\beta$$

and jointly fit columns one and two with equation 8 and columns one and three with equation 9. The value of lnK, the natural logarithm of the molar equilibrium constant, and the values of $C_{b, \alpha}$ and $C_{b, \beta}$, the concentrations of uncomplexed α and β at the radius of the cell bottom, r_b, are fitting parameters common to both equations. The values of ε_α and ε_β, the small baseline error terms, are fitting parameters applicable only to equations 8 and 9 respectively, and are frequently not necessary. M_α and M_β are the values of the molar masses of α and β, respectively; A_α and A_β are defined by:

$$(10) \quad A = (\partial\rho/\partial c)_\mu \omega^2 / 2\, R\, T$$

The thermodynamically more appropriate term $(\partial\rho/\partial c)_\mu$ is essentially equal to $(1 - \bar{v}^o \rho)$, where $\bar{v}^o$ is the partial specific volume of the appropriate solute as its concentration approaches zero. The chapter by Durchschlag (1986) should be consulted for a concise review of this topic. The values of ω, R, T, M_α, and M_β are known; the values of $(\partial\rho/\partial c)_\mu$ may then be obtained experimentally from the concentration distributions of the reactants as will be described later. We are assuming here that $A_{\alpha\beta} M_{\alpha\beta} = A_\alpha M_\alpha + A_\beta M_\beta$. This assumption can be validated by making measurements at different rotor speeds. Since this will result in different mean hydrostatic pressures in the liquid columns, if there is a change in partial specific volume of the reactants which would indicate a significant ΔV of reaction when associating, different values for *ln*K will be obtained at different rotor speeds as well as at different pressures corresponding to different column heights in the cell.

COMPUTER SIMULATIONS

Simulation of a Protein–Nucleic Acid Interaction

Initial evaluation of a method of analysis such as that described in the previous section is best performed by computer simulation. For this purpose, we have used the mathematical modeling system MLAB[§] operating on a 80486 computer for all of the simulations and data analysis presented here. Using data from our ultracentrifugal study on the interaction of the enzyme DNA polymerase-β with the synthetic oligonucleotide pd(T)$_{16}$ (Lewis, *et al.*, 1992), the C matrix was generated using equimolar concentrations of the reactants selected so that none of the resultant absorbances was in excess of 1.30, a rotor speed of 20000 rev min^{-1}, and a temperature of 14°. Values of 11, 13, 15, 17, 19, and 21 were used for *ln*K in order to demonstrate the effects of the value of this parameter on the behavior of the analytical method as described below.

Figure 1 illustrates the molar extinction coeffients of the enzyme and of the oligonucleotide.

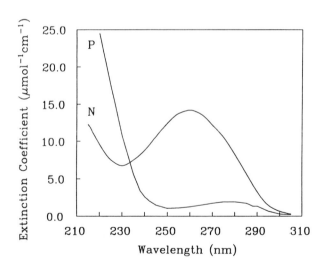

Figure 1. Molar extinction coefficients of DNA polymerase-β and pd(T)$_{16}$ in 50 mM KCl with 50 mM K phosphate, pH 7.4 at 22° measured in a Hewlett–Packard Model 8452A diode array spectrophotometer using a 1 cm optical path length. Data for the protein and the DNA are labeled P and N, respectively.

Wavelengths were selected so that protein absorbance was dominant at the lower wavelengths and nucleic acid absorbance was dominant at the higher wavelengths. The values of the extinction coefficients at 228 nm, 230 nm, 232 nm, 234 nm, 236nm, 238 nm, and 240 nm were obtained by calculating the ratios of the absorbances at these wavelengths to the absorbance at 260 nm for the oligonucleotide and at 280 nm for the enzyme and then multiplying these ratios by the known extinction coefficients of the oligonucleotide and the enzyme, respectively. The E matrix was generated using these values, and the A matrix was then obtained as the product of C and E. The A matrix then had normally distributed random error with a mean of zero and a standard error of 0.01 absorbance units added to every element in the matrix. This produced the equivalents of rather noisy scans, about double the preferred noise level and near the upper limit of what would be considered acceptable data from an XL–A ultracentrifuge. Such a noise level represents a stringent test for the method. A new and noisy C matrix was then obtained as the product of the noisy A matrix and the pseudoinverse E^+; the radius vector was concatenated to this C matrix, and the resultant data matrix was fit using equations 8 and 9 as mathematical models.

Monte Carlo Simulations

A 25–simulation Monte Carlo study using this procedure was performed for each set of generating values. There is little change in the fitting statistics if a greater number of repetitions is performed except for the largest values of lnK. The results of fitting such data are given in Table 1.

TABLE 1. Mean returned values of lnK, the standard error of the mean, and the mean standard error obtained based upon assuming a linear model as functions of the generating values of lnK.

lnK	lnK (mean)	S. E.	S. E. (mean)
11.0	11.011	0.039	0.036
13.0	13.012	0.042	0.039
15.0	15.019	0.066	0.061
17.0	17.015	0.086	0.143
19.0	19.061	0.243	0.386
21.0	21.229	0.612	1.391

With the exception of the results obtained with a generating value of 21.0 for lnK, the mean values of lnK which were obtained by fitting are in excellent agreement with the generating values. The deviations from the generating values are essentially due to the limited number of simulations in the Monte Carlo procedure, and the increase in the deviations at the larger values of lnK suggests the need for a greater number of simulations for these cases. With the same exception noted for the values of lnK, the standard errors of the mean of lnK are quite small. These results indicate that this method may be expected to give accurate estimates provided the data have normally distributed error which is not excessively noisy. The mean values of the standard error returned by the fitting algorithm are based upon assuming that the mathematical model being fit behaves as if it were linear. That this is a reasonable approximation under certain circumstances is demonstrated in Table 1 for values of lnK up to approximately 16. In the next section of this chapter we describe a detailed examination of one data set and describe a simple procedure for determining when the use of this value is valid and when a Monte Carlo study is necessary.

Detailed Evaluation of a Single Simulation

The unperturbed **C** matrix of the simulation which is examined in detail here was generated using equations 8 and 9 with lnK = 15.0, $C_{b, P}$ = 5.0 μM, $C_{b, N}$ = 1.0 μM, and both values of ε = 0 on a radius vector of 6.740 cm to 7.180 cm with increments of 0.002 cm. These reactant concentration at the cell bottom (r_b = 7.200 cm) gives an approximately 1:1 molar ratio of the reactants. The **A** matrix was obtained using **A** = **C E**, noise with a mean of zero and a standard error of 0.01 was added as before. The noisy **A** matrix was then transformed to a noisy **C** matrix using **C** = **AE⁺**. **C** was then concatenated to the radius vector and fit using equations 8 and 9, all as previously described. This gave a value of lnK = 15.045 with a standard error of 0.064. This is the standard deviation returned by the fitting algorithm and, as can be seen in Table 1, is very close to the standard error of the mean from Monte Carlo simulations. The root–mean–square error of the fit was 2.324 x 10⁻⁷, a value consistent with the magnitude of the added noise.

Figure 2 illustrates the absorbances as a function of radial position at wavelengths of 228, 234, and 240 nm, showing the range of gradients obtained.

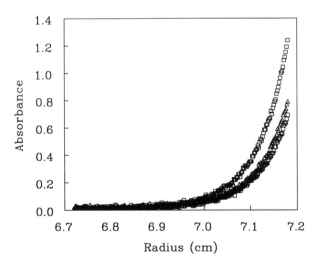

FIGURE 2. Radial distributions of the absorbances with added noise as described in the text at wavelengths of 228 nm at the top to 240 nm at the bottom.

Figure 3 illustrates the fits using equations 8 and 9 to the molar concentration distributions of protein plus complex and nucleic acid plus complex calculated from the absorbances. Figure 4 illustrates the distributions of the residuals about the fitting lines in Figure 3.

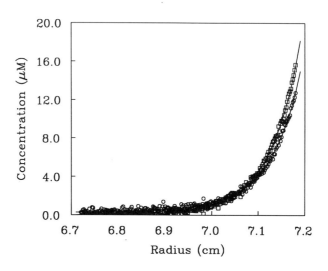

FIGURE 3. Fits of equations 8 and 9 to the molar concentrations of protein plus complex and nucleic acid plus complex as described in the text. The joint fit RMS error = 2.324 x 10^{-7} M.

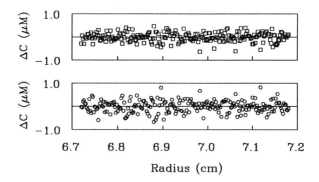

FIGURE 4. Distribution of the residuals about the fitting lines in Figure 3. The upper panel shows the protein plus complex distribution; the lower panel shows the nucleic acid plus complex distribution.

Since the random error added to the absorbances is normally distributed and the pseudoinverse transformation is linear, the error in the concentrations obtained by the transformation would be expected to be normally distributed. It is possible to determine whether or not this is true by plotting the cumulative distribution fraction of the residuals as a function of the values of the residuals. The cumulative distribution fraction is that fraction of the total number of residuals that have a value less than or equal to a given residual value. This distribution can then be compared to the theoretical cumulative distribution fraction calculated from the mean and variance of a given set of residuals. Thus, ϕ_i, the cumulative distribution fraction for the ith value of v in the set of residuals is given by

$$(11) \quad \phi(u_i) = \frac{1}{\sqrt{2\pi}} \int_{-\infty}^{u_i} e^{-t^2/2} \, dt \quad \text{where} \quad u_i = (v_i - \text{mean}(v)) / \sqrt{\text{var}(v)}$$

A satisfactory comparison can usually be made visually. Figure 5 shows the cumulative distribution fractions and the theoretical lines for the data we have been examining. It is is clearly demonstrated that the residuals exhibit normal distribution to three standard deviations from the mean.

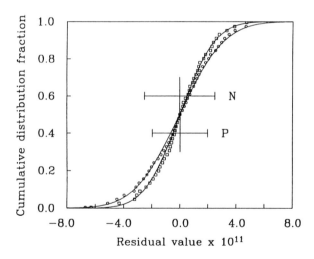

FIGURE 5 Cumulative distribution fraction as a function of the values of the residuals from fitting the protein and nucleic acid distributions, indicated by squares and circles, respectively. The positions of the means are indicated by vertical lines with $\text{mean}_P = 5.2 \times 10^{-10}$ and $\text{mean}_N = -2.0 \times 10^{-9}$; the values of the standard deviations from the mean are indicated by horizontal lines with $SD_P = 1.97 \times 10^{-7}$ and $SD_N = 2.52 \times 10^{-7}$. P and N are used to denote protein plus complex and nucleic acid plus complex, respectively.

Since the standard error of the mean is needed in order to calculate the reciprocal of the variance, the appropriate weight for lnK when this value is fit as a function of temperature in order to determine the values of ΔG^o, ΔH^o, ΔS^o, and ΔC_p^o, it would be useful to know when the estimated value of the standard error obtained during fitting is a sufficiently good approximation and when it is necessary to do Monte Carlo simulations. Table 1 demonstrates that for values of ln K of 15 or less, the estimated standard error is slightly less than the Monte Carlo standard error, but is close enough to be acceptable. For values of lnK of 17 or greater, the estimated standard error is progressively greater than the Monte Carlo standard error and is not close enough to be used. Differences in experimental conditions could readily alter which estimated values of the standard error would be acceptable and which would not. One approach would be to do Monte Carlo simulations routinely. However, the data in Table 1 suggest that an answer might be found by examining the values of the sum of squares as a function of the value of lnK by fixing lnK at various values and letting the other parameters attain their optimal values when fitting the data. A graph of the sum of squares as a function of lnK for the data we have been examining is shown in Figure 6.

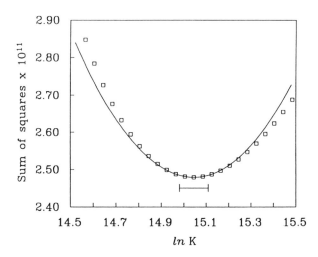

FIGURE 6. Sum of squares as a function of the values of lnK for the data generated with $lnK = 15.0$. The curve is a parabola fit to the sums of squares for values of lnK that lie within ± 3 standard errors obtained in fitting. The center of the parabola was fixed at the fit value of $lnK = 15.045$ and the minimum of the parabola at the corresponding sum of squares. The horizontal line denotes ± 1 standard error (S. E. = 0.064).

With this procedure, if a mathematical model is a linear function of its parameters, and if the data have normally distributed error, and if each datum is properly weighted with the reciprocal of its variance, then the graph of the sum of squares as a function of the parameter of interest will be a parabola that is symmetrical about the optimal value of the parameter. Additionally, the value of the sum of squares at parameter values equal to the optimal parameter value plus and minus one standard error will be equal to the sum of squares at the optimal parameter value plus the square of the root–mean–square error obtained in fitting. These values are 2.48459×10^{-11} and 2.48457×10^{-11}, respectively, for this case. As can be seen, the data are symmetrically parabolic to approximately ± 3 standard errors, although significant deviations occur beyond these values. On the basis of meeting these conditions, we can classify the model and data used here as constituting a pseudolinear system within this range of values; thus, the returned value of the standard error is a very good approximation of the best value. For practical purposes, it is not necessary to obtain the values of the sum of squares for values of $ln\ K$ beyond plus and minus four standard errors or to use increments smaller than one-half a standard error. The necessary computations can be made very rapidly and a Monte Carlo simulation need be made only if the system is not pseudolinear.

For comparative purposes, Figure 7 shows a comparable treatment for data which were generated with a value of $lnK = 21.0$, had noise added as before, and which then gave a fit value of $lnK = 20.876 \pm 1.013$. Only sums of squares for values of lnK lying between $lnK - 1$ S. E. and $lnK + 0.5$ S. E. have been fit and the parabola has been extended onto the regions of values of greater magnitude. It can be seen readily that pseudolinear conditions are not met in this case and that a Monte Carlo simulation analysis is required.

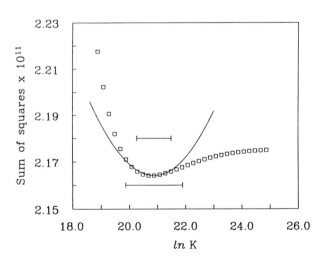

FIGURE 7. Sum of squares as a function of the values of lnK for data generated with $lnK = 21.0$ and fit as described in the text. The lower horizontal line denotes ± 1 standard error returned during fitting (1.013) and the upper horizontal line denotes ± 1 standard error obtained by the Monte Carlo simulation (0.612)

We now consider the effects of error in the extinction coefficients used for analysis upon the results obtained. Various types of error were added to the extinction coefficient matrix before obtaining its pseudoinverse and calculating the molar concentration distributions for fitting using the same set of absorbance data previously used The effects of these various types of added error were examined by performing a 25–simulation Monte Carlo study for each case. The first two studies involved adding normally distributed random error with means of zero and standard errors of five per cent and ten percent to the extinction coefficient matrices. The effects of systematic error were examined by studies where the protein extinction coefficients were increased by ten per cent and the nucleic acid extinction coefficients were decreased by ten per cent in one case, and by reversing this procedure in the other case. These simulated errors in determining extinction coefficients where the wavelength was actually at shorter and longer wavelengths, respectively, than the wavelengths set on the monochromator. The final two studies involved increasing all elements in the extinction coefficient matrix by ten percent in one case and decreasing them by ten per cent in the other. The results of these studies are presented in Table 2.

TABLE 2. The effects of various types of error in the extinction coefficient matrix on the values of lnK and of their standard errors. NDRE = normally distributed random error; P=protein extinction coefficients; N = nucleic acid extinction coefficients.

Error	lnK	S. E.	S. E. of Mean
None	15.0447	0.0643	
+ 5% NDRE	15.0440		0.0579
+10% NDRE	15.0658		0.0908
P+10%, N-10%	16.0589		0.1184
P-10%, N+10%	14.5028		0.0061
P, N +10%	15.1400	0.0643	
P, N -10%	14.9393	0.0643	

Three important conclusions can be reached by examining these data. The first is that the presence of normally distributed random error with a standard error as great as ten percent in the extinction coefficient matrices has a minimally adverse effect upon the lnK values obtained. This means that the random error arising from measurements in the XL–A analytical ultracentrifuge will not have significantly adverse effects upon the results of a properly designed experiment. Second, systematic errors where the extinction coefficients are either greater or less than they should be also have a minimal effect on the results provided the deviations are not greater than about ten per cent. Finally, very serious deviations from optimal values of lnK result when systematic error results from obtaining the extinction coefficients of the reactants using wavelengths which are not identical with those used to obtain the absorbances of the mixture of reactants. The best means for minimizing this problem is to measure the absorbances of reactants and mixture with a procedure where the setting of the monochromator is not moved between measurements.

This will be discussed in more detail in the section on experimental measurements.

Simulation of a Protein–Protein Interaction

It is obvious that this method of multi–wavelength analysis should be equally suitable for the study of protein–protein interactions if the reactants had sufficiently different absorption spectra. While chromophore labels can be chemically attached to proteins, there is a risk that such chromophores might alter the reactivity of the reactants in a significant manner. However, it has recently been demonstrated that it is possible to substitute 5–hydroxy tryptophan for tryptophan in proteins during biosynthesis and that these proteins have normal biological activity (Ross, *et al.*, 1992, Hogue, *et al.*, 1992). In Figure 8, which shows the absorption spectra of tryptophan and 5–hydroxytryptophan, it can be seen that, in contrast to normal tryptophan, 5–hydroxy tryptophan has significant absorption at wavelengths greater than 320 nm, and these spectra sufficiently different that it is possible to simulate data for protein–protein interactions as we have for protein–nucleic acid interactions.

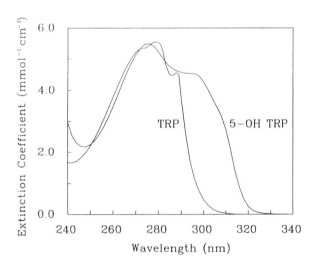

FIGURE 8.   Molar extinction coefficients of normal tryptophan and of 5–hydroxy tryptophan in 50 mM KCl with 50 mM K phosphate, pH 7.4 at 22° measured in a Hewlett–Packard Model 8452A diode array spectrophotometer using a 1 cm optical path length.

The application of the method to protein–protein interactions was tested with a model system where one protein, with a molar mass of 30000, contained a single simulated 5–hydroxy tryptophan, and the other protein, with a molar mass of 40000, contained a single simulated normal tryptophan. Since each protein had a single chromophoric amino acid, we then assumed that the molar extinction coefficients of the proteins would be the same as the molar extinction coefficients of the respective tryptophans. A non-noisy **C** matrix was generated assuming equimolar concentrations of the proteins, $lnK = 12.00$, $(\partial\rho/\partial c)_\mu = 0.280$, a rotor speed of 10000 rev min^{-1} and a temperature of 20°. The **E** matrix was constructed using the appropriate extinction coefficients at wavelengths from 278 nm to 302 nm at 4 nm increments. The **A** matrix was obtained and normally distributed random error with a mean of zero and a standard error of 0.01 absorbance units was added. The radius vector was concatenated to the noisy **C** matrix obtained using the **E$^+$** matrix, and this data matrix was fit using equations 8 and 9. Fitting returned a value of $lnK = 11.990 \pm 0.103$. These fits are shown in Figure 9 and the distribution of the residuals is shown in figure 10.

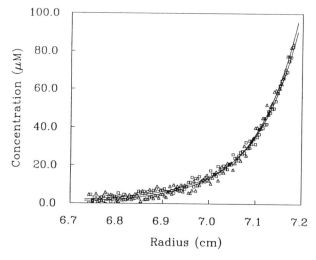

FIGURE 9. Fits of the molar concentrations for the normal tryptophan containing protein (triangles) and the 5-hydroxy tryptophan containing protein (squares). The joint fit RMS error = 1.849 x 10^{-6}.

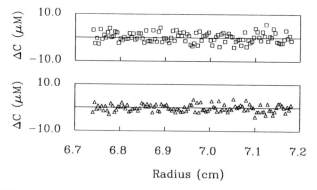

FIGURE 10 Distributions of the residuals of the fits shown in Figure 9.

This system exhibited pseudolinear behavior when the sum of squares was plotted as a function of the values of lnK and the graph of the cumulative distribution function showed that the residuals had a normal distribution. This result, which was expected, indicates that the method is equally applicable to protein-protein interactions. Experimental work currently in progress supports this assertion. Additionally, it is expected that this method of analysis using multi–wavelength scans should be equally applicable to studies of various small chromophoric ligand–macromolecule interactions such as drug–DNA and drug–proteininteractions. The potential applicability is readily evaluated by computer simulations similar to those described here.

EXPERIMENTAL EVALUATION

The experimental application of this method is well demonstrated by data from studies on the interaction of a heat shock factor polypeptide with a cognate DNA in the XL–A analytical ultracentrifuge. The polypeptide (dHSF(33–163)) encompassed the DNA–binding domain of heat shock factor, spanning the 33 to 163 region of the full–length *Drosophila* heat shock factor and had a molar mass of 15259, calculated from the amino acid sequence, and a molar extinction coefficient of 9079 mol^{-1} cm^{-1} at 280 nm. Sequence–specific DNA binding activity was demonstrated by DNase I footprinting. The synthetically prepared DNA was a 13–base pair double–stranded DNA with a –GAA– recognition sequence at its center and had a compositional molar mass of 8011 and a molar extinction coefficient of 90489 mol^{-1} cm^{-1} at 260 nm.

For the data shown here, solutions of polypeptide, DNA, and a 2:1 molar ratio mixture of polypeptide and DNA were simultaneously centrifuged to equilibrium at a rotor speed of 14000 rev min^{-1} over a range of temperatures in a buffer of 100 mM KCl, 10 mM K phosphate, pH 6.3, and 0.1 mM EDTA. The 2:1 molar ratio was used to increase the fraction of the total absorbance attributable to the polypeptide at the lower wavelengths. The final scans at each temperature were taken at wavelengths of 280 nm, 260 nm, and from 248 nm through 230 nm at 2 nm increments. The data presented here were obtained at a temperature of 4.0°.

The first step in the analysis was to obtain the molar extinction coefficients and the values of $(\partial\rho/\partial c)_\mu$ for the polypeptide and the DNA. The concentration distribution of a homogeneous, ideal solute at ultracentrifugal equilibrium in terms of absorbency as a function of radial position is given by:

(12) $A_{r,i} = A_{b,i} \exp(M (\partial\rho/\partial c)_\mu \omega^2/2 R T\} (r^2 - r_b^2))$

where the terms have their usual meaning and A_b is the absorbancy at the radial position of the cell bottom. Six scans of the polypeptide solution at wavelengths of 230 nm through 240 nm were jointly fit for the values of the A_b parameters as functions of wavelength and the value of $(\partial\rho/\partial c)_\mu$ as a global fitting parameter. These wavelengths were used because their scans had the greatest absorbance range and the resultant value of $(\partial\rho/\partial c)_\mu$ was the best attainable. The scans at

242 nm through 248 nm and also at 280 nm were then fit for the values of A_b, with the value of $(\partial\rho/\partial c)_\mu$ fixed. A value of 0.2616 ± 0.0045 was obtained for $(\partial\rho/\partial c)_\mu$ of the polypeptide from the 6–scan joint fit which is shown in figure 11. The same procedure was followed for the DNA solution, except that the wavelengths of the six scans used for the joint fit to obtain the values of $(\partial\rho/\partial c)_\mu$ and A_b were 238 nm through 248 nm. The scans at 230 nm through 236 nm and also at 260 nm were then fit for the values of A_b, with the value of $(\partial\rho/\partial c)_\mu$ fixed as before. A value of 0.4966 ± 0.0066 was obtained for $(\partial\rho/\partial c)_\mu$ of the DNA from the 6–scan joint fit which is shown in figure 12.

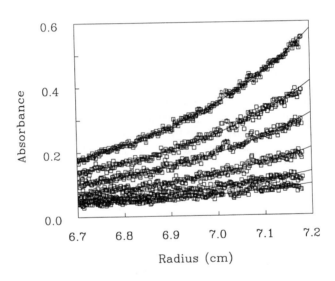

FIGURE 11 Distributions of the absorbances of the polypeptide at equilibrium at 14000 rev min^{-1} and 4.0°. The scanned wavelengths are, from top to bottom, 230 nm, 232 nm, 234 nm, 236 nm, 238 nm, and 240 nm. The curves are for the joint fit as described in the text. The joint fit RMS error = 0.00520 absorbance units.

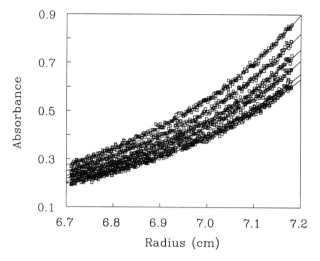

FIGURE 12. Distributions of the absorbances of DNA at equilibrium at 14000 rev min^{-1} and 4.0°. The scanned wavelengths are, from top to bottom, 248 nm, 246 nm, 244 nm, 242 nm, 240 nm, and 238 nm. The curves are for the joint fit as described in the text. The joint fit RMS error = 0.00527 absorbance units.

The ratios of the values of A_b as functions of wavelength to the value of A_b at 280 nm for the polypeptide or A_b at 260 nm for the DNA were then multiplied by the molar extinction coefficient at 280 nm for the polypeptide or at 260 nm for the DNA to obtain the molar extinction coefficients at the other wavelengths used for taking scans. These values were then used for the construction of the **E** matrix from which the **E⁺** matrix was calculated.

Since the wavelength selection of the XL–A monochromator is accurate only to ± 2 nm, it is essential that all of the measurements to be made at a given wavelength be made without changing the monochromator setting. When a single wavelength is selected, the monchromator does not shift between scans of each cell. Thus, while the wavelength selected may be in error by ± 2 nm, the calculated extinction coefficients will be appropriate for the actual wavelength used for scanning all three solutions provided the wavelength is not changed as a result of instrumental flaw. This precludes setting the XL–A to scan at three wavelengths since, at this time, the XL–A program results in scanning each cell at the three wavelengths before scanning the next cell. Since the absorption spectra of proteins and nucleic acids have relatively broad maxima at 280 nm and 260 nm, respectively, an error of ± 2 nm in actual wavelength will have a minimal effect on the values of the other extinction coefficients. The best results are obtained by using extinction coefficients calculated from data obtained with the XL–A as described above rather than using extinction coefficients calculated from spectrophotometric measurements because of this uncertainty of the wavelengths in the XL–A. The effects of both random and systematic error have been demonstrated previously.

Three scans of the equilibrium distribution of the polypeptide, the DNA, and their complex at chemical and centrifugal equilibrium are shown in figure 13 The wavelengths are selected to present distributions at the maximum, the minimum and an average absorbance for the solution.

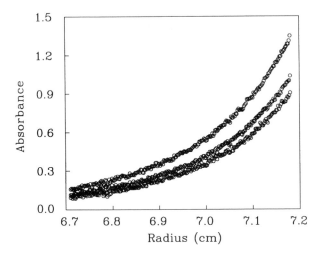

FIGURE 13.   Distributions of the absorbances of a mixture of the polypeptide and DNA in a 2 to 1 molar ratio at equilibrium at 14000 rev min^{-1} and 4.0^{0}.   The scanned wavelengths are,   from top to bottom, 230 nm, 234 nm, and 238 nm.

The molar concentration distributions and the fitting lines resulting from a joint fit using equations 8 and 9 as mathematical models are shown in figure 14. A value for lnK of 15.57 ± 0.55 was obtained. Figure 15 shows the distribution of the residuals about the fitting lines. The distributions are quite adequately uniform with no significant systematic deviations, demonstrating not only the quality of the fit, but also that a 1:1 stoichiometry appears to be an appropriate model for the interaction. This value of lnK agrees well with the value of the natural logarithm of the value of K_a obtained by fluorescence titration.

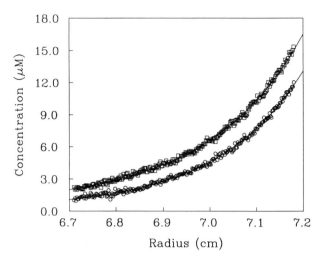

FIGURE 14   Molar concentration distributions for the polypeptide–DNA interaction. The upper distribution is for the polypeptide plus polypeptide–DNA complex; the lower distribution is for the DNA plus polypeptide–DNA complex. The curves are for the joint fit as described in the text. The joint fit RMS error = 0.0590 x 10^{-6} M.

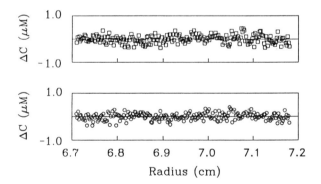

FIGURE 15.   Distributions of the residuals for the polypeptide–DNA interaction shown in Figure 14.

Measurements of hyperchromism or hypochromism for these reactants have not been made. However, it has been demonstrated that these effects are at a maximum at a wavelength of 265 nm and decrease significantly at shorter wavelengths (Jansen, et al., 1976). A simple calculation shows that the value of lnK changes by approximately $\pm$ 0.1 for a $\pm$ 10% alteration in chromicity, and thus is within the limit of reasonable experimental error.

PRACTICAL CONSIDERATIONS

While this new method has demonstrated itself to be quite powerful, it is not without its problems. One problem is the time that it takes to collect and process the quantity of data which is required. Given the nature of the data acquisition system in the XL–A ultracentrifuge, currently there is little that can be done to enhance the rate of acquisition. However, if experimental design permits, the use of 6–channel centerpieces will effect a significant savings in time. Not only can three sets of samples be run simultaneously, thus utilizing running time much more efficiently, but each cell can now contain a complete sample set consisting of the two individual components and the mixture. This will now permit programming three wavelengths at a time since each cell will have internal wavelength consistency. Total scanning time will be longer but a much greater quantity of data can be collected with a significant reduction in the time required for actual attendance on the instrument.

Utilization of the graphical editing facilities in the XL–A software has proved to be the most efficient manner of performing most of the data editing required, and it is rapid and simple to see that the initial and final radial positions in the data sets are the same for all wavelengths for a given component or mixture. While it is desirable for the data sets for the individual components, it is mandatory for the data sets for the mixture of components. While the radial increments for a scan are set using the XL–A software, occasional radial positions will be skipped. Thus, additional editing is required using the analysis software to see that any radial positions and their corresponding absorbances which are missing from the data from the mixture of components are replaced by appropriate interpolation. This is essential for the proper construction of the absorbance matrix. MLAB accomplishes this and all of the other computations previously described in a very rapid and user friendly manner.

One problem which we have not been able to overcome is that there does not appear to be any practical way to generate appropriate weights for the molar concentration data, so we have been forced to perform unweighted fits. The fitting algorithm in MLAB requires weights, and if a weighting vector is not specified, a default weight of one is assigned to each datum. Fortunately, as can be seen from examination of the distributions of the residuals, these distributions appear to be relatively uniform and it is reasonably probable that little significant error results from performing unweighted fits.

A particular problem related to fitting needs to be considered. The spectral properties of the reactants may make it very difficult to obtain a range of wavelengths where each reactant has a domain of absorbance dominance. This can be overcome to some degree for protein–nucleic acid interactions by taking data at shorter wavelengths where the protein absorption becomes more dominant. Additionally, it is possible to also use molar ratios other than 1:1. However, we have found that for strong interactions ($lnK>16.0$), there can be increasing difficulty in accomplishing the fit if the molar ratios of the uncomplexed reactants is greater than 2:1. The need for the 1:1 ratio appears to become more marked when the value of lnK is 17 or greater. One would be well advised to perform a number of simulated experiments covering a wide range of possible conditions in order to optimize the design of the experiments to be performed in the XL–A.

In summary, the reliability of this method is attributable to two factors. The first is the advantages that the application of the pseudoinverse transform confer. The use of scans taken at different wavelengths may be compared to taking replicate measurements with the additional advantage that the contributions of the components to the observed gradient are separated. It must be emphasized again that the values of the molar concentrations that are obtained by the pseudoinverse transform are the optimal values obtainable from the absorbance matrix since the multiplication of the E^+ matrix by each row of the A matrix constitutes linear least–squares curve–fitting of that row of the A matrix. Additionally, since the pseudoinverse transformation is linear, error distribution distortion does not occur and normally distributed error in the A matrix results in normally distributed error in the C matrix. This procedure is significantly superior to the use of the distributions of different average molecular weights as functions of reactant concentrations as a means of obtaining the values of equilibrium constants; such distributions are not the result of linear transformations and non–normal error is unavoidably added. The second factor is that this method utilizes the joint fitting of data matrices of molar concentrations; this imposes very stringent constraints in terms of permitted values of the parameters since both matrices must have identical concentrations of the complex as functions of radial position and these concentrations must be dependent on the radial distributions of the concentrations of the individual reactants in their respective matrices and on the value of the logarithm of the equilibrium constant.

FOOTNOTE

§MLAB, from Civilized Software, 7735 Old Georgetown Road, Bethesda, MD 20814 USA.

Mention of a specific product represents the opinions of the authors and does not constitute an endorsement on the part of their respective Institutes, the National Institutes of Health or the Department of Health and Human Services.

ACKNOWLEDGMENT

The authors thank Dr. Carl Wu, Laboratory of Biochemistry, National Cancer Institute, National Institutes of Health, for the heat shock factor polypeptide and the DNA. The authors also thank their respective Institutes and the National Institutes of Health for the support which made this work possible.

GLOSSARY OF SYMBOLS

A	$(\partial\rho/\partial c)_\mu \, \omega^2 / 2\,R\,T$ or $(1 - \bar{v}^o \rho)\, \omega^2 / 2\,R\,T$
$(\partial\rho/\partial c)_\mu$	density increment at constant chemical potential μ
ρ	solution density in gm cm^{-3}
c	concentration in gm cm^{-3}
ω	rotor angular velocity in radians sec^{-1}
R	gas constant
T	absolute temperature or Total (subscript)
$\bar{v}^o$	limit of solute partial specific volume as solute concentration approaches zero
A	absorbance of solution
$\mathbf{A}$	absorbance matrix
α, β	component designators (subscripts)
b	cell bottom (subscript)
C	molar concentration
$\mathbf{C}$	molar concentration matrix
E	molar extinction coefficient
$\mathbf{E}$	molar extinction coefficient matrix
$\mathbf{E}^+$	Moore–Penrose pseudoinverse of $\mathbf{E}$
K	molar equilibrium constant
λ	wavelength
M	molar mass
N	nucleic acid (subscript)
P	protein or polypeptide (subscript)
ϕ_i	cumulative distribution fraction for the ith residual
r	radial position in cell
RMS	root mean square
S. E.	standard error

REFERENCES

Durchschlag H (1986): Specific volumes of biological macromolecules and some other molecules of biological interest. In: *Thermodynamic Data for Biochemistry and Biotechnology.*, Hinz, H–J, ed. New York: Springer–Verlag.

Hogue CWV, Rasquinha I, Szabo A and MacManus JP (1992): A new intrinsic fluorescent probe for proteins. Biosynthetic incorporation of 5–hydroxy tryptophan into oncomodulin. *FEBS Letters* 310: 269–272

Jansen DE, Kelly RC and von Hippel PH (1976): DNA "Melting" Proteins. II Effects on bacteriophage T4 gene 32–protein on the conformation and stability of nucleic acid structures. *J Biol Chem* 251: 7215–7228

Lewis, MS (1991): Ultracentrifugal analysis of a mixed association. *Biochemistry* 30: 11716–11719

Lewis MS, Kim S–J, Kumar A and Wilson SH (1992): Thermodynamic parameters of oligodeoxynucleotide binding to DNA β–polymerase revealed by equilibrium ultracentrifugation. *Biophys J* 61: A489

Lewis MS, Shrager RI and Kim S–J (1994): Ultracentrifugal analysis of protein–nucleic acid interactions using multi–wavelength scans. *Colloid and Polymer Sci* : in press.

Ross JBA, Senear DF, Waxman E, Kombo BB, Rusinova E, Huang YT, Laws WR and Hasselbacher CA (1992): Spectral enhancement of proteins: Biological incorporation and fluorescence characterization of 5–hydroxy tryptophan in bacteriophage λ cI repressor. *Proc Nat Acad Sci USA* 89: 12023–12027

Strang G (1986) *Introduction to Applied Mathematics*. Wellesley, MA: Wellesley–Cambridge Press. pp 138–139

Part II
SEDIMENTATION VELOCITY

SEDIMENTATION BOUNDARY ANALYSIS OF INTERACTING SYSTEMS: USE OF THE APPARENT SEDIMENTATION COEFFICIENT DISTRIBUTION FUNCTION

Walter F. Stafford, III

INTRODUCTION

New methods for the analysis of sedimentation velocity data have extended the sensitivity of the UV scanning and Rayleigh interferometric optical systems of the analytical ultracentrifuge by 2-3 orders of magnitude. Boundaries with concentrations lower than $10\mu g/ml$ can be visualized readily with the Rayleigh optical system allowing the thermodynamic analysis of interacting systems in a concentration range previously inaccessible to the analytical ultracentrifuge. The increase in sensitivity has been achieved by a combination of analytical (Stafford, 1992; Stafford, 1992; Stafford, 1994) techniques that use the time derivative of the concentration profile and of instrumental techniques (Liu and Stafford, 1992; Yphantis et al., 1994) that employ a rapid acquisition video-based Rayleigh optical system. Use of the time derivative achieves an automatic optical background correction, and the video system allows signal averaging of the sedimentation patterns, resulting in a considerable increase in the signal-to-noise ratio. Sedimenting boundaries are represented as apparent sedimentation coefficient distribution functions, $g(s^*)$ vs. s^*, where s^* is the apparent sedimentation coefficient defined as $s^*=\ln(r/r_m)/\omega^2 t$ and $g(s^*)$ has units proportional to concentration per svedberg. A plot of $g(s^*)$ vs. s^* is geometrically similar to the corresponding plot of dc/dr vs. r obtained with the schlieren optical system. This chapter will describe methods for the analysis of both self-associating and hetero-associating systems using the apparent distribution function.

Because of the relatively high sensitivity and precision that can be attained by combining the time derivative and the signal averaging techniques, it seemed worthwhile to investigate the limitations of the use of $g(s^*)$ as a general approach to the analysis of interacting systems at high dilution. The theoretical derivation of sedimentation coefficient distributions is predicated on the assumption of negligible diffusion and of no interaction between sedimenting species. Therefore, the use of $g(s^*)$ in cases for which diffusion and interaction were present would require acceptance of some degree of approximation. It will be shown that the degree of approximation is quite small and probably acceptable in most practical situations and especially in cases for which thermodynamic information (i.e. the standard free energy of association, ΔG^0) is being sought. The effects of pressure, and how the apparent distribution function can be used to identify such effects, on interacting systems which have a non-zero volume change of association are also discussed.

THEORETICAL BACKGROUND

The apparent sedimentation coefficient distribution function, $g(s^*)$ vs. s^*, can be computed from the time derivative of the concentration profile according to the following relationship, in which the value of $\partial c/\partial t$ has been corrected for the "plateau" contribution by an iterative procedure as described previously (Stafford, 1992; Stafford, 1994).

$$g(s^*) = \left(\frac{\partial c}{\partial t}\right)_{corr} \left(\frac{1}{c_o}\right) \left(\frac{\omega^2 t^2}{\ln(r_m/r)}\right) \left(\frac{r}{r_m}\right)^2 \tag{1}$$

where c is the concentration, $g(s^*)$ has units of concentration per svedberg, r is the radius, r_m is the radius of the meniscus, ω is the angular velocity of the rotor attained after acceleration, c_o is the initial loading concentration, t is the equivalent time of sedimentation, and s^* is defined as

$$s^* = \frac{1}{\omega^2 t} \ln\left(\frac{r}{r_m}\right)$$

Because the angular velocity of the rotor is not constant at the beginning of the run, one must compute t from the following integral

$$t = \frac{1}{\omega^2_f} \int \omega(t)^2 dt \tag{3}$$

where, in this case, $\omega(t)$ is the angular velocity during acceleration and ω_f is the final angular velocity.

We will use the *unnormalized* distribution function, $\hat{g}(s^*)$, defined in the following way to compute the weight average sedimentation coefficient.

$$\hat{g}(s^*) = \left(\frac{\partial c}{\partial t}\right)_{corr} \left(\frac{\omega^2 t^2}{\ln(r_m/r)}\right) \tag{4}$$

The unnormalized distribution function can be related directly to the concentrations in the cell. The weight average sedimentation coefficient, to within a very good approximation (see below), can be shown to be given by

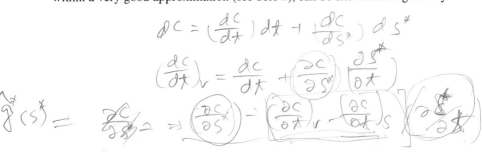

$$\left(\frac{dc}{dt}\right)_r + 2\,w^2 \mathbf{121} \ast \left(\frac{dc}{ds\ast}\right)_r$$

$$s_w = \frac{1}{c_p} \int_{s\ast=0}^{s\ast_p} s\ast \hat{g}(s\ast)ds\ast \tag{5}$$

where the limits of integration refer to integration from the meniscus to the plateau region (the region centrifugal to the boundary), and where c_p is given by

$$c_p = \int_{s\ast=0}^{s\ast_p} \hat{g}(s\ast)ds\ast \tag{6}$$

SIMULATION AND ANALYSIS

Simulated sedimentation velocity profiles were computed by the method of Claverie (Original program kindly supplied by Dr. David J. Cox)(Claverie et al., 1975; Claverie, 1976; Cox and Dale, 1981). Sedimentation patterns for various monomer, n-mer systems of the type $nA=A_n$, and hetero- associating systems of the type $A+B=C$; $C+B=D$ were simulated as a function of time for different values of the equilibrium constant and loading concentration. The computation of the monomer n-mer system was treated as a single component system so that after each step of sedimentation and diffusion the system was allowed to re-equilibrate and a new weight average sedimentation coefficient and gradient average diffusion coefficient were computed and used for the next round of sedimentation and diffusion. Use of the gradient average diffusion coefficient allows the explicit introduction of pressure dependence of the equilibrium constant (Cox and Dale, 1981). The gradient average diffusion coefficient is given by

$$D_{ave} = \frac{D_1 \frac{\partial c_1}{\partial r} + D_n \frac{\partial c_n}{\partial r}}{\frac{\partial c_1}{\partial r} + \frac{\partial c_n}{\partial r}} \tag{7}$$

where D_{ave} is the gradient average diffusion coefficient, D_1 and D_n are the diffusion coefficient of the monomer and n-mer, respectively, and c_1 and c_n are then concentration of monomer and n-mer, respectively.

Simulation of hetero-associating systems was carried out for four separate species (equations 12 and 13). After each step of sedimentation and diffusion, the equilibrium constants and total amount of A and B at each radial position were used to recompute the equilibrium concentrations of A, B, C, and D. These values were used for the next step of sedimentation and diffusion.

Pressure effects were simulated by allowing the equilibrium constant to vary with radius according to equation (22).

The simulated curves of $g(s^*)$ were used to compute a weight average sedimentation coefficient according to equations (5) and (6). Integration was carried out using the trapezoidal rule with a grid spacing of 0.05S. For the monomer n-mer system, s_w was used to compute a value of α using equation (10). The value of α, in turn, was used to compute a value of θ as explained below (equation 12). For the case of the hetero-associating systems, a short FORTRAN program was written to calculate θ from the observed weight average sedimentation coefficient and the computed values of the weight fraction of each species at equilibrium at the plateau concentration. An iterative procedure using equations 19, 20 and 21 was devised.

Monomer-n-mer Systems:

A self-associating monomer-polymer system of the type $nA = A_n$ was simulated. If α denotes the weight fraction of polymer, we have that

$$\theta \equiv k_n c^{n-1} = \frac{\alpha}{(1-\alpha)^n} \qquad (8)$$

where c is the total macromolecular concentration, k is the equilibrium constant on the weight concentration scale, n is the degree of polymerization, and θ is a dimensionless parameter. We also have that the weight average sedimentation coefficient is given by

$$s_w = (1-\alpha)s_1 + \alpha s_n \qquad (9)$$

so that α may be computed from experimental data as a function of concentration by rearranging equation (9) to give

$$\alpha = \frac{s_w - s_1}{s_n - s_1} \qquad (10)$$

where s_1 and s_n are the sedimentation coefficient of the monomer and n-mer, respectively.

Having computed α from s_w, as we will see below, we may compute k_n from the following relationship

$$k_n = \frac{1}{c_p^{n-1}} \left(\frac{\alpha}{(1-\alpha)^n} \right) \qquad (11)$$

where c_p is the plateau concentration.

Figure 1 shows several curves for a monomer-dimer system, and Figure 2 shows curves for a monomer tetramer system for various values of the dimensionless parameter θ, defined by equation 8.

The accuracy of this method for determination of standard free energies appears to be reasonably good. Table I and Table II summarize the analysis and list (column G) the error in ΔG^o expressed in Kcal/mole of polymer that were

θ I S

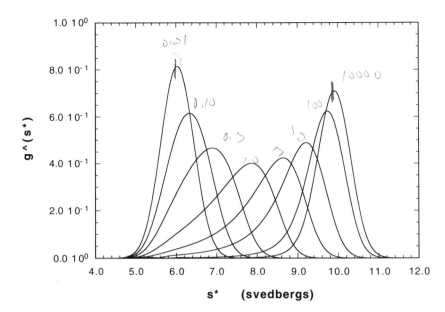

s*   (svedbergs)

Figure 1. Monomer dimer system: S1=6S, D1=1.25F, S2=10S, M₁=440000, speed= 60,000rpm, t= 3000sec. Simulations for various values of $\theta_0 = kc_0 = 0.01, 0.10, 0.3, 1.0, 3.0, 10.0, 100.0, 1000.0$ from left to right.

			TABLE I			
		Monomer dimer S1=6S, D1=1.25F, S2=10S				
A	B	C	D	E	F	G
$\theta_0=kc_0$	$c_p=\int\hat{g}ds$	S_w	α	$\theta=k_{obs}c_p$	k_{obs}	$\delta\Delta G^0$
0.01	0.8668	6.034	0.0085	0.0086	0.0099	0.01
0.10	0.8611	6.310	0.0774	0.0910	0.106	0.03
0.30	0.8525	6.729	0.182	0.272	0.320	0.04
1.0	0.8379	7.455	0.364	0.898	1.07	0.04
3.0	0.8234	8.188	0.547	2.66	3.24	0.04
10.0	0.8105	8.858	0.714	8.76	10.8	0.04
100.	0.7989	9.567	0.892	76.1	95.3	0.03
1000	0.7916	9.865	0.966	848.3	1072.	0.04

Column A is the value of θ at the cell loading concentration, $c_0 = 1.0g/L$.
Column B is the plateau concentration for the g(s*) pattern used
Column C is the weight average sedimentation coefficient. Equations 4 and 5.
Column D is the weight fraction of polymer computed from s_w.
Column E is the value of θ computed from α.
Column F is the observed value of $k = \theta_{obs}/c_p$
Column G is the error in ΔG^0 expressed in Kcal/mole of complex

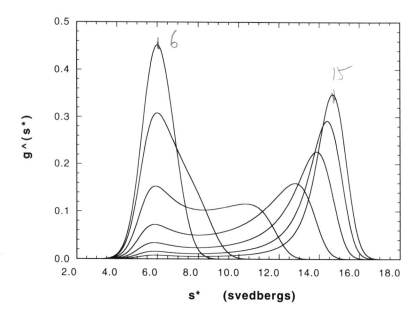

Figure 2. Monomer tetramer system: s_1=6S, D_1=1.25F, s_4=15S; M_1=440000, speed=40000rpm, t=4320sec. Curves are for the values of θ listed in Table II

TABLE II						
monomer tetramer system: s_1=6S, D_1=1.25F, s_4=15S.						
A	B	C	D	E	F	G
$\theta=kc^3$	$c_p=\int\hat{g}ds$	s_w	α	$k_{obs}c_p^3$	k_{obs}	$\delta\Delta G^o$
0.01	0.9120	6.069	0.0077	0.00079	0.0087	0.08
0.10	0.9049	6.569	0.0632	0.0820	0.0906	0.06
1.0	0.8818	8.236	0.248	0.775	0.879	0.08
10.	0.8530	10.40	0.489	7.172	8.41	0.10
100	0.8309	12.14	0.682	66.69	80.2	0.13
1000	0.8167	13.30	0.811	635.6	778.2	0.15
10000	0.8081	14.01	0.890	6079.	7522	0.17
Columns are the same as for Table I						

obtained for this example. In the case of the monomer-tetramer system, and also higher degrees of association (not shown here) there is a systematic but tolerable error especially if it is expressed per mole of monomer. Expressed on a monomer basis, it would correspond more closely to the error in the value of ΔG^o for inter-subunit contacts, the exact value depending on the arrangement of subunits in the complex.

Hetero-associating Systems:

Hetero-associating systems of the following general type were simulated and then analyzed.

$$A + B = C \qquad (12)$$
$$C + B = D \qquad (13)$$

Starting with amounts of A and B in molar ratio 1:2 or 1:1, sedimentation patterns were simulated for various values of the equilibrium constants defined by the following equations

$$k_1 = \frac{c_c}{c_a c_b} \qquad (14)$$

$$k_2 = \frac{c_d}{c_c c_b} \qquad (15)$$

where c_i (i=a,b,c,d) is the concentration of species i in g/L.

If one writes that the total weight concentration c_t is given by

$$c_t = c_{a,tot} + c_{b,tot} \qquad (16a)$$

then

$$f_{a,o} + f_{b,o} = 1 \qquad (16b)$$

where $f_{a,o}$ and $f_{b,o}$ are the weight fraction of total A and B, respectively, in the mixture.

Now, we can write

$$\theta_1 \equiv k_1 c_t = \frac{f_c}{f_a f_b} \qquad (17a)$$

and

$$\theta_2 \equiv k_2 c_t = \frac{f_d}{f_c f_b} \qquad (17b)$$

Conservation of mass requires that

$$f_{o,a} = f_a + f_c \frac{M_a}{M_c} + f_d \frac{M_a}{M_d} \qquad (18a)$$

$$f_{o,b} = f_b + f_c \frac{M_b}{M_c} + 2 f_d \frac{M_b}{M_d} \qquad (18b)$$

where the M_i (i=a,b,c,d) the the molecular weights of species i.

Substituting and rearranging we have

$$f_a = f_{o,a} / [\ 1 + \theta_1 f_b \frac{M_a}{M_c} + \theta_1 \theta_2 f_b^2 \frac{M_a}{M_d}\ ] \qquad (19a)$$

$$f_b = f_{o,b} / [\ 1 + \theta_1 f_a \frac{M_b}{M_c} + 2\theta_1 \theta_2 f_a f_b \frac{M_b}{M_d}\ ] \qquad (19b)$$

$$f_c = \theta_1 f_a f_b \qquad (19c)$$

$$f_d = \theta_2 f_b f_c = \theta_1 \theta_2 f_a f_b^2 \qquad (19d)$$

These equations can be solved for f_a, f_b, f_c and f_d as a function of θ_1 and θ_2 by iteration starting with an initial guess for f_b. Equations 19a and 19b were solved iteratively for f_a and f_b which in turn were substituted into equations 19c and 19d to obtain f_c and f_d.

The weight average sedimentation coefficient for this system is given by

$$s_w = f_a s_a + f_b s_b + f_c s_c + f_d s_d \qquad (20)$$

where $f_a + f_b + f_c + f_d = 1$; and f_a, f_b, f_c, and f_d are the weight fractions of each species at the plateau concentration. The value of s_w is a unique function of the total macromolecular concentration for a given starting ratio of total A to total B and values of k_1 and k_2.

The Case A+B=C; $[B]_o = [A]_o$

First, we will examine the case represented by equation 12 describing the binding of just one mole of B to A so that $\theta_2 = 0$. Once a value of $f_{a,o}$ has been chosen, this is a single parameter system, the parameter being θ_1. The values of f_a, f_b and f_c are uniquely determined by the value of the dimensionless variable θ_1. Figure 3 shows a series of plots of $g(s^*)$ vs. s^* for various values of θ_1. One set of these curves expressed in terms of the reduced variable θ_1 will not describe all possible interacting systems because A and B can have any possible molecular weights and sedimentation coefficients. However, one set of curves will suffice for any particular system. Table III summarizes results of the analysis of the curves shown in Figure 3. Again, the accuracy of the values of K_{obs} obtained from the analysis is quite good.

The Case A+B=C; C+B=D; $[B]_o = [2A]_o$

For the more complex case of A+B=C; C+B=D, we will restrict ourselves to a simplified version of the problem by treating the case for which the intrinsic molar binding constant, K_{int}, for binding of each molecule of B, is the same at each site on A and independent of whether or not a molecule of B is already bound at the other site. In this case,

$$K_{int} = 0.5 K_1 = 2.0 K_2 \qquad (21)$$

where K_1 and K_2 are the macroscopic association constants for the reactions .

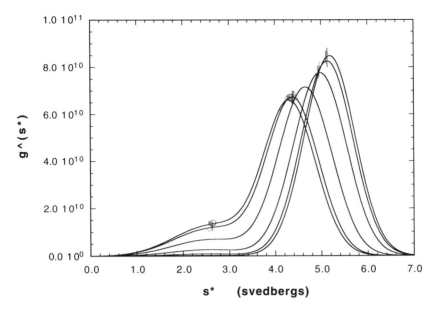

Figure 3. System A+B=C: Curves from left to right correspond to values of θ listed in Table III. $s_A = 2.6S$, $s_B = 4.3S$, and $s_C = 5.2S$; $M_A = 26000$, $M_B = 90000$, $M_C = 116000$. Speed=60000 rpm, t= 6000sec.

TABLE III						
Hetero-associating system A+B=C: $[A]_0=[B]_0=1 \times 10^{-7}M$ $s_1=2.6S$, $s_2=4.3S$, $s_3=5.2S$; $M_A=26000$, $M_B=90000$						
A	B	C	D	E	F	G
θ_0	$c_p=\int \hat{g} ds$ $\times 10^3$	s_w	$[A]_p$ $\times 10^5$	θ_{obs}	K_{obs}	$\delta \Delta G^0$
0.01	11.26	3.931	9.707	0.009707	1.00×10^5	0.00
0.10	11.25	4.023	9.698	0.09601	0.99×10^6	0.01
1	11.22	4.397	9.672	0.9479	0.98×10^7	0.01
10	11.18	4.847	9.638	9.541	0.99×10^8	0.01
100	11.16	5.077	9.621	98.13	1.02×10^9	0.01
1000	11.15	5.162	9.612	961.2	1×10^{10}	0.00
Column A: $\theta_0=K[A]_0=\theta$ at the initial loading concentration Column D: $[A]_p$ was computed from column B using $M_c=116000g/mole$. Column E: Value of θ required to give s_w for $[A]_p$. Column F: Value of $K_{obs} = \theta_p/[A]_p$. Column G: Error in ΔG^0 expressed in Kcal per mole of C.						

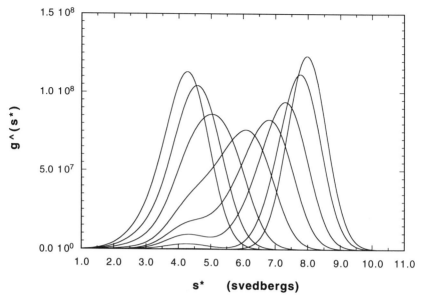

Figure 4. A+B=C; C+B=D for θ ranging from 0.01 to 1000 (see Table IV) from
left to right for the case $[B]_0=2[A]_0$, and $K_{int} = 0.5K_1 = 2K_2$. $S_A= 3.4S$,
$S_B= 4.3S$, $S_C= 6.0S$, $S_D= 8.0S$, $M_A= 52000$, $M_B= 90000$, $M_C= 142000$,
$M_D= 232000$. Speed=60000rpm; t=3600 sec. The y-axis units are arbitrary.

TABLE IV						
Hetero-associating system A+B=C; C+B=D: $[A]_0=1 \times 10^{-7}M$ $s_1=3.4S$, $s_2=4.3S$, $s_3=6.0S$, $s_4=8.0S$; $M_A=52000, M_B=90000$						
A	B	C	D	E	F	G
θ_0	$c_p=\int \hat{g}ds$ $\times 10^3$	s_w	$[A]_p$ $\times 10^8$	θ_p	$K_{int,obs}$	$\delta\Delta G^0$
0.010	20.62	4.132	8.886	0.00712	0.79×10^5	0.13
0.100	20.44	4.462	8.806	0.0907	1.03×10^6	0.02
0.333	20.18	4.912	8.696	0.274	3.15×10^6	0.03
1.0	19.78	5.627	8.525	0.904	1.07×10^7	0.04
3	19.35	6.311	8.341	2.65	3.18×10^7	0.03
10	19.00	7.458	8.188	8.68	1.06×10^8	0.03
100	18.62	7.612	8.022	83.4	1.04×10^9	0.02
1000	18.47	7.868	7.957	796.	1.00×10^{10}	0.00
Column A: $\theta_0 = \theta$ at the initial loading concentration.						
Column D: The total molar concentration of A in the plateau from column B.						
Column E: The value of θ_p obtained from s_w and $[A]_p$.						
Column F: The value of K_{int} obtained by dividing θ_{obs} by $[A]_p$.						
Column G: Magnitude of the error in ΔG^0 in Kcal per mole of complex, D.						

represented by equations 12 and 13.

It is convenient to reparameterize the system in terms of another dimensionless variable, call it θ, such that

$$\theta \equiv K_{int} [A]_0 = \left((\theta_1 \theta_2) \left(\frac{M_a M_b^2}{M_d^3} \right) \right)^{1/2} \tag{21}$$

For any given set of initial concentrations, molecular weights, intrinsic binding constant, and sedimentation coefficients for A, B, C and D, the composition, and, therefore, the weight average sedimentation coefficient, is uniquely determined by the value of θ.

Figure 4 shows a series of curves for the system having molecular weights 52000 and 90000 for A and B, respectively. The initial concentrations are in the stoichiometric ratio of 1:2 and the successive macroscopic molar equilibrium constants are in the ratio 4:1. The curves are normalized so that the different curves correspond to different values of K_{int} for the same loading concentration of $23.2\mu g/ml$. Table IV summarizes the results of analysis of the simulated curves for this system. Again the errors in determining ΔG^o, although systematic, are quite small and probably within the usual experimental error.

EFFECT OF PRESSURE

This section describes some of the ways that pressure effects can be revealed in g(s*) patterns. The effect of pressure on interacting systems for which there is a volume change on association can be dramatic if the molar volume change is sufficiently large(Cann, 1970; Harrington and Kegeles, 1973). Supramolecular assemblies composed of many subunits, each contributing a small volume change on association, can exhibit strong pressure sensitivity since the total volume change per mole of complex can become quite large(Potschka and Schuster, 1987).

The radial dependence of the equilibrium constant due to the pressure gradient in the ultracentrifuge cell is given by the following relationship (Kegeles et al., 1967; Cann, 1970; Harrington and Kegeles, 1973)

$$K(r)=K(r_m)EXP\left(- \frac{\omega^2(r^2-r_m^2)\rho\Delta V}{2RT} \right) \tag{22}$$

where K(r) is the value of K at radius r, $K(r_m)$ is the value of K at the meniscus $(P=\approx 1atm)$, ρ is the solution density, ΔV is the molar volume change in cc/mole of complex, R is the gas constant and T is the absolute temperature.

For the systems simulated here, the molar volume change would have to be larger than about ± 100 cc/mole at 60 000 rpm before the effects would be of sufficient magnitude that they would have to be taken into account. Figure 5 shows g(s*) patterns for the system A+B=C for $\theta=1$, $K=1x 10^7 M^{-1}$ and $[A]_0=[B]_0=1 x 10^{-7}M$ with volume changes of 0.0, +100 and -100 cc/mole.

Values of s_w obtained were 4.33, 4.40, and 4.46S, respectively, and differ by only a few per cent. This difference would not introduce serious error into the computation of the equilibrium constant for this system.

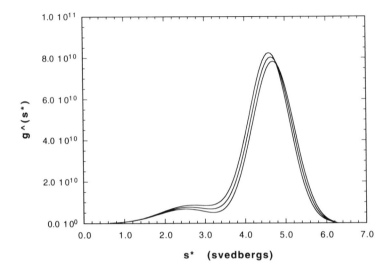

Figure 5. Hetero-associating system A+B=C with a volume change of association. Patterns for ΔV = +100, 0, and -100 cc per mole of C from left to right. K=1x $10^7 M^{-1}$ and $[A]_0=[B]_0=1 \times 10^{-7}M$. Speed=60000rpm and t=6000sec. Same system as shown in Figure 3. See text for explanation.

Pressure dependence for the system A+B=C; C+B=D is shown in Figure 6 for ΔV = +100, 0.0, and -100 cc/mole. Even for the seemingly quite large differences in the patterns, the estimated value of the equilibrium constant is not in serious error. Table V gives the values of s_w computed from g(s*) as well as the magnitude of the error in ΔG^0 and shows that the effect of pressure for this system could be disregarded for most practical purposes. Larger volume changes would have to be explicitly taken into account.

In general, for pressure dependent systems, the gradient in the region centrifugal to the boundary will not be zero and the integration carried out to compute s_w will not be strictly correct; however, if the volume change is not too large, the approximation introduced by stopping the integration at an arbitrary place in the region centrifugal to the boundary will not be sufficiently large to obviate the use of s_w computed this way.

A characteristic of pressure dependent associations that is evident in the simulated example shown in Figures 7 and 8 is the potential development of negative concentration gradients either within the boundary or in the region centrifugal to it. Whether or not negative gradients will form in any particular system will depend on the sign and magnitude of the volume change as well as the magnitude of the diffusion coefficient (Johnson et al., 1973). Negative

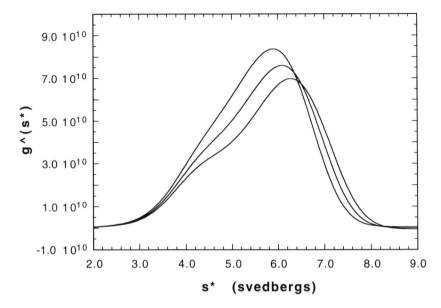

Figure 6. System A+B=C; C+B=D for the case θ=1.0 with ΔV= +100, 0, -100 cc/mole of D from left to right. See Table V for the analysis.

Table V						
Effects of Pressure Dependence for the system A+B=C; C+B=D. $[A]_0=1 \times 10^{-7}M$ and $K_{int}=1 \times 10^7 M^{-1}$						
ΔV cc/mole	$\int \hat{g} ds$ $\div 10^{-3} g/L$	s_w	$[A]_p$ $\div 10^{-8} M$	θ_{obs}	$K_{int,obs}$ $\times 10^{-7} M$	$\delta \Delta G^o$ Kcal/mol
+100.	21.23	5.530	9.151	0.7778	0.85	0.09
0.	19.77	5.628	8.522	0.9033	1.06	0.03
-100.	18.32	5.726	7.810	1.054	1.35	0.16

concentration gradients can be sustained only in the presence of a stabilizing gradient of an additional substance. Often redistribution of buffer components can be sufficient to stabilize the boundary against convection if they are present at high enough concentration. Otherwise, it may be necessary to add an inert background component like sucrose to provide the necessary stabilization.

The time dependence of the apparent distribution function can be used as a convenient device to detect pressure dependence. When sedimentation data are presented in terms of g(s*) vs. s*, the development of the pattern is characteristic of the sign and magnitude of the volume change. One feature of g(s*) patterns for systems that are neither pressure dependent nor kinetically controlled is that they are relatively broad at early times and become narrower and taller with increasing time with little or no drift in the positions of the peaks. However, for pressure dependent systems, the peak positions may not only drift to higher or lower values of s*, but also may decrease with time.

Four examples at two extreme values of θ are show in Figure 7 for the system A+B=C; C+B=D for θ=100 with a volume change of -1000cc/mole (Figure 7A) and 0.0 cc/mole (Figure 7B) and for θ=0.01 with a volume change of 0.0cc/mole (Figure 7C) and +1000cc/mole (Figure 7D). Three examples at a value of θ=1.0 are shown in Figure 8A,B,C&D for the system A+B=C; C+B=D with volume changes of +1000, 0 and -1000cc/mole, respectively.

In the case of a positive volume change (θ=100, Figure 7A and θ=1.0, Figure 8A), the maxima in the g(s*) plots tend to lower values of s* with increasing time reflecting increasing dissociation as the polymer system experiences higher and higher pressures while an overall negative gradient tends to develop in the region centrifugal to the boundary.

In the case of a negative volume change, for a weakly associating system, θ=0.01 (Figure 7D), the curves start to increase at earliest times but as the boundary progresses down the cell, they eventually decrease with time. The plateau concentration for this system decreases much more rapidly than expected based on the peak position of the g(s*) pattern. In fact, for this system, the average value of s_w computed by following the decrease in the plateau concentration is about 15S, explaining the decrease in the area of the peak of g(s*) with time. A plot of $\ln(c_p)$ vs t shows considerable negative curvature as well. A slight negative gradient develops in the boundary and a relatively large positive concentration gradient is seen in the region centrifugal to the boundary during the entire simulation (Not shown).

For the case θ=1.0, for which there are significant amounts of all species present, and a positive volume change of 1000cc/mole (Figure 8A), there is a negative gradient in the region centrifugal to the boundary throughout the simulation. As for the case shown in Figure 7A, the peak in g(s*) tends to drift to lower values of s* while it becomes sharper. Again the case for the positive volume change shows a positive gradient in the region centrifugal to the boundary and the position of the tallest peak in the g(s*) plot drifts to higher values of s* reflecting increasing association with increasing pressure. A slight negative gradient develops in the boundary between the smaller and larger peaks.

Incidentally, when there is a significant gradient in the region centrifugal to the boundary for any reason, it may not be possible to a obtain a reliable value of s_w from either g(s*) or from a second moment analysis. and, therefore, other methods of analysis must be employed. Visual comparison of simulated g(s*) plots to the experimental data can be a helpful aid in understanding the behavior of these systems.

Pressure dependence of several experimental systems was investigated by Kegeles (Kegeles, 1970; Kegeles and Johnson, 1970; Kegeles and Johnson, 1970). These papers should be consulted for examples. From a comparison of simulated sedimentation profiles for various volume changes of association, it was inferred that a significant volume change on association must be occurring for the case of the self-association of α-chymotrypsin. Volume changes associated with polymerization of myosin, for example, have been reported to be on the order of +380 cc/mole of monomer (Josephs and Harrington, 1968) and

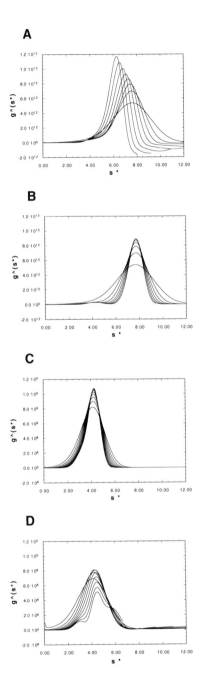

Figure 7. Time dependence of g(s*) patterns for the system A+B=C; C+B=D; Times are 600sec apart at 60,000 rpm. (A) $\theta=100$, $\Delta V=+1000cc/mole$ (B) $\theta=100$, $\Delta V=0cc/mole$, (C) $\theta=0.01$, $\Delta V=0cc/mole$.and (D) $\theta=0.01$, $\Delta V=-1000cc/mole$. Broader curves are for earlier times; see text for explanation.

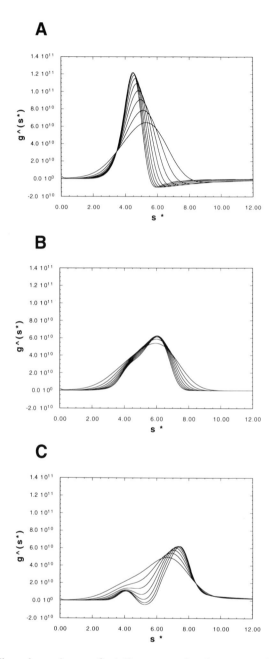

Figure 8. Time dependence of g(s*) patterns for the system A+B=C; C+B=D; θ=1.0. Values of s_i and M_i are the same as for Figure 4. Times are 600sec apart, at 60,000 rpm. (A) ΔV=+1000cc/mole, (B) ΔV=0cc/mole, and (C) ΔV=-1000cc/mole. See text for explanation. The broader curves correspond to the earlier times.

have been shown to produce dramatic effects on the sedimentation patterns observed for the polymerization of monomers to form filaments at pH 8.3.

Pressure effects are best demonstrated by overlayering the solution with oil or solvent or by pressurizing with nitrogen gas to produce different pressures at the same speed, rather than by simply changing the speed, since other types of associating systems, namely kinetically controlled interactions, may also exhibit speed dependence (Harrington and Kegeles, 1973).

CONCLUSIONS:

This chapter has discussed the use of the apparent sedimentation coefficient distribution function, $g(s^*)$ vs. s^*, as a tool for studying interacting systems especially at low concentrations. The apparent distribution function can be computed from the time derivative of sedimentation concentration curves. The relatively high precision afforded by combining use of the time derivative with signal averaging allows the analysis of systems at total concentrations of a few micrograms per milliliter with the Rayleigh optical system.

The simulations presented here were carried out in the absence of added random noise. Observed root mean square noise levels for a single time derivative pattern are typically on the order of ± 0.004 fringes corresponding to about $\pm 0.1 \mu g/ml$ (Yphantis et al., 1994). When patterns are combined by averaging from 10 to 40 Rayleigh photographs spanning a time interval of a few minutes, the noise level can be reduced by a factor of from 3 to 7 further attenuating the root mean square noise level to ± 0.03 to $\pm 0.014 \mu g/ml$.

The weight average sedimentation coefficient can be estimated from the $g(s^*)$ patterns by simple integration according to equations 5 and 6 and if one knows the sedimentation coefficient of each species as well as the stoichiometry, one may obtain an accurate estimate of the equilibrium constants and standard free energies describing the system.

Moreover, the apparent distribution function offers a convenient way to present sedimentation data so that pressure effects may be identified. In the absence of convection, the time dependence of the $g(s^*)$ patterns can be quite characteristic of the sign and magnitude of the volume change for the reaction. Pressure effects may make analysis of associating systems by sedimentation analysis difficult unless one is aware of and can recognize the characteristics of pressure dependent sedimentation. In most cases, the effects of pressure can be minimized or practically eliminated by running at lower speeds. The pressure effect is proportional to $\omega^2 \Delta V$ and will be the same for equal values of this product (see equation 22).

---------------   Glossary of Symbols   ---------------

$[A]_0$	total molar loading concentration of species A. (moles-L^{-1})
$[B]_0$	total molar loading concentration of species B. (moles-L^{-1})
$[A]_p$	total molar concentration of species A in the plateau. (moles-L^{-1})
c	weight-concentration (g-liter^{-1})
c_0	cell loading weight-concentration (g-liter^{-1})

c_p weight-concentration in the plateau region (g-liter^{-1})

D_i diffusion coefficient of species i.

f_i the weight fraction of species i.

$f_{o,i}$ the total weight fraction of species i.

g normalized sedimentation coefficient distribution function. (svedbergs^{-1})

$\hat{g}$ unnormalized sedimentation coefficient distribution function (g-L^{-1}-svedberg^{-1})

k equilibrium constant on the mass scale (L/g)

K macroscopic molar equilibrium binding constant (L-mole^{-1})

K_{int} intrinsic molar equilibrium constant (L-mole^{-1})

M_i molecular weight of species i (g-mole^{-1})

r radius (cm)

r_m radius of the meniscus (cm)

R the gas constant

s_i sedimentation coefficient of species i (svedbergs)

s_w weight average sedimentation coefficient (svedbergs)

s^* the apparent sedimentation coefficient (svedbergs=10^{-13}sec^{-1})

T absolute temperature (^{o}K)

t time (sec)

α weight fraction of polymer for monomer-n-mer system

ΔV molar volume change of association (cc/mole of complex)

ΔG^o standard Gibbs free energy (Kcal-mole^{-1})

$\delta \Delta G^o$ error in the standard Gibbs free energy (Kcal-mole^{-1})

θ a dimensionless parameter equal to the product of the binding constant and the total protein concentration on the same concentration scale.

ρ solution density (g/cc)

ω angular velocity (radians-second^{-1})

ACKNOWLEDGEMENT

This work was supported in part by NIH grant U01 CA-51880.

REFERENCES

Cann, J. R. (1970). Interacting Macromolecules. New York, Academic Press.

Claverie, J.-M. (1976). "Sedimentation of generalized systems of interacting particles III. Concentration dependent sedimentation and extension to other transport methods." Bioplymers 15, 843-857.

Claverie, J. M., H. Dreux and R. Cohen (1975). "Sedimentation of generalized systems of interacting particles. I. Solution of systems of complete Lamm equations." Biopolymers 14, 1685-1700.

Cox, D. J. and R. S. Dale (1981). "Simulation of Transport Experiments for Interacting Systems." in Protein-Protein Interactions Ed. C. Frieden and Nichol, L.W. New York, John Wiley & Sons. 173-212.

Harrington, W. F. and G. Kegeles (1973). "Pressure Effects in Ultracentrifugation of Interacting Systems." in Methods in Enzymology, 27D Eds. C. H. W. Hirs and S. E. Timasheff. New York, Academic Press. 306-345.

Johnson, M. L., D. A. Yphantis and G. H. Weiss (1973). "Instability in Pressure-Dependent Sedimentation of Monomer-Polymers Systems." Biopolymers 12, 2477-2490.

Josephs, R. and W. F. Harrington (1968). "On the stability of myosin filaments." Biochemistry 7, 2834-47.

Kegeles, G. (1970). "Effect of Pressure on Sedimentation Velociy Patterns." Arch. Bioch. Bioph. 141, 68-72.

Kegeles, G. and M. L. Johnson (1970). "Effect of Pressure on Sedimentation Velocity Patterns. II. Myosin at pH 8.3." Arch. Bioch. Bioph. 141,

Kegeles, G. and M. L. Johnson (1970). "Effects of Pressure on Sedimentation Velocity Patterns. I. Alpha Chymotrypsin at pH 6.2." Arch. Bioch. Bioph. 141, 59-62.

Kegeles, G., L. Rhodes and J. L. Bethune (1967). "Sedimentation Behavior of Chemically Reacting Systems." Proc. Natl. Acad. Sci. 58, 45-51.

Liu, S. and W. F. Stafford (1992). "A Real-Time Video-Based Rayleigh Optical System for an Analytical Ultracentrifuge Allowing Imaging of the Entire Centrifuge Cell." Biophys. J. 61, A476, #2745.

Potschka, M. and T. M. Schuster (1987). "Determination of Reaction Volumes and Polymer Distribution Characteristics of Tobacco Mosaic Virus Coat Protein." Anal. Biochem. 161, 70-79.

Stafford, W. F. (1992). "Boundary Analysis in Sedimentation Transport Experiments: A Procedure for Obtaining Sedimentation Coefficient Distributions Using the Time Derivative of the Concentration Profile." Anal. Biochem. 203, 295-301.

Stafford, W. F. (1992). "Sedimenation Boundary Analysis: An averaging Method for Increasing the Precision of the Rayeigh Optical System by Nearly Two Orders of Magnitude." Biophys. J. 61, A476,(#2746).

Stafford, W. F. (1994). "Methods of Boundary Analysis in Sedimentation Velocity Experiments." in Numerical Computer Methods, Part B., Methods in Enzymology in press Eds. L. Brand and M. L. Johnson. Orlando, Academic Press.

Yphantis, D. A., W. F. Stafford, S. Liu, P. H. Olsen, J. W. Lary, D. B. Hayes, T. P. Moody, T. M. Ridgeway, D. A. Lyons and T. M. Laue (1994). "On line Data Acquistion for the Rayleigh Interference Optical System of the Analytical Ultracentrifuge." in MODERN ANALYTICAL ULTRACENTRIFUGATION: Acquisition and Interpretation of Data for Biological and Synthetic Polymer Systems Eds. T. M. Schuster and T. M. Laue. Boston, Birkhäuser Boston, Inc.

STUDIES OF MACROMOLECULAR INTERACTION BY SEDIMENTATION VELOCITY

James C. Lee and Surendran Rajendran

INTRODUCTION

As our knowledge expands and new systems are investigated, it becomes clear that elucidating precise biochemical regulatory mechanisms requires detailed understanding of macromolecular assembly processes. For example, the regulation of gene expression involves an intricate network of protein-protein and protein-nucleic acid interactions and the mechanism of some allosteric enzymes is linked to subunit assembly. In order to define these mechanisms one needs information on the identities of proteins in the complex, the affinities of these proteins for each other and the effects of regulators on the formation of these complexes. A direct way of studying macromolecular assembly is to monitor the resulting changes in mass as a consequence of the formation of these macromolecular complexes. One of the methods that enables one to directly monitor the mass of macromolecules is the transport technique. Among the transport methods sedimentation analysis is the technique of choice because of the sound fundamental principles on which the method is based and because of its resolving power. Excellent reviews on the applications of sedimentation equilibrium in studying macromolecular self-associations and heteropolymers formation are included in Part I of this volume. In this chapter the focus is on applications of sedimentation velocity. One of the advantages of sedimentation velocity over that of equilibrium is its speed of analysis, e.g., a run can be completed within an hour whereas equilibrium experiments may take much longer. Hence, if a biological sample is unstable the study on that system may have to be conducted using sedimentation velocity. However, a larger amount of sample will be required for velocity analysis than for equilibrium measurements, e.g., the calf brain tubulin and rabbit muscle phosphofructokinase systems require a few mg of protein to define a curve of sedimentation coefficient vs concentration.

In studying any system that involves macromolecular assembly the primary goal is to establish the mode of association. Basically that means determination of stoichiometry and the equilibrium constants that govern the reaction. Furthermore, information on the hydrodynamic properties of the associating species will be useful in elucidating the quaternary structure of the complex, e.g., to define the configuration of subunit arrangements be it head-to-tail or side-by-side. Depending on the specific three-dimensional orientation of these subunits, the hydrodynamic properties of two complexes with the same number of subunits will be quite different although the masses will be the same. As sedimentation velocity data contain information on both size and shape of the macromolecular species, this technique can provide information on the quaternary structure.

Macromolecular assembly processes invariably are linked to ligand-binding. The ligand with the simplest structure is H^+ and the most complex is a biological macromolecule, be it proteins, nucleic acids, glyco-conjugates or lipid components. Hence, it is important to be cognizant of the potential complexity of ligand-facilitated or - mediated assembly processes reviewed by J. Cann later in Part III of this volume and only the practical aspects of identifying this process will be dealt with in this chapter, which will specifically concentrate on protein self assembly. In this review the focus is on: a) data acquisition and analysis; b) model fitting and c) simulation of sedimentation patterns based on the conclusion derived from (a) and (b) so that the simulated data can be directly compared with the primary experimental data acquired in (a). By completing the cycle of (a) to (c) successfully, the investigator can then accept the derived model with its corresponding quantative parameters with reasonable confidence.

A.   Data Acquisition and Analysis:

In the sedimentation velocity method, the ultracentrifuge is operated at higher speeds compared to the equilibrium method, resulting in the sedimentation of macromolecules towards the bottom of the cell. This migration can be followed by any one of the following optical methods depending on the sensitivity required and convenience: (i) schlieren optical system which can detect concentrations greater than 1 mg/ml, (ii) interference optical system with a sensitivity extended to 0.1 mg/ml and (iii) UV scanning system which can monitor protein concentrations down to a few $\mu g/ml$. This migration of macromolecules results in a region containing only solvent molecules, in addition to the region in the cell where the concentration of solute is uniform. Between the solvent and the solution of uniform concentration known as the plateau region there is a transition zone in which the concentration varies with distance from the axis of rotation. The velocity technique essentially follows the movement of this transition zone, or boundary, as a function of time. Typical tracings of the concentration gradient as a function of distance in the ultracentrifuge cell at different intervals of time are shown in Fig. 1. The rate of movement of the boundary as a function of protein concentration contains information for defining the mode of association and the equilibrium constant that governs the reaction. Traditionally, these

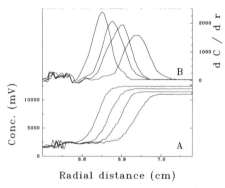

FIGURE 1. Velocity sedimentation profiles of rabbit pyruvate kinase at a concentration of 0.14 mg/ml. (A) The photoelectric scanner tracings of optical density at 230 nm in terms of mV of photomultiplier signal versus radial distance. (B) The respective concentration gradient plots. The rotor speed and temperature were 44000 rpm and 20°C.

boundaries are expressed as derivative curves of dc/dr $\underline{vs}$ r, where c and r are concentration and radial distance, respectively, as shown in Fig. 1. The shape of the peak and the number of peaks provide an excellent visual presentation of the association behavior of the system.

The mechanisms of protein self-association can be broadly classified into three types: (i) Formation of stable aggregates characterized by infinitely slow rate of reequilibration relative to the time of the experiment. Accordingly the sedimentation velocity patterns will resolve into two or more boundaries, therefore, two or more peaks corresponding to reactants and products will appear; conventional analysis of the patterns yields their mobilities and concentrations in the initial equilibrium mixture. A typical example of this is the slow but reversible dissociation of globulin arachin (Johnson and Shooter, 1950). (ii) A system in which the rate of reequilibration is rapid with respect to the time of the sedimentation experiment (Gilbert, 1955; Gilbert and Gilbert, 1973). In this case peaks in the transport patterns cannot be placed into correspondence with individual reactants and products. Typical examples of this are the tetramerization of ß-lactoglobulin (Timasheff and Townend, 1960; Townend et al., 1960a; Townend and Timasheff, 1960; Townend et al., 1960b; Timasheff and Townend, 1961; Kumosinski and Timasheff, 1966; Basch and Timasheff, 1967), the polymerization of calf brain tubulin (Frigon and Timasheff, 1975 a,b) and the subunit assembly of rabbit muscle phosphofructokinase (Hesterberg and Lee, 1981). (iii) A system in which the rate of reequilibration is rapid, but the self-association is mediated by ligand(s) (Cann and Goad, 1972; 1973; Cann, 1970; Cann, 1994, this volume).

Though there are various procedures recommended for distinguishing Case (i) from Case (ii) or (iii), the most unambiguous interpretation is based on a fractionation test of the sample employing the Yphantis-Waugh partition cell (Yphantis and Waugh, 1956). As the sedimenting boundary progresses down the cell, fractions from the slow moving peak can be collected. These fractions can be subjected to sedimentation experiments separately after concentrating the samples to the initial concentrations followed by reequilibration against the same buffer. If bimodality is due to an association-dissociation reaction, the fractions would exhibit a sedimentation profile identical to that of the unfractionated sample, whereas for heterogeneity the sedimentation profiles are not identical.

Another test for the presence of rapid re-equilibration is to conduct the experiments as a function of rotor speed. If a self associating system were in a slow equilibrium or if noninteracting or denatured components of the system were present, then the sedimentation pattern would be expected to change as a function of the angular velocity, ω, which in turn is directly related to the rotor speed.

After establishing that the system is indeed undergoing rapid association-dissociation, the sedimentation patterns should be determined as a function of protein concentration. The specific shape of the derivative curves can be employed to qualitatively diagnose the mode of association.

i. Gilbert system - formation of a polymer with n subunits.

Let us examine the reaction scheme of

$$nM \overset{K}{\rightleftharpoons} M_n \tag{1}$$

where M and M_n are the monomeric and polymeric species; n is the degree of polymerization and

$$K = C_p / C_m^n \qquad \text{(2)}$$

where C_p and C_m are concentrations of the polymer and monomer, respectively.

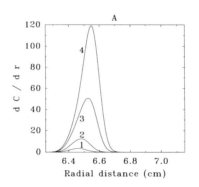

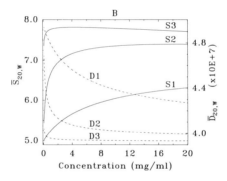

FIGURE 2.   A.   Simulated velocity sedimentation derivative profiles for a theoretical dimerization scheme. The rotor speed, sedimentation time, and K_2 value are 60000 rpm, 40 min and 0.1 (ml/mg) respectively. The loading protein concentrations in mg/ml are 0.5, 2.0, 9.0 and 20 respectively.   The sedimentation coefficient of monomer is 5.0 S. The number by the side of a curve corresponds to the loading protein concentration in increasing order.
          B.   Simulation of concentration dependence of weight average sedimentation (—) and diffusion (‒) coefficients for a theoretical dimerization scheme. The K_2 values in ml/mg used for the simulation are 0.1, 3.0 and 300. The number by the side of each curve represnets the K_2 value in increasing order.

          For n = 2, the derivative curves of the sedimentation boundaries characteristically show a single peak throughout the concentration range employed in the study, as shown in Fig. 2. The sedimentation coefficient exhibits a distinctive dependence on concentration. As shown in curve 3 in Fig. 2B, starting at high protein concentration, with decreasing concentration the sedimentation coefficient increases as a consequence of a reduction of the hydrodynamic nonideality effects, then the s value passes through a maximum. With further decrease in concentration the value of the sedimentation coefficient decreases due to a progressive dissociation into monomers under the influence of mass action. Extrapolation of the value of the sedimentation coefficient to infinite dilution leads to the determination of the sedimentation coefficient of the monomeric kinetic unit of association.

For $n \geq 3$, the derivative curve of the sedimentation boundary will show a single symmetrical peak at low protein concentration. As the concentration of the polymer increases with increasing total protein concentrations, a shoulder develops on the leading edge of the peak. Upon further increase in protein concentration bimodality develops. In addition to resolution of the reaction boundary into two peaks for a system with $n \geq 3$, there is another distinctive feature in the sedimentation profile. At lower concentrations, the single peak grows in area as the protein concentration is increased. Upon further increase of protein concentration bimodality sets in and the fast moving peak grows in area, while the area of the slower moving peak remains constant, as shown in Fig. 3. The constancy of the area of the slower peak with increasing total macromolecule concentration is diagnostic for a higher order polymerization reaction, as shown in Equ. 1.

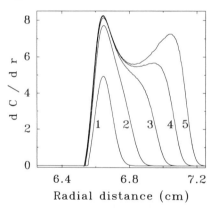

FIGURE 3. Simulated sedimentation velocity derivative profiles for a tetramerization scheme. The rotor speed and sedimentation time are 60000 rpm and 40 min., respectively. K_4 and s_1^0 are 0.01 (ml/mg)^3 and 8.0 S, respectively. The protein concentrations are 1, 2, 3, 4, and 5 mg/ml.

ii. Isodesmic - indefinite self-association.

Another mode of self-association is indefinite self-association with a scheme of

$$
\begin{array}{lll}
M_1 + M_1 & \rightleftharpoons M_2 & K_2 = m_2/m_1^2 \\
M_1 + M_2 & \rightleftharpoons M_3 & K_3 = m_3/m_2 m_1 \\
\quad : \quad : & \quad : & \\
M_1 + M_{i-1} & \rightleftharpoons M_i & K_i = m_i/m_{i-1} m_1
\end{array}
\tag{3}
$$

where M_i denotes the ith aggregate, m_i is the molal concentration of the ith species, and K_i is the equilibrium constant for successive bond formation between monomer and ith-mer. If all K_i's are equal, the extent of the association is considered as linear indefinite and the reaction is sometimes referred to as isodesmic (Holloway and Cox, 1974). The derivative curves of the sedimentation boundaries for such a scheme of association do not exhibit bimodality throughout the concentration range. Instead, peaks skewed towards the leading edge will be observed. The degree of skewness is proportional to the association constant. The stronger the affinity the more asymmetric is the peak.

iii. Ligand induced self-association.

Velocity sedimentation technique is highly sensitive to changes in the state of association of macromolecules induced by ligands. The mechanism of these ligand induced macromolecular associations can be one of two major categories, namely, ligand mediated or ligand facilitated. In the case of ligand-mediated association obligatory binding of ligand to macromolecule should precede association, whereas in the case of a ligand-facilitated mechanism the ligand binds to the aggregated state of the macromolecule. The distinction between the two mechanism can be made by judicious application of ligand binding studies as a function of protein concentration in conjunction with macromolecular assembly as a function of ligand concentration (Timasheff et al., 1991).

Ligand induced self-association may also exhibit bimodal sedimentation profiles, hence, bimodality of a sedimentation boundary indicates either the presence of cooperativity in the macromolecular self-association or the induction of self-association by a strong binding of a ligand or both effects. The former is known as a Gilbert system which was earlier described and the latter is known as a Cann-Goad system. In the Cann-Goad system the bimodality arises due to a concentration gradient of unbound free ligand across the sedimentation boundary caused by the strong stoichiometric ligand-protein interaction (Cann and Goad, 1970; 1972; 1973). Let us consider the ligand- mediated polymerization reaction,

$$mM + nX \rightleftharpoons M_m X_n \qquad \text{(4)}$$

in which the macromolecule, M, associates into a polymer, M_m, medicated by the binding of n ligand molecules or ions, X, to form the complex, $M_m X_n$. Resolution of the reaction boundary or the reaction zone into two peaks is dependent upon generation of stable concentration gradients of unbound ligand across the sedimentation boundary or zone by reequilibration during differential transport of M and $M_m X_n$. The areas under the two peaks do not correspond to the initial equilibrium composition. The sedimentation coefficients of the peaks are not the absolute sedimentation coefficients of the individual species. Furthermore, the sedimentation coefficients cannot be determined by extrapolation of the values to infinite dilution of macromolecule at constant ligand concentration. This is because of constant reequilibration of monomers to polymer within the boundary to maintain reaction equilibrium throughout the boundary. This serves essentially as a ligand pump which constantly removes unbound ligand from the region centripetal to the boundary to the cell bottom and thus generates a free ligand concentration gradient across the boundary. Similarly, for a strong ligand induced dissociation a free ligand concentration gradient would be generated across the boundary with the direction of the gradient reversed (Cann 1970; Cann and Goad, 1970; Cann, 1973; Cann, 1994; Part I this volume).

The bimodal sedimentation boundary of a Cann-Goad system emerges from the meniscus as a single peak which then resolves into a bimodal one after sedimentating a distance from the meniscus. In contrast, in a Gilbert system the boundary emerges from the meniscus as a bimodal one. Furthermore, a Cann-Goad mechanism is dependent on the maintenance of a stable free ligand concentration gradient across the boundary and is thus strongly dependent on rotor speed. At low rotor speed, the ligand concentration gradient tends to diffuse out, resulting in poorly resolved bimodal sedimentation boundaries. Differentiation of a mechanism of ligand induced self-association reaction as a Cann-Goad system or a Gilbert system can also be made by velocity sedimentation as a function of protein concentration. If the ligand optical absorption does not overlap with the protein absorption, with the aid of the UV/VIS scanner the total concentration of ligand across the entire cell can be monitored. In the case of a Cann-Goad system, with respect to bimodal protein peak, a unimodal ligand boundary would be observed whose peak corresponds to the fast moving protein polymer peak and a strong depletion of free ligand centripetal to the boundary is seen. At a constant ligand concentration, the Gilbert type of cooperative self-association shows a unimodal sedimentation boundary at low protein concentration and a bimodal boundary at high protein concentration. No strong depletion of free ligand is observed. Also, the critical protein concentration at which the fast moving peak emerges, decreases with increase in ligand concentration.

Theoretical moving boundary sedimentation patterns were generated for the ligand dimerization and tetramerization reactions which are dealt with by Cann furtheron in Part III, this volume. The vinblastine induced self-association of tubulin is the best documented case of a Cann-Goad system (Lee et al., 1975; Na and Timasheff, 1980).

B.   Model Fitting

Generally, a series of experiments of sedimentation velocity as a function of protein concentration is performed. From the sedimentation patterns one can tentatively select a model system for the association of the proteins, as described in Section A. Though the characteristic features of sedimentation patterns like, asymmetry, bimodality of the boundary and the minima between the peaks are distinctive indicators of the mode of association, selection and elucidation of a model for a self-association system is not a simple straightforward one. Also the self-association need not be following a single distinct model, it can be a mixture of models, such as a combination of isodemic and finite polymer formation examplified by the Mg^{2+} -induced self-association of brain tubulin (Frigon and Timasheff, 1975).

For an associating protein system in a rapid, dynamic equilibrium, the rate of sedimentation of the protein boundary is defined by the velocity of the square root of the second moment of the boundary and corresponds to the weight-average sedimentation coefficient, $\bar{s}$ (Schachman, 1959), since

$$\bar{s} = \frac{\sum_i s_i c_i}{\sum_i c_i} \tag{5}$$

where s_i and c_i are the sedimentation coefficient and concentration of the ith species, respectively. In expressing each sedimentation coefficient, s_i, as a function of the total protein concentration, c, then

$$\bar{s} = \frac{\sum_i s_i^0 (1-g_i c) K_i c_1^i}{\sum_i K_i c_1^i} \tag{6}$$

where s_i^0 is the sedimentation coefficient of the ith species at infinite dilution, g, is the nonideality coefficient, $c = \Sigma K_i c_1^i$, K_i is the association constant in units of ml/mg between any i-mer and the monomer, and c_1 is the monomer concentration in units of mg/ml.

Theoretical value of s as a function of c_i^i, s_i^0, g_i and K_i can be fitted to the observed weight-average sedimentation coefficients as a function of c; thus, the best fitted curve can yield information on the thermodynamic parameters which govern the self-association reaction. The fitting procedure requires good initial estimation of the parameters such as s^0 for monomer and polymer, g_i and K_i. Estimation of s^0 for monomer can be accomplished by extrapolation of $\bar{s}$ values to infinite dilution in an experimental condition that favors the dissociated state. On the contrary, estimation of s^0 for polymer should employ experimental condition that favors the associated state. In both cases the purpose is to minimize the concentration range through which the extrapolation has to be conducted, as illustrated by Fig. 2. Estimation for s_1^0 is very difficult for the data sets with K_2 of 3 and 300 (ml/mg). Significant error can be introduced by extrapolation of a curve with a steep slope. However, the data set with K_2 of 0.1 (ml/mg) provides a significant protein concentration range over which the value of s is essentially a linear function of protein concentration. A linear extrapolation of the data to infinite dilution should yield a reliable estimate of s_1^0. For the same rationale, s^0 for the polymer can be more accurately estimated for the data set with K_2 of 300 (ml/mg) in Fig. 2 by a linear extrapolation of the data at high protein concentration. In this case a value of 7.9 S, the exact value for a dimer, can be obtained.

For the purpose of using s_i^0 as an assay of molecular weight M_i, the following empirical equation

$$\frac{M_i}{M_1} = \left(\frac{s_i^0}{s_1^0}\right)^{3/2} \tag{7}$$

is used where M_1 and s_1^0 are the molecular weight and sedimentation coefficient of the monomer at infinite dilution, respectively. This relationship assumes an identical frictional ratio (or shape factor) for all species (Cann, 1970; Nichol et al., 1964). However, if distinct evidence regarding the morphology of the end polymer is available, employing the Kirkwood theory (Kirkwood, 1954), the sedimentation coefficient can be deduced. A classical example of the application of this theory is the calculation for a double ring structure of tubulin polymer, as seen in the election microscope (Frigon and Timasheff, 1975a).

Numerical fitting should be performed for all the possible pathway models available to form the end product. The change in sedimentation profiles as a function of protein concentration, as discussed in Section A, will provide a good qualitative estimation of the mode of association, which can also be inferred from the dependence of $\bar{s}_{20,w}$ on total protein concentration. If the plot of $\bar{s}_{20,w}$ vs total protein concentration shows a simple parabolic curve, the association is a simple stepwise one without cooperativity, i.e., isodesmic association. A sigmoidal curve is obtained for cooperative association, i.e., the

Gilbert mode of association. The degree of inflection of the sigmoidal curve and the number of inflection points reflect the degree of cooperativity and the minimum number of cooperative steps present, respectively.

As a rule, in fitting the experimental data with theoretical models, the simplest model with the least number of variables that can best fit the data should be the choice. The derived parameters of stoichiometry and equilibrium constants can in turn be used for computer simulation of sedimentation patterns employing suitable simulation methods described in Section C. The overall agreement between the experimental and calculated patterns indicate the possible validity of the model chosen. This approach has been successfully employed for the self-association of tubulin in the presence of Mg^{2+} (Frigon and Timasheff, 1975 a.b) and vinblastine (Na and Timasheff, 1980).

C.   Simulation of Sedimentation Profiles

If during data analysis sedimentation boundaries are expressed and analyzed in the form of weight average sedimentation coefficients some of the fine features, such as the shape of the peak, are lost. These features reflect significantly the particular model of association, as discussed earlier. Hence, a complete analysis of the association behavior of a system should include simulation of sedimentation profiles in the form of derivative curves so that the simulated data can be quantitatively compared with the experimental results.

During transport, the solute concentration at each point in the moving boundary changes with time, hence, the chemical equilibrium for the self-association system is perturbed. If the solute relaxes to the new equilibrium dictated by the altered total solute concentration and if the reequilibrating process occurs more rapidly than the local changes in solute concentration resulting from the transport process, the solute will be very close to local chemical equilibrium everywhere in the system throughout the experiment. In that case the solute can be treated as a single component (Fujita, 1962). The local sedimentation coefficient depends on the relative amounts of monomer and aggregates present, and the population of the solute species depends in turn on the total concentration. Since the solute concentration varies across the boundary, the local sedimentation coefficient also varies. Different parts of the boundary migrate at different rates, and the shape of the solute profile is distorted, resulting in distinctive boundary shapes characteristics for different self-associating solutes.

The simulation procedure requires a set of data of $\hat{s}$ and $\bar{D}$, the weight average diffusion coefficient, as a function of C based on the mode of association defined earlier in Section B.

The monomer (C_1) concentration is related to the total solute concentration, (C_T):

$$C_T = \sum_i K_i C_1^i \qquad (8)$$

where $K_1 = 1$ and the K_i is product of successive stoichiometric equilibrium association constants. C_1 is determined by a trial and error procedure described in Section B. This C_1 is inserted to Equ. 6 to calculate $\bar{s}$. The corresponding $\bar{D}$ is calculated by

$$\bar{D} = \frac{\sum_i i D_i K_i C^{i-1}}{\sum_i i K_i C^{i-1}} \qquad (9)$$

where D_i is the diffusion coefficient of the i-mer. The hydrodynamic concentration dependence of these transport coefficients is supplied to the simulation routine, based on either previous literature or educated guess. This value remains the same irrespective of the polymeric status of the species computed. The individual transport coefficients (s_i and D_i) are computed using the following standard equations:

$$s_i = \frac{i M_1 (1-\bar{v}\rho)}{N f_i} \qquad (10)$$

$$D_i = \frac{RT}{N f_i} \qquad (11)$$

$$f_i = 6\pi\eta R_i^s (f/fo)_i \qquad (12)$$

$$R_i^s = \left(\frac{3i M_1 \bar{v}}{4\pi N} \right)^{1/3} \qquad (13)$$

where M_1 is the molecular weight of the monomer, f/fo is the frictional ratio of each species, $\bar{v}$ is partial specific volume, T is temperature in Kelvin units, ρ is density of solvent, η is viscosity of solvent, R is gas constant, N is

Avogadro's number, f_i is the frictional coefficient and R^s is the Stokes radius of individual species. Alternately our program can calculate the transport coefficient of individual species, from monomer coefficients assuming equal frictional ratios for all species by Equ. 7 for $\bar{s}_i^0$ and

$$D_i = D_1(i)^{-1/3} \qquad (14)$$

For complete systems, the computation of $\bar{s}$ and $\bar{D}$ would be quite long, taking quite lengthy computing time. This difficulty can be dealt with by assembling tables of $\bar{s}$ and $\bar{D}$ for a few hundred values of C_T before the simulation begins. The values of $\bar{s}$ and $\bar{D}$ needed by the simulation are then extracted from the tables by an interpolation routine.

Simulated sedimentation profiles not only can be used to ascertain the mode of association, they can also quantitatively refine the parameters of association constants derived from analyzing the $\bar{s}$ vs C relationship. This analysis is possible because of the enhanced resolving power of analyzing the derivative curves of a reaction boundary. The simulated results of a system undergoing a tetramerization scheme of $4M \rightleftharpoons M_4$ are shown in Fig. 4A and B to illustrate the point. Fig. 4A shows the relations of $\bar{s}$ vs C and $\bar{D}$ vs C. $\bar{s}_1^0$ assumes a value of 5.0 S, g is zero and the values of K_4 are 32 and 64 $(ml/mg)^3$, i.e., a two fold difference in association constant. Due to the relatively high association constants, there is very little indication of cooperativity in the relation between $\bar{s}$ and C. The curves show a significant increase in the values of $\bar{s}$ between C of 0 to 2 mg/ml, then they approach the same maximum value of $\bar{s}$. The maximum differences between the two curves are observed for the range of C between 0.5 and 2 mg/ml. The difference is about 0.5 S at 0.5 mg/ml. In practice, using the Model E scanner system, a precision of ± 0.2 S is observed, although considerably better precision is possible using the Beckman XLA analytical ultracentrifuge. Hence, in order to differentiate between a two fold difference in the association constant one requires the highest precision possible and a significant number of data points. A casual inspection of the simulated curves of $\bar{s}$ vs C would lead to a conclusion that, in practice, it might be difficult to differentiate between these two cases with confidence. Fig. 4B shows the simulated sedimentation profiles of this associating system at 0.5 mg/ml after 40 min of sedimentation at 60000 rpm. There is an obvious difference in these profiles. In the case of $K_4 = 64$ $(ml/mg)^3$, the area encompassed by the slow moving peak is smaller than that of $K_4 = 32$ $(ml/mg)^3$, whereas the corresponding area for the fast moving peak is greater. Notice that the minimum between the two peaks remains at the same position, as predicted by the Gilbert theory (Gilbert, 1955), since it is a function only of the stoichiometry, which remains as 4. Hence, quantitative comparison between the experimental and simulated sedimentation profiles can increase the accuracy in defining the mode of association.

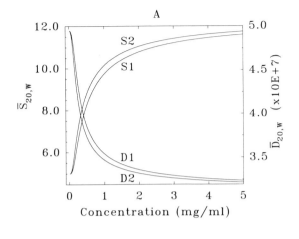

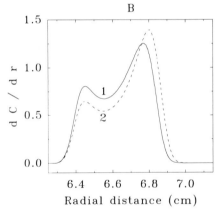

FIGURE 4.   A.   Simulation of concentration dependence of $\bar{s}$ and $\bar{D}$ for a tetramerization scheme. The values for K_4 are 32 and 64 $(ml/mg)^3$ for curves 1 and 2, respectively. All other conditions are the same as in Fig. 3.

B.   Simulated sedimentation velocity derivative profiles for the tetramerization scheme as in (A). The protein concentration is 0.5 mg/ml.

Among the simulation methods the distorted grid method of Cox has been widely used. It describes the sedimentation of solutes involved in self association equilibria, provided that the self-association reaction relaxes toward equilibrium very rapidly (Cox, 1978; Cox and Dale, 1981). The ultracentrifuge cell is divided into an array of n boxes, each Δ x deep and to the midpoint of box i the weight concentration $\bar{c}_i$ is assigned. The simulation proceeds by treating the concentration array alternately with expressions that simulate diffusion without sedimentation for a short time and then sedimentation without diffusion for an equal time.

This method was applied to the study of magnesium-induced self-association of calf brain tubulin (Frigon and Timasheff, 1975a). This simulation was helpful in confirming that calf brain tubulin initially undergoes linear indefinite type of progressive association to be followed by a ring formation of n = 26. Later the method was successfully applied to simulate indefinite self-association of tubulin in the presence of high concentration of vinblastine (Na and Timasheff, 1980).

D.   Prospective

A judicious application of sedimentation velocity can provide significant insights to the mode of macromolecular assembly. The speed of analysis renders this approach as the technique of choice for biological systems that are not stable. The derivative curves of the reaction boundary provide additional resolving power to define the mode of association. Thus, the technology and theory to study macromolecular assembly are available. With the advent of the biotechnology in molecular cloning and expression of biomacromolecules more systems are becoming available for study. More interesting and complex questions on heteropolymer formations are being addressed. Future development of the technology will include improvement in the sensitivity of signal detection so that the concentration range of study can be extended to the next regime of ng/ml. The availability of the new generation of analytical ultracentrifuge in coincidence with the explosion in biological systems to be investigated indicates that new insights in the regulatory mechanisms of macromolecular assembly will be forthcoming.

ACKNOWLEDGEMENTS

Supported in part by grants GM-45579 and DK-21489 from the National Institutes of Health and grants H-013 and H-1238 from the Robert A. Welch Foundation. We thank Shirley Broz for preparing the manuscript.

GLOSSARY OF SYMBOLS

K	association constant
n	stoichiometry
C_p	concentration of the polymeric species
C_m	concentration of the monomeric species
m_i	molal concentration of the ith species
s_i^0	sedimentation coefficient of the ith species
$\bar{s}$	weight average sedimentation coefficient
$\bar{D}$	weight average diffusion coefficient
g	nonideality coefficient
M_i	molecular weight of the ith species
$\bar{v}$	partial specific volume
f_i	frictional coefficient of the ith species

f/f_0 frictional ratio
ρ density of solvent
η viscosity of solvent
R   gas constant
N   Avogadro's number
R^s Stokes radius
ω angular velocity

REFERENCES

Basch, J.J. and Timasheff, S.N. (1967): Hydrogen ion equilibria of the genetic variants of bovine ß-lactoglobulin. *Arch. Biochem. Biophysics* 118: 37-47.

Cann, J.R. (1970): *Interacting Macromolecules* - The Theory and Practice of Their Electrophoresis, Ultracentrifugation, and Chromatography. 93-151 New York: Academic Press.

Cann, J.R. and Goad, W.B. (1970): Bimodal sedimenting zones due to ligand-mediated interactions. *Science* 170: 441-445.

Cann, J.R. and Goad, W.B. (1972): Theory of sedimentation for ligand-mediated dimerization. *Arch. Biochem. Biophys.* 153: 603-609.

Cann, J.R. (1973): Theory of zone sedimentation for non-cooperative ligand-mediated interactions. *Biophys. Chem.* 1: 1-10.

Cann, J.R. and Goad, W.B. (1973): Measurements of protein interactions medicated by small molecules using sedimentation velocity. *Methods in Enzymology* 27: 296-306.

Cox, D.J. (1978): Calculation of simulated velocity profiles for self-associating solutes. *Methods in Enzymology* 48: 212-242.

Cox, D.J. and Dale, R.S. (1981): Simulations of transport experiments for interacting systems. In: *Protein-Protein Interactions* 173-211, New York: Wiley Interscience Publications.

Frigon, R.P. and Timasheff, S.N. (1975a): Magnesium-induced self-association of calf brain tubulin. I. Stoichiometry. *Biochemistry* 14: 4559-4566.

Frigon, R.P. and Timasheff, S.N. (1975b): Magnesium-induced self-association of calf brain tubulin. II. Thermodynamics. *Biochemistry* 14: 4567-4572.

Fujita, H. (1962): *Mathematical theory of Sedimentation Analysis*. New York: Academic Press.

Gilbert, G.A. (1955): *Discuss. Faraday Soc.* 20: 68-71.

Gilbert, L.M. and Gilbert, A.G. (1973): Sedimentation velocity measurement of protein association. *Methods in Enzymology* 27: 273-296.

Hesterberg, L.K. and Lee, J.C. (1981): Self-association of rabbit muscle phosphofructokinase at pH 7.0: Stoichiometry. *Biochemistry* 20:

2974-2980.

Holloway, R.R. and Cox, D.J. (1974): Computer simulation of sedimentation in the ultracentrifuge. VII. Solutes undergoing indefinite self-association. *Arch. Biochem. Biophys.* 160: 595-602.

Johnson, P. and Shooter, E.M. (1950): The globulin of the groundnut. I. Investigation of arachin as a dissociation system. *Biochem. Biophys. Acta* 5: 361.

Kirkwood, J.G. (1954): The general theory of irreversible processes in solutions of macromolecules. *J. Polym. Sci.* 12: 1-14.

Kumosinski, T.F. and Timasheff, S.N. (1966): Molecular interactions in ß-lactoglobulin. X. The stoichiometry of the ß-lactoglobulin mixed tetramerization. *J. Am. Chem. Soc.* 88: 5635-5642.

Lee. J.C., Harrison, D. and Timasheff, S.N. (1975): Interaction of vinblastine with calf brain microtubule protein. *J. Biol. Chem.* 250: 9276-9282.

Na, G.C. and Timasheff, S.N. (1980): Stoichiometry of the vinblastine - induced self-association of calf brain tubulin. *Biochemistry* 19: 1347-1354.

Nichol, L.W., Bethune, J.L., Kegeles, G. and Hess, E.L. (1964): Interacting Protein Systems. In: *The Proteins,* H. Neurath, ed., Vol II 2nd Ed. 305-403. New York: Academic Press.

Schachman, H.K. (1959): *Ultracentrifugation in Biochemistry.* New York, Academic Press.

Timasheff, S.N. and Townend, R. (1960): Molecular interactions in ß-lactoglobulin. I. The electrophoretic heterogeneity of ß-lactoglobulin close to its isoelectric point. *J. Am. Chem. Soc.* 82: 3157-3161.

Timasheff, S.N. and Townend, R. (1961): Molecular interactions in ß-lactoglobulin. V. The association of the genetic species of ß-lactoglobulin below the isoelectric point. *J. Am. Chem. Soc.* 83: 464-469.

Timasheff, S.N., Andreu, J.M. and Na, G.C. (1991): Physical and spectroscopic methods for the evaluation of the interactions of antimitotic agents with tubulin. *Pharmac. Ther.* 52: 191-210.

Townend, R., Winterbottom, R.J. and Timasheff, S.N. (1960a): Molecular interactions in ß-lactoglobulin. II. Ultracentrifugal and electrophoretic studies of the association of ß-lactoglobulin below its isoelectric point. *J. Am. Chem. Soc.* 82: 3161-3168.

Townend, R. and Timasheff, S.N. (1960): Molecular interactions in ß-lactoglobulin. III. Light scattering investigation of the stoichiometry of the association between pH 3.7 and 5.2. *J. Am. Chem. Soc.* 82: 3168-3174.

Townend, R., Weinberger, L. and Timasheff, S.N. (1960b): Molecular interactions in ß-lactoglobulin. IV. The dissociation of ß-lactoglobulin below pH 3.5. *J. Am. Chem. Soc.* 82: 3175-3179.

Yphantis, D.A. and Waugh, D.F. (1956): Ultracentrifugal characterization by direct measurement of activity. II. Experimental. *J. Phys. Chem.* 60: 630-635.

MEASURING SEDIMENTATION, DIFFUSION, AND MOLECULAR WEIGHTS OF SMALL MOLECULES BY DIRECT FITTING OF SEDIMENTATION VELOCITY CONCENTRATION PROFILES

John S. Philo

INTRODUCTION

Sedimentation velocity experiments have traditionally been used for samples with relatively high sedimentation coefficients and low diffusion. Such samples give sharp boundaries from which it is relatively easy to extract the sedimentation coefficient, and which permit the separation of multicomponent samples into distinct boundaries. However, many proteins of interest for therapeutic purposes, such as cytokines and growth factors, have molecular masses of only 10-40 kDa. Even at 60000 rpm, such small molecules give very broad boundaries which are difficult to analyze by existing techniques.

For example, the approach used in the program XLAVEL (Beckman Instruments) involves numerically differentiating concentration profiles to give $\partial c/\partial r$. The derivative data is then treated like Schlieren data, *e.g.* the natural log of the position of the maximum in $\partial c/\partial r$ may be plotted versus $\omega^2 t$ and then fitted to a straight line to determine s. However, the broad boundaries from small molecules give particularly small values of $\partial c/\partial r$. This makes the derivative data particularly sensitive to dirt or scratches on the windows, so typically considerable smoothing of the derivative data is required to even see a peak. In our hands, for small proteins this method gives significantly different s values from the same data depending on the degree of smoothing and the person manually defining the position of the peaks in $\partial c/\partial r$.

Another approach that might seem more suitable for small molecules is to determine the boundary position from the second moment of the concentration profiles (Goldberg, 1953) and/or the diffusion coefficients from the boundary spreading (Muramatsu and Minton, 1988), as is done in the program VELGAMMA (Beckman Instruments). However, these methods require data at times where the meniscus is clear and there is also a plateau region at the bottom of the cell. For small molecules, by the time the meniscus is clear a plateau may no longer exist, due to the large movement of the leading edge of the boundary, as well as the typically 1-2 mm region where there is back diffusion of material accumulated at the bottom of the cell. (This is particularly true if aluminum centerpieces are not used and rotor speed is limited to 42,000 rpm). At best, there is often only a short time in the run during which this method may be applied, and the small boundary movement during this time can limit the accuracy of the analysis.

A more fundamental flaw with both of these approaches is that neither directly fits the raw experimental data. Rather, one or two properties are extracted from each scan, and these values are in turn fitted to an appropriate function of time during the run. These two-step approaches make it virtually impossible to correctly evaluate the confidence limits for the hydrodynamic parameters, and difficult to assess how strongly the results may be affected by factors such as the exact values chosen for the concentration in the plateau region or manual selection of peaks in $\partial c / \partial r$.

Moreover, neither of these approaches is useful when the sample contains more than one species, i.e. exactly the situation where sedimentation velocity can be often be particularly useful. All these considerations, as well as the availability of considerable computing power at low cost, have led us to develop a method where multiple raw data sets of concentration vs. radius, taken at various times during the run, are simultaneously fitted to appropriate approximate solutions of the Lamm equation, with s, D, and the loading concentration as fitting variables. As a bonus, the determination of s and D from the same experiment allows calculation of the molecular weight, and as shown below, this can be surprisingly accurate. In addition, with this approach it is possible to analyze samples containing multiple non-interacting species.

METHODS

Numerical Methods. In order to directly fit sedimentation velocity concentration profiles by non-linear least squares techniques, an appropriate theoretical fitting function must be chosen. Since there is no general, closed form solution of the Lamm equation, an approximate solution is needed which is accurate enough to represent the data, but not too time consuming to compute. We have chosen initially to ignore the concentration dependence of sedimentation coefficients, since for modern absorbance or refractometric optical systems one can easily work at concentrations of <1 mg/ml where the concentration dependence is generally negligible.   Fujita (1975) has summarized a number of approximate solutions of the Faxén type (unbounded at the bottom of the cell), appropriate for different types of cells during different times during the run. For data from the conventional sector-shaped velocity cell, we therefore use equation 2.94 from Fujita (1975):

$$c = \frac{c_o e^{-\tau}}{2}\left\{1 - \mathrm{erf}\left[\frac{1-\left(xe^{-\tau}\right)^{1/2}}{\left[\varepsilon\left(1-e^{-\tau}\right)\right]^{1/2}}\right] + \frac{\left[2\varepsilon\sinh\left(\tau/2\right)\right]^{1/2}}{\sqrt{\pi}x^{1/4}\left[1+\left(xe^{-\tau}\right)^{1/4}\right]}\exp\left(-\frac{\left[1-\left(xe^{-\tau}\right)^{1/2}\right]^2}{\varepsilon\left(1-e^{-\tau}\right)}\right)\right\}$$

where c_0 is the loading concentration, r_0 is the meniscus position, erf() is the error function, $\tau \equiv 2s\omega^2 t$, $x \equiv (r/r_0)^2$, and $\varepsilon \equiv 2D/s\omega^2 r_0^2$. This solution does not account for restricted diffusion near the meniscus, and thus may not be appropriate at times very early in the run. It also cannot be applied near the bottom of the cell where solutes accumulate.

For synthetic boundary cells, equation 2.127 of Fujita (1975) is used:

$$c = \frac{c_o e^{-\tau}}{2}\left[1 - \mathrm{erf}\left(\frac{\tau - \ln x}{2\sqrt{\varepsilon\tau}}\right)\right]$$

where the terms are as defined above except that r_0 is now the initial position of the boundary. This equation accounts for diffusion of solutes above the initial position of the boundary, but again does not treat accumulation of solutes at the cell bottom.

A program, SVEDBERG, was written using Microsoft Visual Basic for Windows which incorporates these fitting functions into a non-linear least squares fitting routine. This program directly reads up to nine raw data files from the Beckman XL-A and extracts the elapsed time, $\omega^2 t$, and rotor speed from the data header. Baseline files may also be subtracted from the data to correct for poor matching of window absorbance or poor matching of solvent absorbance in sample and reference channels. The user then selects whether the data is from a conventional or synthetic boundary cell, and whether 1, 2, or 3 non-interacting species are to be included in the fit. The program then allows the user to graphically set the position of the meniscus (or the initial position of the boundary for synthetic boundary data) and an inner and outer radius that define the range of data to be included in the fit (constant for all data sets).

For each species, the program can either fit the values of s, D, and c_0, or hold them fixed at values input by the user. For synthetic boundary data the initial position of the boundary is also generally fitted, while for the conventional cell the meniscus position usually is held fixed (but may be fitted as desired). In addition, a zero offset (common to all data sets) may be included in the fit for those situations where the absorbance of the reference channel is not well matched to the sample.

The fitting algorithm uses a modified Gauss-Newton method developed in-house which is functionally identical to the "preferred method" described by Johnson & Faunt (1992; see also this volume). When desired, SVEDBERG can also evaluate the confidence interval for each fitted parameter (which is generally asymmetric about the "best" value) again using the "preferred method" described by Johnson & Faunt. Evaluating these fitting functions is relatively slow, due primarily to the required numerical approximations of the error function. Therefore the derivatives of the fitting function, which are needed for the Gauss-Newton method, are calculated algebraically rather than by difference methods to avoid extra evaluations of the error function. For the conventional ultracentrifuge cell, in calculating these derivatives the last term in the large brackets in the fitting function (above) is ignored because it is quite small compared with other terms. (This approximation is equivalent to treating $\partial c/\partial r$ as a Gaussian function.)

Numerical simulations of sedimentation velocity experiments were carried out using the finite element Claverie method (Claverie *et al.*, 1975), using program code kindly provided by David Cox and Walter Stafford. The sample was divided into 400 radial segments, and the time integration step size was 1 s.

Experimental. Centrifugation experiments were carried out with a Beckman XL-A ultracentrifuge using either conventional aluminum centerpieces or aluminum-filled epon synthetic boundary cells. Bovine serum albumin "monomer" was purchased from Sigma (#A1900). Recombinant transforming growth factor α and brain-derived neurotrophic factor were expressed in *E. coli*, refolded, oxidized, and purified to homogeneity by sequential column chromatography. All samples were made up in, or dialyzed into, Dulbecco's phosphate buffered saline (PBS) (Gibco). The light scattering/size exclusion analysis of the BSA sample was kindly done by Jie Wen (Protein Chemistry, Amgen) using methods described previously (Philo *et al.*, 1993).

RESULTS and DISCUSSION

Single Species. We have tested this method with a number of proteins with $M_r \sim$ 6,000-70,000. Figure 1 shows results for brain-derived neurotrophic factor (BDNF) in a conventional velocity cell at 60000 rpm. In solution BDNF is a very tightly bound homodimer with a sequence molecular weight of 27,274 per dimer. The derivative data shown in the inset illustrate the difficulty of defining the boundary position from the peak even after considerable smoothing. Using our method, the fit to 9 scans (for clarity only 5 are shown) returns an s value of 2.513 S with a 95% confidence interval of 2.509 to 2.517, a value for D of 8.36 [8.24, 8.47] $\times$ 10^{-7} cm^2/s, a loading concentration of 0.4542 [.4529, .4554] absorbance units (AU) (equivalent to 0.23 mg/ml), and a zero offset of -0.0015 [-.0026, -.0005] AU, with an rms residual of .00704 AU. As also shown in Figure 1, the theoretical curves (solid lines) agree very well with the experimental data, and the residuals from the fit (not shown) are approximately randomly distributed.

If this analysis is correct, we should be able to calculate the molecular weight from the relation

$$M = \frac{sRT}{D(1-\bar{v}\rho)}$$

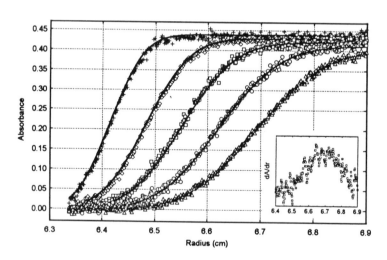

FIGURE 1. Sedimentation velocity data (symbols) and fitted data (curves) for 0.23 mg/ml BDNF in PBS at 20°C and 60000 rpm. Nine data sets were used in the fitting, but for the sake of clarity only the odd numbered sets are shown. The inset shows $\partial A/\partial r$ for the last data set after a 13-point smoothing.

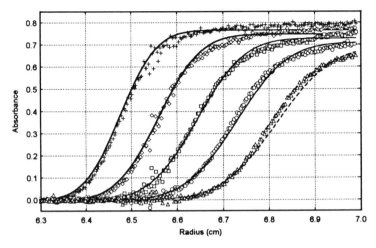

FIGURE 2. Sedimentation velocity data and single species fit for 1 mg/ml BSA "monomer" (Sigma) in PBS at 20°C and 60000 rpm. For the sake of clarity only 5 of the 9 data sets are shown.

where R is the gas constant, $\bar{v}$ is the partial specific volume, and ρ is the solvent density. Using the measured $\rho = 1.00394$ g/ml and $\bar{v} = 0.7271$ ml/g calculated from the amino acid composition (Laue *et al.*, 1992), the fitted values of s and D imply a molecular mass of 27.27 [26.87, 27.71] kDa, in excellent agreement with the sequence molecular weight.

Based on ~1 year of experience to date, for proteins of 15-70 kDa this analysis typically gives molecular weights within ±2% of sequence values *if the sample truly contains only one species.* This accuracy compares quite favorably with that of sedimentation equilibrium, and may be limited in part by inaccuracies in the partial specific volume or density.

Another important point to note is that this fitting function seems to represent the data well even at times prior to complete clearing of the meniscus, even though it does not explicitly account for the restricted diffusion. We generally find it acceptable to include data after the concentration at the meniscus is <20% of the loading concentration. On the other extreme, this function seems to work well to at least the time when the leading edge of the boundary reaches the cell bottom.

Multiple Species. In contrast to the excellent fit of the BDNF data, Figure 2 shows a single species fit for a sample of commercial BSA "monomer", which clearly shows systematic deviations. This fit returns s = 4.53 S and $D = 7.94 \times 10^{-7}$ cm^2/s, which implies $M = 53$ kDa, well below the formula weight of 66,268. As shown in Figure 3, the fit is substantially improved if we include a second species, which reduces the variance by a factor of 1.9. For this fit, the first species (90.2%) has $s =$ 4.439 [4.429, 4.449] S, $D = 6.11$ [5.97, 6.26] $\times 10^{-7}$ cm^2/s, and $M =$ 67.6 [65.9, 69.2] kDa, and therefore is an excellent match for the BSA monomer. The second species, contributing $(9.8 \pm 1)\%$ of the total absorbance, has $s = 6.72$ [6.63, 6.80] S, $D = 5.35$ [3.88, 7.37] $\times 10^{-7}$ cm^2/s, and $M = 117$ [85, 163] kDa. Not only does the apparent M suggest a BSA dimer, but (more significantly) the ratio of sedimentation coefficients for the two species is 1.51, nearly a perfect match for the ratio of 1.50 expected for a dimer of 2 hard spheres (van Holde, 1975).

To further validate these results, we analyzed this same material by size exclusion chromatography with light scattering detection. The chromatograms indeed show the presence of 9% dimer plus ~1% of an unresolved peak which appears to be a mixture of trimer and tetramer. If a third species with fixed s and D values corresponding to those expected for a BSA trimer or tetramer is also included, the sedimentation analysis returns about a 2% contribution, but the improvement in the fit is

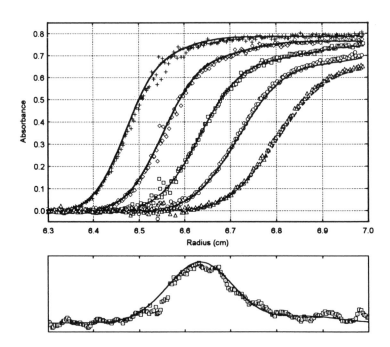

FIGURE 3. (upper) Sedimentation velocity data and two-species fit for BSA data from Fig. 2. (lower) $\partial A/\partial r$ for the last data set after a 13-point smoothing (symbols), and computed derivative from the two-species fit (curve).

marginally significant. The derivative data in the lower panel of Fig. 3 shows that although a second peak for the dimer is resolved in theory, in practice 9 to 10% dimer would not be detected in a derivative analysis.

These BSA results also illustrate a number of general features of this method of analysis: (1) Treating a heterogeneous sample as a single species will usually give s values close to those of the predominant species, but D values which are too large. (The boundary spreading from the large D can partially mimic the spread due to an additional species.) (2) In a multi-species analysis the s and c_0 values are likely to be more reliable than are the D values. (3) The method is fairly insensitive to scratches or dirt on the cell windows, as present near 6.55 cm in Fig. 2. (4) The ability to resolve species with markedly different s values may be limited by the fact that this method requires that the rotor speed is constant and that the slowest sedimenting species is mostly clear of the meniscus, at which time faster sedimenting species may have reached the cell bottom. For example, it would be difficult to get accurate information about the presence of tetramers in the BSA sample, because they are still present in only the first 2 of the 9 data sets. In this case we could, in fact, have included data earlier in the run to get more contribution from faster sedimenting species, but this would compromise our ability to resolve monomers from dimers since this separation is much better later in the run. Such considerations suggest it may never be practical to determine s and D values for 3 species with this method.

Claverie Simulations. We have also tested this methodology with data simulated by the Claverie method (Claverie *et al.*, 1975). Figure 4 shows a Claverie simulation for $s = 2$ S and $D = 1 \times 10^{-6}$ cm^2/s ($M \approx 18$ kDa) at a loading concentration of 1 AU and 60000 rpm. Overlaid is the fit to these simulated data, which returns $s = 2.019$ [2.018, 2.020] S, $D = 1.001$ [1.000, 1.002] $\times 10^{-7}$ cm^2/s, $c_0 = 1.0009$ [1.007, 1.0010], and a zero offset of -0.00031 [-0.0004, -.0002]. The residuals, shown in the right panel, have an rms value of only 6×10^{-6} c_0 and the maximum deviation is only ~0.3% c_0. Thus the simulation suggests that the approximations in the fitting function are good enough to allow determination of s, D, and M with an accuracy of 1% or better.

However, this excellent agreement with the Claverie simulations only holds true if the simulation places the initial position of the boundary away from the meniscus (i.e. allowing diffusion inward from the initial boundary, effectively treating the conventional velocity cell like a synthetic boundary cell). A Claverie simulation that does not allow

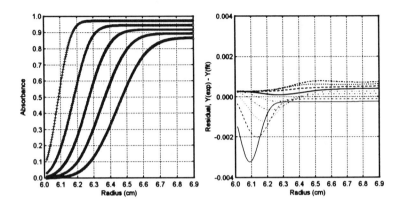

FIGURE 4. (left) Claverie simulation and single species fit for $s = 2$ S and D = 1×10^{-6} cm^2/s at 20°C and 60000 rpm. Five of 9 data sets are shown. (right) Residuals for all 9 data sets. The largest deviation is for the earliest time.

diffusion above the meniscus produces narrower boundaries, especially early in the run. In this case the fit gives the correct s, but has much larger residuals and underestimates the true D by about 6%. This is perhaps not surprising, since the approximate solution we are using does not explicitly account for restricted diffusion. Nonetheless, as noted above, the *experimental* data are well represented by this function, even very early in the run, and the fits give the correct D and M. At present the source of this apparent conflict between experiment and simulation remains unclear. It is likely that the experimental data show early boundaries broader than predicted by the simulation because diffusion is occurring prior to the time the rotor reaches full velocity, while the simulation assumes instantaneous acceleration (and in fact it is our usual practice to pause at 3000 rpm long enough to take an initial scan). It is also possible that part of the conflict arises from the numerical approximations used in the Claverie method.

Claverie simulations have also been used to test whether it would be possible to resolve two species of much lower molecular weight than BSA. Simulations like that in Fig. 4 were run for $c_0 = 0.5$ AU of a monomeric species with s = 1.9 S, $D = 1.1 \times 10^{-6}$ cm^2/s (a ~16 kDa protein) and $c_0 = 0.5$ AU of its dimer, s = 2.85 S, $D = 8.25 \times 10^{-7}$ cm^2/s, at 60000 rpm. Random (Gaussian) noise with an rms amplitude of 0.006 AU, corresponding to typical noise in our XL-A, was then added to the simulated data for a more realistic test. The resulting test data, shown in

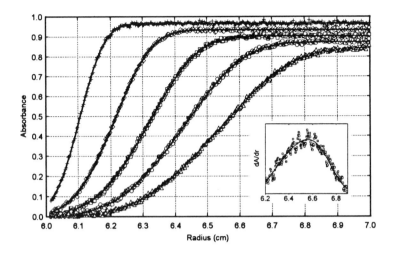

FIGURE 5. Claverie simulation and two-species fit for equal loading concentrations of a 1.9 S, $D = 8.25 \times 10^{-7}$ cm^2/s monomer and its dimer at 20 °C and 60000 rpm. Random noise with rms amplitude 0.006 AU was added to the simulated data. Five of the 9 data sets are shown. The inset shows a 13-point smoothed $\partial A/\partial r$ for the last data set and $\partial A/\partial r$ predicted from the fit.

Figure 5, show no resolution of the two species by conventional criteria, and as shown in the inset there is not even a distinct shoulder in the theoretical $\partial A/\partial r$. These data were first tested to see if fitting could correctly deduce the proportions of the two species if the s and D values of the two species were known independently. A zero offset was also included in the fitting, since this is often necessary for experimental data. This fit found .4846 [.4821, .4871] AU of monomer and .5164 [.5144, .5183] of dimer, an error of ~3% from the true values.

If instead the s and D values are also fitted, .5147 [.4982, .5313] AU of s = 1.936 [1.918, 1.953] S, $D = 1.11$ [10.76, 11.44] $\times 10^{-6}$ cm^2/s and .4861 [.4699, .5034] AU of s = 2.876 [2.863, 2.889] S, $D = 8.18$ [8.04, 8.32] $\times 10^{-7}$ cm^2/s are found (curves shown in Fig. 5). This agreement is remarkably good, but it is also deceptive. An examination of the parameter cross-correlation coefficients (Johnson & Faunt, 1992) shows that the monomer c_0 has a -0.999 correlation with the dimer c_0, and the monomer s has a 0.9815 and -0.9842 correlation with these loading concentrations, respectively. These high cross-correlations mean that we can almost completely compensate for a change in any of these parameters by an adjustment of the others, *i.e.*they are not truly independently determined. (SVEDBERG alerts the user whenever there are cross-correlations > 0.97.) Therefore, for species this small, one would need independent information about s or D for one of the species in order to have confidence in the results.

Comparison to the Attri & Lewis Method. With these results in hand, it is also useful to compare our approach to a somewhat similar one developed independently by Attri & Lewis (1992). Their approach uses a sigmoid function to fit the concentration profiles, a function chosen because it is both rapid to compute and accurately locates the radial position of the square root of the second moment of the concentration data, but which has no direct theoretical basis as a solution of the Lamm equation. The sigmoid function is first fitted to each data set, and the derived position and width parameters are then fitted in turn to an appropriate function of time to derive s and D. This approach appears to be reasonably accurate for single species. A sigmoid function does not reproduce the shape of experimental sedimentation boundaries or Claverie simulations as well as does the function used in our method, especially for broad boundaries like those in Figs. 1 and 4. The D values derived from their approach are therefore probably less accurate, especially when D is large. Attri & Lewis demonstrated some capability of resolving two components, but quantitative results for two-species fits were not presented. Further, like all two-step fitting approaches, their method makes it difficult to accurately assess the errors in the fitted parameters. This is particularly true for D, since the value of s is also explicitly used in the data transformation for determining D. Nonetheless, the Attri & Lewis method is clearly a viable approach and well suited to situations where one is primarily interested in rapidly computing sedimentation coefficients and obtaining at least an estimate of D.

Synthetic Boundary Data. To date we have little experience with applying this methodology to the synthetic boundary cell, but one example is shown in Figure 6, a measurement of a very small (5678 Da) protein, transforming growth factor α (TGFα) at 42000 rpm. The figure shows data at radii beyond the actual fitting region (which stopped at 7 cm) to illustrate a major drawback of this cell: the region where the data are influenced by back diffusion from the cell bottom grows so rapidly that very little boundary movement can be obtained. We also find that the earliest scans must usually be rejected because they show systematic

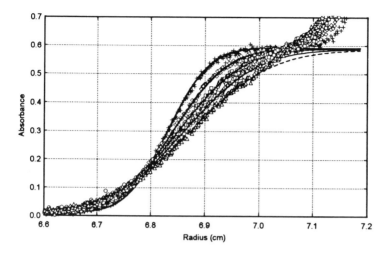

FIGURE 6. Synthetic boundary data and fits for TGFα at 20°C and 42000 rpm.

deviations from the expected shape. This is probably due to inhomogeneities in the initial boundary, which later become less significant as they are averaged out by diffusion. The fit to these data gives s = 0.849 [.825, .872] S, D = 1.43 [1.41, 1.45] × 10^{-6} cm^2/s, the initial position of the boundary as 6.8166 [6.8156, 6.8176] cm, and gives reasonably non-systematic residuals. These results imply M = 5.05 [4.91, 5.18] kDa. This is about 11% below the true value, and this error almost certainly reflects an overestimate of D because the initial boundary was not sharp. We have also tested the synthetic boundary fitting function against Claverie simulations, and again find excellent agreeement and that the approximations are good enough to give an accuracy of much better than 1%. These simulations also further emphasize the problem of back diffusion from the cell bottom, since a simulation of the data in Figure 5 shows that back diffusion contributes >1% of the total concentration at radii >7 cm.

While SVEDBERG also implements multi-species fits for the synthetic boundary cell, the smaller separation obtainable probably means that, at best, this may be useful for high molecular weight species. Overall, we have not found the use of the synthetic boundary cell to be very advantageous.

Limitations and Future Developments. It is important to emphasize that for multiple species this approach can only give meaningful results *if they* *are non-interacting.* Thus it is generally inappropriate for self-associating systems, unless the kinetics of redistribution among species is slow compared to the time scale of the experiment. Sample concentrations must also be kept low enough that the Johnston-Ogston effect is negligible.

As mentioned previously, the fact that the fitting functions assume a fixed rotor speed makes it difficult to measure samples containing widely different s values. In some cases one is only interested in the more slowly sedimenting species, but one or more larger species affect the data early in the run. In such cases we have found that including a species with variable s but a fixed very large value of D can be useful to model aggregates with a broad distribution of sizes.

SVEDBERG uses data from the same radial range for all data sets. This means that a large number of data points are included from the plateau region of early data sets, which can exacerbate the problem just discussed. Moreover, we sometimes observe sloping plateau regions in the data from our XL-A. This slope in the plateau remains constant over several data sets, and therefore it cannot be due to the presence of other species. (The origin of this problem is still unclear). Since the fitting functions cannot produce a sloping plateau, such data will be poorly fit and will give an overestimate of D. Therefore it may be worthwhile in some cases to confine the data range for each data set to the vicinity of the boundary.

Another practical limit to the accuracy of this analysis is the accuracy of sample temperature control. The temperature dependence of the viscosity of aqueous samples will produce a change in the apparent s and D of ~2.2% per °C. While the steady-state temperature control in the XL-A appears to be quite good, during acceleration to 60000 rpm the titanium rotor heats by ~.8°C, so some temperature change during velocity runs seems unavoidable. Lastly, this analysis does not account for the movement of the boundary *during the course of a scan*, which could be significant for sharp boundaries and slow scan speeds.

While these fitting functions are relatively slow to compute, this is not a severe limitation. The two-species BSA fit in Fig. 3 (2189 data points, 6 parameters) takes about 2.5 min to converge using a 33 MHz Intel 486 CPU, while single species fits take < 1 min. The calculation of rigorous confidence limits for 2 species can take ~20 min, but this is usually only done for final results. Numerical calculations using Visual Basic are not particularly rapid, and a true 32-bit compiler would probably speed these computations 3-4 fold.

It is possible that the resolution and reliability of multi-species fitting could be improved by the inclusion of more data sets. (SVEDBERG's present limitation to 9 sets is entirely arbitrary). However, we have tried eliminating some of the data at intermediate run times, and this seems to have little effect other than slightly increasing the uncertainty in the parameters. Fundamentally, the resolution of the analysis is limited by the physical separation actually achieved, so as long as data covering the full time span appropriate for this method are included, varying the number of data sets will not have a dramatic effect.

In the future it will also be interesting to try this approach using more accurate approximate solutions of the Lamm equation. As computation speeds continue to increase, it may be practical on a microcomputer to use Claverie simulations as the fitting function, and in principle such simulations could allow for changing the rotor speed during the run.

[Note added in proof] It has come to our attention that an analytical approach similar to this one was reported some years ago. Holladay (1980) analyzed multiple sedimentation velocity scans by fitting c_0, s, and the ratio s/D, using an approximate solution to the Lamm equation (Holladay, 1979) which does account for restricted diffusion at the meniscus (and is therefore probably superior at early times in the run). Multiple-species fits were not reported. The fitting function used by Holladay is more complex than that employed here (Fujita, 1975), and would probably require ~3 times longer to compute. In future work we hope to compare this fitting function to the one used here.

GLOSSARY OF SYMBOLS

AU	absorbance units
c_0	loading concentration of solute
erf(y)	the error function of y, $\dfrac{2}{\sqrt{\pi}} \int_0^y e^{-y^2} dy$
D	diffusion coefficient
M	molecular mass
R	gas constant
r_0	meniscus or initial boundary position
s	sedimentation coefficient
t	elapsed time of centrifugation
$\bar{v}$	solute partial specific volume
ρ	solvent density
ω	angular velocity

REFERENCES

Attri AK and Lewis MS (1992): A fitting function for the analysis of sedimentation velocity concentration distributions. In: *Analytical Ultracentrifugation in Biochemistry and Polymer Science*, Harding SE, Rowe AJ and Horton JC, eds. Cambridge: The Royal Society of Chemistry.

Claverie J-M, Dreux H and Cohen R (1975): Sedimentation of generalized systems of interacting particles. I. Solution of systems of complete Lamm equations. *Biopolymers* 14, 1685.

Fujita H. (1975): *Foundations of Ultracentrifugal Analysis*. New York: John Wiley & Sons, pp. 64-81.

Goldberg RJ (1953): Sedimentation in the ultracentrifuge. *J. Phys. Chem.* 57: 194-202.

Holladay LA (1979): An approximate solution to the Lamm equation. *Biophys. Chem.* 10: 187-190.

Holladay LA (1980): Simultaneous rapid estimation of sedimentation coefficient and molecular weight. *Biophys. Chem.* 11: 303-308.

Johnson ML and Faunt LM (1992): Parameter estimation by least-squares methods. *Methods Enzymol.* 210: 1-37.

Johnson ML and Straume M (1993): Comments on the analysis of sedimentation equilibrium experiments. In: *Modern Analytical Ultracentrifugation*, Schuster TM and Laue TM, eds. Boston: Birkhauser Publishing Inc., this volume, Chapter 3.

Laue TM, Shah BD, Ridgeway TM and Pelletier SL (1992): Computer-aided interpretation of analytical sedimentation data for proteins. In: *Analytical Ultracentrifugation in Biochemistry and Polymer Science*, Harding SE, Rowe AJ and Horton JC, eds. Cambridge: The Royal Society of Chemistry.

Muramatsu N and Minton AP (1988): An automated method for rapid determination of diffusion coefficients via measurements of boundary spreading. *Anal. Biochem.* 168: 345-351.

Philo JS, Rosenfeld R, Arakawa T, Wen J and Narhi LO (1993): Refolding of brain-derived neurotrophic factor from guanidine hydrochloride: Kinetic trapping in a collapsed form which is incompetent for dimerization. *Biochemistry* 32: 10812-10818.

van Holde KE (1975): Sedimentation analysis of proteins. In: *The Proteins*, 3rd ed., vol. 1, Neurath H and Hill R, eds. New York: Academic Press.

COMPUTER SIMULATION OF THE SEDIMENTATION OF LIGAND-MEDIATED AND KINETICALLY CONTROLLED MACROMOLECULAR INTERACTIONS

John R. Cann

INTRODUCTION

Ultracentrifugal characterization of macromolecular interactions was initiated shortly after construction of the first analytical ultracentrifuge by Svedberg and his co-workers in 1925-26. Thus, following the immediate demonstration that proteins are distinct molecular entities with well defined mass and shape, not ill defined colloids, Svedberg (Svedberg and Pedersen, 1940) anticipated modern concepts of the subunit structure of proteins. Additionally, the prophetic studies on ligand-mediated association-dissociation of hemocyanins (Marimoto and Kegeles, 1971; Kegeles and Cann, 1978; Roxby et al. 1974; Miller and Van Holde, 1974) were initiated (Svedberg and Pedersen, 1940). Self-association of proteins and their interactions with each other and with low molecular weight ligands are central to current biological thought especially as they pertain to the mechanism of regulatory processes. Over the years one seminal finding for the development of biochemistry and molecular biology followed another. Early on, Heidelberger and Pedersen (1937) demonstrated the existence of soluble, protein antigen-antibody complexes in the antigen excess zone of the precipitin reaction. Subsequent ultracentrifuge studies of the soluble complexes by Singer et al. (Singer, 1965) provided direct confirmation of the framework theory of antigen-antibody precipitate and led, in conjunction with electrophoretic analysis, to thermodynamic characterization of antigen- antibody reactions via application of the Goldberg theory (Goldberg, 1952). In another vein, Meselson and Stahl (9) employed density gradient centrifugation to demonstrate for the first time that DNA replication is a semiconservative process, which is a first principle of molecular biology and genetics. Contemporaneously, there was also a fundamental development in the biophysical chemistry of proteins; namely, formulation by Gilbert (1955, 1959) of an asymptotic theory of velocity sedimentation of reversibly associating-dissociating macromolecules

$$mM \rightleftharpoons M_m \qquad (1)$$

under the assumptions of rapid reequilibration and negligible diffusion. The most dramatic prediction of the Gilbert theory is that, whereas only a single weight average sedimenting peak will be observed for macromolecular dimerization (m=2), for higher-order association (m≥3) the reaction boundary will resolve into two peaks despite instantaneous reequilibration during differential transport of monomer and polymer. The two peaks, moreover, cannot be identified with separated species. Experimental examples of these two stoichiometrically and ultracentrifugally distinct classes of self-associating macromolecules include the reversible dissociation of hemoglobin at high salt concentrations (Kirshner and

Tanford, 1964) and the low-temperature tetramerization of β-lactoglobulin A (Brown and Timasheff, 1959).

Seeing that reversible complex formation between different macromolecules is one of the most important of biological interactions, it is particularly significant that Gilbert and Jenkins (1956 and 1959) set forth an asymptotic theory of sedimentation and electrophoresis of such interactions. A generalization of the asymptotic theory was applied by Gilbert and Gilbert (1965) to the simulation of the qualitative features of the moving-boundary electrophoretic patterns shown by a specifically interacting mixture of bivalent antibody with univalent antigen.

While the Gilbert and Gilbert-Jenkins theories are extremely useful in this respect, detailed predictions about a given interaction depend upon computer solution of partial differential conservation equations which are complete in the sense that they include the diffusion term. At least five different numerical procedures are available for computation of quantitatively accurate boundary profiles (Bethune and Kegeles, 1961; Bethune, 1970; McNeil et al., 1970; Dishon et al., 1966; Goad, 1970; Claverie et al., 1975; Cox, 1978; Todd and Haschemeyer, 1983) and have been applied to a variety of macromolecular interactions (Oberhauser et al., 1965; Kegeles et al., 1967; Kegeles and Johnson, 1970; Cann, 1970; Cann and Goad, 1970; Gilbert and Gilbert, 1973; Cann and Kegeles, 1974; Cohen and Claverie, 1975; Timasheff et al., 1976; Cann, 1982a; Werner et al., 1989). In this review we will consider the sedimentation and other manifestations of ligand-mediated and kinetically controlled interactions. Comparison of theory with experiment will demonstrate the insightful power of computer simulation as an aid in the development and interpretation of experiments designed for quantitative characterization of macromolecular interactions.

A few words are in order concerning the meaning of "kinetically controlled interactions". Consider, for example, the sedimentation behavior for the simple macromolecular dimerization represented by Reaction 1 with $m = 2$. In the limit where equilibration is so rapid as to be considered instantaneous as far as mass transport is concerned, the sedimentation pattern will show only a single peak with weight averaged sedimentation and diffusion coefficients. Such fast reactions imply diffusion controlled kinetics characterized by a small activation energy interpreted as the energy of activation for viscous flow of the solvent. In the other limit in which the rates of reaction are so slow that negligible interconversion occurs during the course of sedimentation, the pattern will show two peaks corresponding to separated monomer and dimer. *Reactions which proceed at intermediate rates are referred to as kinetically controlled and characteristically exhibit bimodal reaction boundaries when the half-time of dissociation is greater than about 200 s* (Cann and Kegeles, 1974). The unique feature of kinetically controlled dimerization is that the reaction boundary may be trimodal for half-times comparable to the time of sedimentation depending upon the values of other relevant parameters (Belford and Belford, 1962; Oberhauser *et al.*, 1965).

In the present context, these concepts were formulated quantitatively for the illustrative ligand- mediated dimerization

$$2M + nX \; \rightleftharpoons \; M_2X_n \tag{2}$$

in which a macromolecule, M, associates reversibly into a dimer, with the mediation of a small ligand molecule or ion, X, of which a fixed number, n, are bound into the complex. The set of Lamm equations for sedimentation in a sector-shaped ultracentrifuge cell takes the form

$$\frac{\partial (C_1 + 2C_2)}{\partial t} = \frac{1}{r} \frac{\partial}{\partial r} \left[(D_1 \frac{\partial C_1}{\partial r} - C_1 s_1 \omega^2 r) \, r + 2 \, (D_2 \frac{\partial C_2}{\partial r} - C_2 s_2 \omega^2 r) \, r \right]$$

$$\tag{3}$$

$$\frac{\partial (nC_2 + C_3)}{\partial t} = \frac{1}{r} \frac{\partial}{\partial r} \left[n \, (D_2 \frac{\partial C_2}{\partial r} - C_2 s_2 \omega^2 r) \, r + (D_3 \frac{\partial C_3}{\partial r} - C_3 s_3 \omega^2 r) \, r \right]$$

in which C designates molar concentration; D, diffusion coefficient; s, sedimentation coefficient; ω, angular velocity; r, radial distance and t, time. The subscripts 1, 2 and 3 designate M, M_2X_n and X respectively. These equations conserve constituent macromolecule and constituent small molecule during diffusion and concerted transport in the centrifugal field. A third equation expresses the effect of Reaction 2 on the concentrations. For rates of reaction so fast that local equilibrium at every instant can be assumed

$$C_2 = KC_1^2 C_2^n \; , \tag{4}$$

where K is the equilibrium constant of Reaction 2. In the case of a kinetically controlled interaction, Eqn. 4 is replaced by a set of chemical rate equations. Computer solution of the simultaneous Equations. 3 and 4 (Goad, 1970; Cann and Goad, 1972; Cann and Oates, 1973; Cann and Kegeles, 1974) gives the simulated velocity sedimentation pattern of the interacting system, displayed as a plot of the concentration gradient of constituent macromolecule, usually accompanied by the concentration of unbound ligand, against position in the ultracentrifuge cell for a given time of sedimentation.

In order to reduce the complexities of solving a set of Lamm equations, resort may be had to the rectilinear approximate

$$\frac{\partial C_i}{\partial t} = D_i \frac{\partial^2 C_i}{\partial x^2} - V_i \frac{\partial C_i}{\partial x} \tag{5}$$

for each of the i interacting species (Cann and Kegeles, 1974; Werner et al., 1989) under the assumption of constant diffusion coefficient and driven velocity $V_i = s_i \omega^2 \bar{x}$ where $\omega^2 \bar{x}$ is the constant field strength and $\bar{x}$, and average position. These equations are solved in a frame of reference moving with the average macromolecular velocity so as to minimize truncation error (Cann, 1987) and to approximately center the front in the coordinate system (Cann and Goad, 1965).

RESULTS AND DISCUSSION

Ligand-Mediated Dimerization. Formulation of the theory of analytical sedimentation of macromolecules subject to ligand-mediated dimerization evolved from experimental/theoretical studies on the electrophoretic behavior of proteins that interact with a component of the electrophoresis buffer (Cann and Goad, 1965). The fundamental concept was and remains that, in contradistinction to the behavior of simple dimerizations (Reaction 1, m=2) which always give sedimentation patterns showing a single peak when equilibrium is rapid, ligand-mediated dimerization (Reaction 2) can give rise to a well-resolved bimodal reaction boundary despite rapid equilibration. Verification is provided by Figure 1, which compares a theoretical (Cann, 1970) with an experimental (Na and Timasheff, 1980) sedimentation pattern for rapid, ligand-mediated dimerization.Resolution of the two peaks in the bimodal reaction boundary depends upon the generation and maintenance of a concentration gradient of unbound ligand along the centrifuge cell by reequilibration during differential transport of macromonomer and dimer. The peaks correspond to different equilibrium mixtures and not simply to separated monomer and dimer, so that in general their relative areas do not faithfully reflect the initial equilibrium mixture. Under some conditions the sedimentation coefficient of the slower peak is greater than that of the monomer since the depicted mixture contains dimer, whereas the sedimentation coefficient of the faster peak is always less than that of the dimer since the corresponding mixture contains monomer. There are conditions, of course, under which resolution collapses. Thus, at sufficiently high ligand

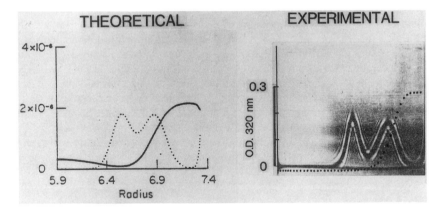

FIGURE 1. Comparison of theoretical and experimental bimodal velocity sedimentation patterns exhibited by rapidly equilibrating protein dimerization mediated by the binding of a single ligand molecule. **Theoretical:** Simulated for Reaction 2, n=1; ·····, protein gradient profile; ——, distribution of ligand along the centrifuge cell (Cann, 1970). **Experimental:** dimerization induced by binding of one vinblastine molecule per tubulin molecule; ——, Schlieren picture; ·····, concentration distribution of vinblastine (Na and Timasheff, 1980).

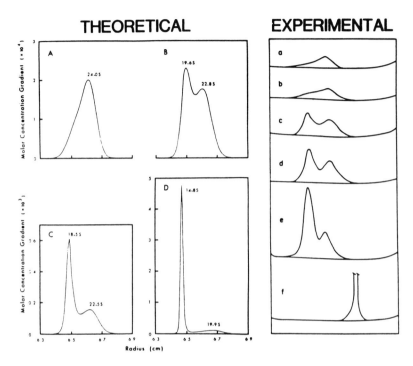

FIGURE 2. Dependence of boundary shape on protein concentration at constant ligand concentration. **Theoretical** sedimentation patterns simulated   for dimerization of protein in the presence of 1.15×10^{-4} $\underline{M}$ ligand (Reaction 2, n=6): A, 0.375×10^{-4} $\underline{M}$ protein; B, 0.5×10^{-4}; C, 0.65×10^{-4}; D, 1.5×10^{-4} (Cann and Goad, 1972). **Experimental** patterns for dimerization of tubulin induced by 2×10^{-5} $\underline{M}$ vinblastine: a, 2.71 mg tubulin/ml; b, 3.15; c, 4.87; d, 6.76; e,9.89. Pattern f, isodesmic selfassociation of tubulin induced by very high concentration of vinblastine (Lee et al., 1975).

concentration (or low centrifugal field) ligand-mediated dimerization will give patterns that show a single peak. In the limit where for any reason the concentration of unbound ligand along the centrifuge cell is not significantly perturbed by the reaction during differential sedimentation of the macromolecular species, the system effectively approaches the case of simple dimerization.

Our theoretical studies on ligand-mediated dimerization predicted (Cann and Goad, 1972) the unique sedimentation behavior displayed in Figure 2 along with experimental verification (Lee et al., 1975). At a given concentration of protein and constituent ligand, the protein is largely dimerized. As the concentration of protein is progressively increased at constant constituent concentration of ligand the pattern resolves into a bimodal reaction boundary, the peak composed largely of monomer growing at the expense of the peak composed largely of dimer. This behavior is a consequence of the law of mass action. When the concentration of protein is

J. R. Cann

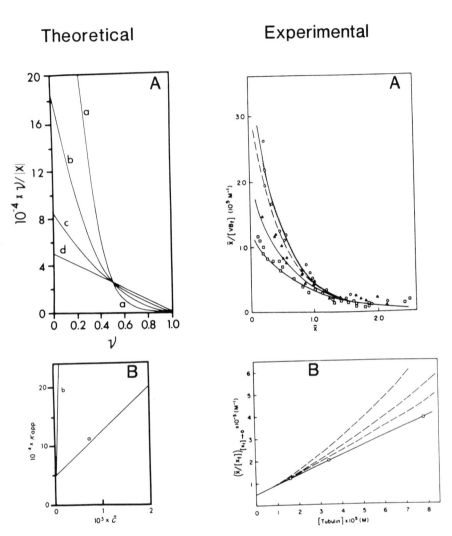

FIGURE 3. Dependence upon macromolecular concentration of Scatchard plots of ligand-binding data for ligand-induced macromolecular selfassociation. **Theoretical** data generated for a range of concentrations in accordance with Reaction Set 6: **A.** Scatchard plots computed for (a) 7.27×10^{-5} $\underline{M}$, (b) 1.8×10^{-5}, (c) 4.55×10^{-6}, (d) limit of infinite dilution; **B.** Linear extrapolation of apparent binding constant to infinite dilution (Cann and Hinman, 1976). **Experimental** binding of vinblastine to tubulin: **A.** Scatchard plots for 1.6×10^{-5}, 3.4×10^{-5}, 7.8×10^{-5} $\underline{M}$ tubulin; **B.** Linear extrapolation to infinite dilution (Na and Timasheff, 1986).

increased there is a concomitant decrease in the concentration of unbound ligand, which in turn shifts the dimerization reaction back toward monomer.

Ligand-mediated selfassociation also has important consequences for interpretation of ligand-binding isotherms (Cann and Hinman, 1976; Cann, 1978a). Consider, for example, the model system

$$M + X \rightleftharpoons MX, \quad k_0 \tag{6a}$$

$$M + MX \rightleftharpoons M_2X, \quad K_a \tag{6b}$$

for which binding data have been generated for a range of constituent concentration of macromolecule, $\overline{C}$, with the following assignments: intrinsic binding constant $k_0 = 5 \times 10^4$ $\underline{M}^{-1}$ and dimerization constant $K_a = 1.5 \times 10^5$ $\underline{M}^{-1}$. The simulated data are displayed in Figure 3A (Theoretical Panel) as Scatchard plots of $\nu/[x]$ vs ν, where ν is the mean number of moles of ligand bound per mole of constituent macromolecule at the equilibrium concentration of unbound ligand, $[x]$. The plots exhibit two distinguishing features. Firstly, the plots are concave toward the abscissa, reminiscent of nonassociating systems characterized either by inherent heterogeneity of binding sites with respect to their intrinsic affinity for ligand or by binding to multiple sites with negative cooperativity. Secondly, in contrast to heterogeneity and negative cooperativity, however, the extent of binding is dependent upon macromolecule concentration, the intercept with the ordinate

increasing with increasing $\overline{C}$. Analytically, the intercept is $K_{app} = k_0 + k_0 K_a \overline{C}$, where extrapolation of the apparent ligand-binding constant, K_{app}, to infinite dilution of macromolecule (Figure 3B, Theoretical Panel) erases the role of dimerization in determining the extent of binding at finite concentrations and yields k_0. The value of K_a is then given by the ratio of slope to intercept. Guidelines have been delineated for interpretation of experimental binding data once it is known that the ligand induces association of the protein or other macromolecule (Cann, 1978a). Na and Timasheff (1986) applied these guideline to the vinblastine-tubulin system thereby determining the values of the equilibrium constant for specific binding of vinblastine to the first of two sites on the tubulin molecule and the tubulin self-association constant. Their data, which are displayed in Figures 3A and 3B (Experimental Panel), while for the more complex ligand induced and facilitated selfassociation, exhibit virtually the same features as predicted theoretically for our model system.

The class of interactions considered thus far are for instantaneous establishment of equilibrium, but many interactions of timely interest are kinetically controlled; e.g., hybridization of Esterase-5 in *Drosophilia pseudobscura* (Cann, 1987), protein-DNA interactions (Cann, 1989) and the hexamer-dodecamer reaction of New England lobster hemocyanin induced by Ca^{+2}-binding (Kegeles and Tai, 1973; Tai and Kegeles, 1975). This Ca^{+2}-mediated dimerization gives bimodal sedimentation patterns similar to those in Figure 1 (Morimoto and Kegeles, 1971). In fact, the system behaves in every way as a rapidly reequilibrating interaction, although the interaction does not equilibrate instantaneously, the half-times of reaction being of the order of 40-100 sec depending upon conditions. This finding prompted a theoretical investigation (Cann and Kegeles, 1974) into the effect of chemical kinetics upon the shape of the reaction boundary for the model ligand-mediated reaction set

$$\text{M} + \text{X} \underset{k_2}{\overset{k_1}{\rightleftharpoons}} \text{MX} \tag{7a}$$

$$2\text{MX} \underset{k_4}{\overset{k_3}{\rightleftharpoons}} \text{M}_2\text{X}_2 \tag{7b}$$

where the k's are specific rate constants. It is noteworthy that because the rates of Reaction 7a were taken to be very fast in conformity with the upper range of values observed for binding of small molecules by macromolecules, we applied the theory of relaxation kinetics in formulating the computational procedure for this particular reaction set. Simulated sedimentation patterns are displayed in Figure 4. Those shown in Figure 4A for instantaneous establishment of equilibrium serve as a point of reference for kinetically controlled interactions. The family of patterns in Figure 4B is for kinetic-control at a ligand concentration for which the sedimenting reaction boundary is unimodal when equilibrium is instantaneous. Given that Reaction 7a equilibrates rapidly, the monomer-dimer interconversion (Reaction 7b) becomes rate controlling, the definitive kinetic parameter being the half-time of dissociation of the dimer. A half-time as long as 60 sec. simply causes a slight distortion of the boundary, as illustrated in Figure 2B of Cann and Kegeles (1974). Upon further increase in the half-time (Figure 4B), resolution into two peaks ensues and becomes increasingly sharp as the limit is approached asymptotically where negligible interconversion occurs during the course of sedimentation. It is evident that conclusions concerning the sedimentation behavior of ligand-mediated interactions in the limit of instantaneous equilibration are valid for kinetically controlled interactions characterized by half-times as long as 60 sec., as confirmed by Figure 4C. The agreement between this theoretical result and the experimental findings on New England lobsters hemocyanin is gratifying.

As detailed previously (Cann, 1978b), the results of these several theoretical and experimental investigations have added to our store of fundamental understanding required for unambiguous interpretation of the sedimentation patterns shown by ligand-mediated interactions and for characterization of the interaction in thermodynamic and kinetic terms. But, analytical sedimentation alone does not suffice. It is imperative that at least two independent biophysical methods be brought to bear in order to establish the exact nature of the interaction. These results also have important implications for the many analytical and preparative applications of velocity sedimentation in biochemistry and molecular biology. Thus, an inherently homogeneous macromolecule can give sedimentation patterns showing two well-resolved peaks due to ligand-mediated association-dissociation even when equilibration is rapid. It cannot be overemphasized that unequivocal proof of inherent heterogeneity is afforded only by isolation of the various components via fractionation in a partition cell, preparative ultracentrifugation, or gel filtration.

Ligand-Mediated Heterogeneous Association. The results of the foregoing studies suggested that the possible involvement of ligands in certain heterogeneous

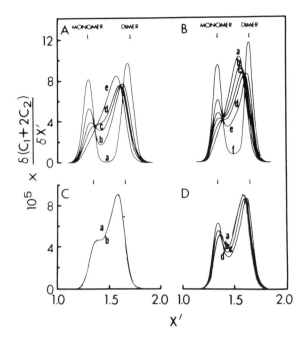

FIGURE 4. Effect of chemical kinetics upon the shape of sedimenting reaction boundaries simulated for ligand-mediated dimerization (Reaction Set 7): A. Instantaneous reequilibration during differential transport of the several species: a, initial concentration of unbound ligand, $CX° = 10^{-7}$ M; b, 5×10^{-7}; c, 7.5×10^{-7}; d, 10^{-6}; e, 2×10^{-6}. Time of sedimentation, $t_s = 2 \times 10^3$ s. B. Kinetically controlled interaction for $CX° = 2 \times 10^{-6}$ M: a, $k_4 = 3.6 \times 10^{-2}$ s^{-1} (half-time, $t_{1/2} = 19$ s); b, 6×10^{-3}; c, 3×10^{-3}; d, 2×10^{-3}; e, 10^{-3}; f, 10^{-4} ($t_{1/2} = 6.9 \times 10^3$ s). $t_s = 1.5 \times 10^3$ s. C. Comparison of the sedimentation pattern for instantaneous reequilibration [curve a] with kinetically controlled interaction, $k_4 = 1.2 \times 10^{-2}$ s^{-1} ($t_{1/2} = 58$ s) [curve b]; $CX° = 10^{-6}$ M; $t_s = 1.5 \times 10^3$ s. D. Comparison of rapidly equilibrating with kinetically controlled reactions for $CX° = 7.5 \times 10^{-7}$ M: a, instantaneous reequilibration; b, $k_4 = 1.2 \times 10^{-2}$ s^{-1}; c, 3×10^{-3}; d, 1.5×10^{-3}. $t_s = 1.5 \times 10^3$ s. (Cann and Kegeles, 1974)

association-dissociation reactions might also play a dominant role in determining the sedimentation behavior of such systems. Examples of the structural role of ions and small ligand molecules in stabilizing subunit protein and protein assemblies, each evidently by involving a single kind of ligand, have been referenced (Cann, 1982b), and it is conceivable that the stabilization of some protein assemblies might depend upon two different ligands. Accordingly, analytical sedimentation patterns were computer simulated for model heterogeneous associations mediated by either a single kind of ligand or two different ligands acting in a stepwise fashion.

In the first model, complex formation between three dissimilar protein molecules A, B and D is mediated in part by a small ligand molecule X as schematized by the rapidly reversible reaction set

$$A + B \; \rightleftharpoons \; AB, \qquad\qquad K_1 \qquad\qquad (8a)$$

$$AB + D + X \; \rightleftharpoons \; ABDX, \qquad\qquad K_2 \qquad\qquad (8b)$$

The salient features of the computed patterns (Figure 5) is their trimodality consisting of a boundary of A and two more rapidly sedimenting peaks constituting a bimodal reaction boundary. The faster of the two peaks corresponds to an equilibrium mixture rich in ABDX and, thus, sediments with a velocity between AB and ABDX; the slower one being rich in AB sediments between B and AB. Resolution of the bimodal boundary is dependent, through the agency of mass action, upon generation of a concentration gradient of ligand along the centrifuge cell.

In the other model, two dissimilar proteins A and B assemble into a complex with the mediation of two different ligands X and Y acting in a stepwise fashion,

$$A + B + X \; \rightleftharpoons \; ABX, \qquad\qquad K_1 \qquad\qquad (9a)$$

$$2ABX + Y \; \rightleftharpoons \; (ABX)_2Y, \qquad\qquad K_2 \qquad\qquad (9b)$$

in which X is obligatory for complex formation between A and B, and Y is obligatory for dimerization of the complex ABX. This model can give sedimentation patterns exhibiting four peaks (Figure 6A): a boundary of A and a trimodal reaction boundary. Resolution of the reaction boundary into three peaks is dependent upon generation of concentration gradients of both ligands along the centrifuge cell, and each peak corresponds to a different equilibrium composition and not to an individual protein species.

Finally, it is instructive to return to the preceding consideration of selfassociation, but this time mediated by two different ligands acting in a stepwise fashion in accordance with the tetramerization schema

$$2M + X \; \rightleftharpoons \; M_2X, \qquad\qquad K_1 \qquad\qquad (10a)$$

$$2M_2X + Y \; \rightleftharpoons \; (M_2X_2)Y, \qquad\qquad K_2 . \qquad\qquad (10b)$$

As illustrated in Figure 6B this schema can give sedimentation patterns consisting of a well resolved trimodal reaction boundary; the slowest peak sedimenting slightly faster than monomer, the central peak slower than dimer, and the fastest peak slower than tetramer. This result and those for dimerization mediated by a single ligand molecule could conceivably have import for active enzyme sedimentation in solutions containing inhibitors, cofactors or allosteric affectors, which might promote association of the enzyme.

The results described above for various model interactions provide guidelines for interpretation of sedimentation patterns in the case of ligand-mediated association reactions and broaden the base for biophysical investigations into the architectural and regulatory roles played by protein associations *in vivo*. It is of first importance that the sedimentation velocities of peaks comprising a reaction boundary are not characteristic parameters of products or reactants, and their areas cannot be used to calculate association constants. On the other hand, mechanistic insights can be obtained by systematically varying the proportions of protein

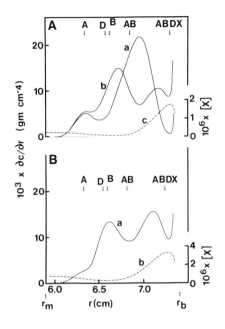

FIGURE 5. Representative sedimentation patterns simulated for ligand-mediated heterogeneous association, Reaction Set 8:

Panel A. Pattern a, control without ligand mediation (analogous to Reaction Set 8 without participation of X): b, pattern computed for Reaction Set 8 with constituent concentrations $A > B > C$; curve c, concentration profile of unbound ligand.

Panel B. Pattern a, computed for $A = B = C$; curve b, profile of unbound ligand (Cann, 1982b).

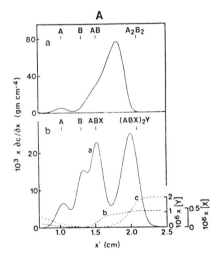

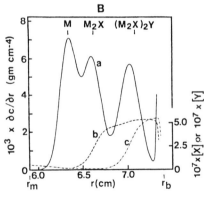

FIGURE 6. Representative sedimentation patterns simulated for macromolecular association mediated in stepwise fashion by two different ligands: A. Heterogeneous association schematized by Reaction Set 9. Upper Panel: Control without ligand mediation (analogous to Reaction Set 9 without participation of X and Y). Lower Panel: Curve a, pattern for ligand mediation; curve b, concentration profile of X; curve c, profile of Y. B. Selfassociation mediated by two different ligands, the tetramerization Reaction Set 10 (Cann, 1982b).

reactants and the concentration of ligand(s) which promotes their association (Cann, 1982b).

A particularly provocative result is that rapidly equilibrating ligand-mediated associations can give multimodal sedimentation patterns. So can inherent heterogeneity and other interactions such as the simple association monomer $\rightleftharpoons$ trimer $\rightleftharpoons$ nonamer (Bethume and Grills, 1967) and sufficiently slow reaction rates. It is essential, therefore, that appeal be made to additional biophysical methods, including fractionation, in order to distinguish between heterogeneity and interaction and to establish the exact nature of an associating system including precise characterization in terms of thermodynamic and other parameters.

Ligand-Mediated Isomerization. The enzyme aspartate transcarbamoylase (ATCase) from *Escherichia coli* catalyzes the carbamoylation of aspartate by carbamoyl phosphate. This is the committed step in the biosynthesis of pyrimidines in the bacteria; and the enzyme is inhibited by cytidine triphosphate, an end product of the pyrimidine pathway, thus establishing the regulatory pattern of feedback inhibition. Because ATCase consists of separable catalytic and regulatory subunits and because its three-dimensional X-ray structure has been determined, it is the enzyme of choice for studies on the mechanism of allosteric regulation. Ultracentrifugation showed that a 3.5% decrease in sedimentation coefficient of the enzyme accompanies the binding of saturating amounts of either carbamoyl phosphate and the aspartate analog succinate, or the inhibitory bisubstrate analog N-(phosphonacetyl)- L-aspartate (PALA), indicating that ATCase expands upon binding substrates. X-ray studies revealed that changes in both quaternary and tertiary structures accompanies the binding of PALA.

These ligand-promoted conformational changes have been characterized qualitatively in terms of unliganded T and the swollen, liganded R states of the enzyme by analysis of boundary spreading in sedimentation velocity experiments on partially liganded ATCase (Werner and Schachman, 1989). Strikingly different results were obtained with two different active-site ligands. With substoichiometric amounts of the tightly bound PALA the dynamic equilibrium between the T and R conformations appeared to be uncoupled. The sedimentation patterns exhibited single broad boundaries approximating those for a mixture of noninteracting components with slightly different sedimentation coefficients. These results in conjunction with experiments involving mutant ATCase demonstrated that the ligand-promoted allosteric transition T $\rightarrow$ R was concerted and that the presence of molecules of intermediate conformation was negligible. In contrast to PALA, approximately equal populations of T and R-state molecules produced by a large excess of the weakly bound ligand succinate exhibited boundary spreading corresponding to a single species with intermediate hydrodynamic properties, reminiscent of rapidly equilibrating macromolecular isomerization A $\rightleftharpoons$ B (Cann, 1970).

Theoretical explanation of these results was provided by simulation of sedimentation velocity patterns for the rapidly equilibrating model of allosteric interactions

$$T + nL \rightleftharpoons RL_n \,, \tag{11}$$

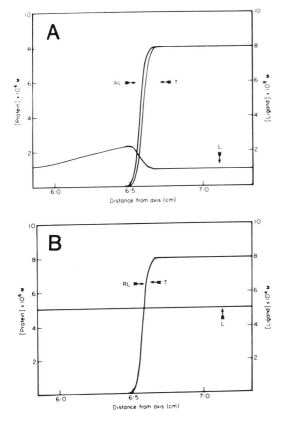

FIGURE 7. Simulation of sedimentation patterns for aspartate transcarbamoylase partially saturated with active-site ligands (Reaction 11 with n=1): A. The high-affinity PALA. B. The low-affinity succinate in the presence of carbamoyl phosphate (Werner et al., 1989).

where n is the number of ligand molecules, L, bound to the enzyme in the R conformation (Werner et al., 1989). For the calculation considered here a value of n = 1 was assumed. In order to faithfully simulate the sedimentation velocity experiments, the T-state molecules (11.7 S) migrated faster than the swollen RL_n complex (11.3 S) and the sedimentation of free L, because of its low molecular weight, was negligible.

Calculations for 1.6×10^{-5} $\underline{M}$ enzyme in the presence of substoichiometric amounts of the strongly binding PALA assumed a dissociation constant of 10^{-8} $\underline{M}$, which is the concentration of free ligand when 50% of the enzyme is in the R conformation. The sedimentation pattern simulated for these conditions is displayed in Figure 7A. It is immediately apparent that the broad boundary observed experimentally is due to separation of T and RL in the steep ligand gradient generated by reequilibration during differential sedimentation of T and RL such that the equilibrium would appear to be uncoupled. In this regard, take note that the concentration of free L is higher in the supernatant region than the initial concentration because of the continuous dissociation of small amounts of L from the slower migrating RL during the course of sedimentation.

The simulated behavior of a system in which the ligand binds weakly to the enzyme is in marked contrast to that of the PALA interaction. When a subsaturating amount of the weak ligand succinate interacts with 1.6×10^{-5} $\underline{M}$ ATCase, dissociation constant 5×10^{-4} $\underline{M}$, the equilibrium concentration of unbound L for 50% of ATCase in the R state is about two orders of magnitude greater than the concentration of enzyme. Thus, as evidenced by the simulated sedimentation pattern displayed in Figure 7B, the concentration of unbound ligand along the centrifuge cell can not be significantly perturbed by reequilibration during differential sedimentation of T and RL. Consequently, the system effectively collapses to the simple isomerization reaction $T \rightleftharpoons R$, which behaves as a single macromolecular species with concentration averaged diffusion and sedimentation coefficients thereby accounting for the experimental result. These results attest once again to the power of computer simulations.

CODA

The several foregoing considerations dealt with boundary sedimentation, but in conclusion, note should be taken of the many important applications of zonal sedimentation in biochemistry and molecular biology some of which involve interacting systems. Early on (Cann and Goad, 1970), computer simulations showed that ligand-mediated association-dissociation reactions can give rise to band sedimentation patterns exhibiting bimodal bands despite instantaneous establishment of equilibrium. Rapid ligand-mediated dimerization and isomerization reactions can also give bimodal zones during velocity sedimentation through a density gradient (Cann, 1973). Mention should also be made of the simulation studies on active enzyme sedimentation, the results of which validated a far wider range of experimental conditions for this method than previously recognized (Cohen and Claverie, 1975). In another vein, zone patterns for kinetically controlled dissociation of either a large enzyme complex or ribosome-enzyme complex (C) to give a residual complex or ribosome (A) stripped of a particular enzyme (B), which now exist in free solution, were simulated (Cann and Oates, 1973) for the reaction

$$C \underset{k_2}{\overset{k_1}{\rightleftharpoons}} A + B \tag{12}$$

with the assignment $V_C = V_A > V_B$. As illustrated by the patterns presented in Figure 6 of Cann and Oates (1973), both the profile of total material and of enzymic activity are typically bimodal. The faster sedimenting peak is a mixture of C, A and B in the pattern of total material and a mixture of C and B in the pattern of enzyme activity, and in both cases the slower peak is comprised largely of B. Zone patterns similar to the simulated ones have been observed experimentally with enzymically active complexes between aminoacyltransferase I and proteinaceous cytoplasmic particles (Shelton et al., 1970) and between aminoacyl-tRNA synthetases and ribosomes (Roberts and Olsen, 1976). In both cases, fractionation experiments established that the two peaks in the pattern of enzymic activity do not correspond to isozymes of different molecular weights, but instead constitute a bimodal reaction zone arising from dissociation of the complex during the course of sedimentation.

GLOSSARY OF SYMBOLS

M	Monomeric macromolecule
M_m	Macromolecular m-mer
X	Small ligand molecule
Y	Second small ligand molecule
C_i	Molar concentration of species i
D_i	Diffusion coefficient of species i
s_i	Sedimentation coefficient of species i
w	Angular velocity
r	Radial distance
t	Time
K	Equilibrium constant
V_i	Driven velocity of species i in rectilinear coordinate system
x	Position in rectilinear coordinate system
x'	Position in moving rectilinear coordinate system
k_0	Intrinsic ligand binding constant
v	Mean moles of ligand bound per mole of constituent macromolecule
C	Constituent molar concentration
$CX°$	Initial molar concentration of unbound ligand
k	Specific rate constant
$t_{1/2}$	Half-time of reaction
t_s	Time of sedimentation
T	Tense conformation of allosteric enzyme
R	Relaxed conformation of allosteric enzyme
L	Allosteric effector

REFERENCES

Belford GG and Belford RL (1962): Sedimentation in Chemically Reacting Systems. II. Numerical Calculations for Dimerization. *J. Chem. Phys.* 37: 1926-1932.

Bethune JL (1970): Directed Transport of Monomer-Single Polymer Systems. A Comparison of the Countercurrent Analog and Asymptotic Approaches. *J. Phys. Chem.* 74: 3837-3845.

Bethune JL and Grill PJ (1967): The Effect of Intermediates upon the Transport Properties of Polymerizing Systems. I. Monomer, Trimer and Nonamer. *Biochemistry* 6: 796-800.

Bethune JL and Kegeles G (1961): Countercurrent Distribution of Chemically Reacting Systems. III. Analogs of Moving Boundary Electrophoresis and Sedimentation. *J. Phys. Chem.* 65: 1761-1764.

Brown RA and Timasheff SN (1959): Applications of Moving Boundary Electrophoresis to Protein Systems. In: *Electrophoresis. Theory, Methods and Applications, Bier M ed.*, Chapter 8 New York: Academic Press.

Cann JR (1970): *Interacting Macromolecules. The Theory and Practice of Their Electrophoresis, Ultracentrifugation, and Chromatography*, Chapter 4, New York: Academic Press.

Cann JR (1973): Theory of Zone Sedimentation for Non-Cooperative Ligand-Mediated Interactions. *Biophys. Chem.* 1: 1-10.

Cann JR (1978a): Ligand-Binding by Associating Systems. *Methods Enzymol.* 48: 299-307.

Cann JR (1978b): Measurements of Protein Interactions Mediated by Small Molecules Using Sedimentation Velocity. *Methods Enzymol.* 48: 242-248.

Cann JR (1982a): Theory of sedimentation for Antigen-Antibody Reactions: Effect of Antibody Heterogeneity on the Shape of the Pattern. *Mol. Immun.* 19: 505-514.

Cann JR (1982b): Theory of sedimentation for Ligand-Mediated Heterogeneous Association-Dissociation Reactions. *Biophys. Chem.* 19: 41-49.

Cann JR (1987): Theory of Electrophoresis of Hybridizing Enzymes with Kinetic Control: Implications for Population Genetics of Electrophoretic Markers. *J. Theor. Biol.* 127: 461-477.

Cann JR (1989): Phenomenological Theory of Gel Electrophoresis of Protein-Nucleic Acid Complexes. *J. Biol. Chem.* 264: 17032-17040.

Cann JR and Goad WB (1965): Theory of Moving Boundary Electrophoresis of Reversibly Interacting Systems. *J. Biol. Chem.* 240: 148-155.

Cann JR and Goad WB (1970): Bimodal Sedimenting Zones Due to Ligand Mediated Interactions. *Science* 170: 441-445.

Cann JR and Goad WB (1972): Theory of Sedimentation for Ligand-Mediated Dimerization. *Arch. Biochem. Biophys.* 153: 603-609.

Cann JR and Hinman ND (1976): Hummel-Dryer Gel Chromatographic Procedure as Applied to Ligand-Mediated Association. *Biochemistry* 15: 4614-4622.

Cann JR and Kegeles G (1974): Theory of Sedimentation for Kinetically Controlled Dimerization Reactions. *Biochemistry* 13: 1868-1874.

Cann JR and Oates DC (1973): Theory of Electrophoresis and Sedimentation for Some Kinetically Controlled Interactions. *Biochemistry* 12: 1112-1119.

Claverie J-M, Dreux H and Cohen R (1975): Sedimentation of Generalized Systems of Interacting Particles. I. Solution of Complete Lamm Equations. *Biopolymers* 14: 1685-1700.

Cohen R and Claverie J-M (1975): Sedimentation of Generalized Systems of Interacting Particles. II. Active Enzyme Centrifugation-Theory and Extensions of its Validity Range. *Biopolymers* 14: 1701-1716.

Cox DJ (1978): Calculation of Simulated Sedimentation Velocity Profiles for Self-Association Solutes. *Methods Enzymol.* 48: 212-242.

Dishon M, Weiss GH and Yphantis DA (1966): Numerical Solutions of the Lamm Equation. I. Numerical Procedure. *Biopolymers* 4: 449-455.

Gilbert GA (1955): *Disc. Faraday Soc.* 20: 68-71.

Gilbert GA (1959): Sedimentation and electrophoresis of interacting substances I. Idealized boundary shape for a single substance aggregating reversibly. *Proc. Roy. Soc. London* A250: 377-388.

Gilbert GA and Jenkins RC Ll (1956): Boundary Problems in the Sedimentation and Electrophoresis of Complex Systems in Rapid Reversible Equilibrium. *Nature* 177: 853-854.

Gilbert GA and Jenkins RC Ll (1959): Sedimentation and electrophoresis of interacting substances II. Asymptotic boundary shape for two substances interacting reversibly. *Proc. Roy. Soc. London* A253: 420-437.

Gilbert LM and Gilbert GA (1965): Generalized treatment of Reversibly Reacting Systems in Transport Experiments, Illustrated by an Antigen-Antibody Reaction. *Biochem. J.* 97: 7c-9c.

Gilbert LM and Gilbert GA (1973): Sedimentation Velocity Measurement of Protein Association. *Methods Enzymol.* 27: 273-296.

Goad WB (1970): Numerical Methods. In Cann JR (1970): *Interacting Macromolecules. The Theory and Practice of Their Electrophoresis, Ultracentrifugation, and Chromatography.* Chapter 5, New York: Academic Press.

Goldberg RJ (1952): A Theory of Antibody-Antigen Reactions. I. Theory for Reactions of Multivalent Antigen with Bivalent and Univalent Antibody. *J. Am. Chem. Soc.* 74: 5715-5725.

Heidelberger M and Pedersen KO (1937): The Molecular Weight of Antibodies. *J. Exper. Med.* 65: 393-414.

Kegeles G and Cann JR (1978): Kinetically Controlled Mass Transport of Associating-Dissociating Macromolecules. *Methods Enzymol.* 48: 248-270.

Kegeles G and Tai M-S (1973): Rate Constants for the Hexamer-Dodecamer Reaction of Lobster Hemocyanin. *Biophys. Chem.* 1: 46-50.

Kegeles G and Johnson M (1970): Effects of Pressure on Sedimentation Velocity Patterns. *Arch. Biochem. Biophys.* 141: 59-62.

Kegeles G, Rhodes L and Bethune JL (1967): Sedimentation Behavior of Chemically Reacting Systems. *Proc. Natl. Acad. Sci. USA* 58: 45-51.

Kirshner AG and Tanford C (1964): The Dissociation of Hemoglobin by Inorganic Salts. *Biochemistry* 3: 291-296.

Lee JC, Harrison D and Timasheff SN (1975): Interaction of Vinblastine with Calf Brain Microtubule Protein. *J. Biol. Chem.* 250: 9276-9282.

McNeil BJ, Nichol LW and Bethune JL (1970): Directed Transport of Monomer-Dimer-Trimer Systems. Comparison of Asymptotic and Countercurrent Distribution Approaches. *J. Phys. Chem.* 74: 3846-3852.

Meselson M and Stahl FW (1958): The Replication of DNA in Escherichia Coli. *Proc. Nat. Acad. Sci. USA* 44: 671-682.

Miller K and Van Holde KE (1974): Oxygen Binding by *Callianassa californiensis* Hemocyanin. *Biochemistry* 13: 1668-1674.

Morimoto K and Kegeles G (1971): Subunit Interactions of Lobster Hemocyanin 1. Ultracentrifuge Studies. *Arch. Biochem. Biophys.* 142: 247-257.

Na GC and Timasheff SN (1980): Thermodynamic Linkage between Tubulin Self-Association and the Binding of Vinblastine. *Biochemistry* 19: 1355-1365.

Na GC and Timasheff SN (1986): Interaction of Vinblastine with Calf Brain Tubulin: Multiple Equilibria. *Biochemistry* 25: 6214-6222.

Oberhauser DF, Bethune JL and Kegeles G (!965): Countercurrent Distribution of Chemical Reacting Systems. IV. Kinetically Controlled Dimerization in a Boundary. *Biochemistry* 4: 1878-1884.

Roberts WK and Olsen ML (1976): Studies on the Formation and Stability of Aminoacyl-tRNA Synthetase Complexes from Ehrlich Ascites Cells. *Biochim. Biophys. Acta.* 454: 480-492.

Roxby R, Miller K, Blair DP and Van Holde KE (1974): Subunits and Association Equilibria of *Callianassa californiensis* Hemocyanin. *Biochemistry* 13: 1662-1668.

Shelton E, Kuff EL, Maxwell ES and Harrington JT (1970): Cytoplasmic Particles and Aminoacyl Transferase I Activity. *J. Cell. Biol.* 45: 1-8.

Singer SJ (1965): Structure and Function of Antigen and Antibody Proteins. In: *The Proteins. Composition, Structure, and Function*, Neurath H, ed. Second Edition, Vol. III Chapter 15 New York: Academic Press.

Svedberg T and Pedersen KO (1940): *The Ultracentrifuge*, London: Oxford University Press.

Tai M-S and Kegeles G (1975): Mechanism of the Hexamer-Dodecamer Reaction of Lobster Hemocyanin. *Biophys. Chem.* 3: 307-315.

Timasheff SN, Frigon RP and Lee JC (1976): A solution physical-chemical examination of the self-association of Tubulin. *Fed. Proc., Fed. Am. Soc. Exp. Biol.* 35: 1886-1891.

Todd GP and Haschemeyer RH (1983): Generalized Finite Element Solution to One-Dimensional Flux Problems. *Biophys. Chem.* 17: 321-336.

Werner WE and Schachman HK (1989): Analysis of the Ligand-promoted Global Conformational Change in Aspartate Transcarbamoylase. *J. Mol. Biol.* 206: 221-230.

Werner WE, Cann JR and Schachman HK (1989): Boundary Spreading in Sedimentation Velocity Experiments on Partially Liganded Aspartate Transcarbamoylase. A Ligand-mediated Isomerization. *J. Mol. Biol.* 206: 231-237.

REFINING HYDRODYNAMIC SHAPES OF PROTEINS: THE COMBINATION OF DATA FROM ANALYTICAL ULTRACENTRIFUGATION AND TIME-RESOLVED FLUORESCENCE ANISOTROPY DECAY[1]

Evan Waxman, William R. Laws, Thomas M. Laue, and
J.B. Alexander Ross

INTRODUCTION

Classical physical biochemical techniques are currently experiencing a renaissance as the result of several technological advances made over the past decade. One major reason for this renaissance is the need to understand the structures and functional characteristics of wild-type and mutant proteins that are now readily available through recombinant-DNA methods. Mutations of interest include alterations at specific functional sites, truncations, and switched domains. Another major reason for this renaissance is the availability of cheap, accessible computing power to facilitate data reduction. It is now possible and appropriate to combine the data from different physical techniques to obtain information which would not be obtainable from any single technique alone. This chapter examines a way in which the combination of analytical ultracentrifugation and time-resolved fluorescence anisotropy data permits knowledge of the hydrodynamic shape of a protein to be refined.

BACKGROUND

Analytical ultracentrifugation is a classical method for examining the global structure of a protein in solution by assessment of its hydrodynamic properties. Sedimentation equilibrium and sedimentation velocity measurements provide the molecular weight and the

---

[1]Supported by grants GM-39750 and HL-29019 from the National Institutes of Health, and by grant DIR-9002027 from the National Science Foundation.

translational friction coefficient of the protein. The translational friction coefficient, f, is a parameter that depends on the size, shape, and flexibility of the molecule.   It can be obtained from the Einstein-Sutherland equation

$$f = kT/D \tag{1}$$

or the Svedberg equation

$$f = \frac{M_r(1-\bar{\upsilon}\rho)}{N_A \, s} \tag{2}$$

where k is Boltzmann's constant, T is the absolute temperature, D is translational diffusion coefficient, N_A is Avagadro's number, and M_r, is the molecular weight, $\bar{\upsilon}$ the partial specific volume, ρ the solvent density, and s the sedimentation coeffcient of the protein.   With knowledge of the molecular weight, f is often used to assess the asymmetry of a protein by comparing its value with that expected for a rigid, spherical molecule of equal volume, f_o. From Stokes' law

$$f_o = 6\pi\eta_o R_{sphere} \tag{3}$$

where η_0 is the standard state solvent viscosity and R_{sphere} is the radius of the equivalent sphere.   The measure of asymmetry is generally expressed in terms of the friction ratio f/f_o. Using f/f_o, protein shapes are often modelled as oblate (discus shaped) or prolate (cigar shaped) ellipsoids of revolution. However, as discussed by Teller et al. (1979), this calculation involves assumptions about hydration, and unfortunately the calculation often results in absurd conclusions (also see Laue et al., 1992). In addition, because f is a single parameter, it is impossible to distinguish between the oblate and prolate models used when proteins are considered as hydrodynamic ellipsoids of revolution described by two molecular axes. Moreover, it is impossible to distinguish between a flexible, asymmetric molecule and a rigid molecule of lower asymmetry.

Small and Isenberg (1977) carefully considered the problem of rotatory and linear diffusion of a rigid body, and they have suggested that one approach for placing limits on the family of ellipsoids is to include information about rotational diffusion from time-resolved fluorescence anisotropy experiments. This approach has also been suggested by Stafford and Szent-Györgyi (1978).

In this chapter, using data that we have obtained on the clotting enzyme human factor VIIa and its complex with a soluble truncation mutant of human tissue factor, sTF (Waxman et al., 1993), we show that time-resolved fluorescence anisotropy measurements can be used to place limits on the family of ellipsoids predicted from the f values obtained by analytical ultracentrifugation. In addition, we show how the combination of the information content of time-resolved fluorescence anisotropy and analytical ultracentrifugation is useful for distinguishing between changes in f due to asymmetry and those due to increased rigidity of the macromolecule. What follows is a description of the strategy we used for evaluating the VIIa system concluded by discussion of important experimental requirements.

STRATEGY

Figure 1 provides an outline of the flow of information used to place limits on the semiaxes of a general ellipsoid. Equilibrium ultracentrifugation experiments are performed to assess possible self-association of the system and to obtain M_r, and sedimentation velocity experiments are performed to evaluate the deviation of the molecules from rigid spheres. Using equations derived by Perrin (1934), a family of ellipsoids is then generated which is consistent with the centrifugation data for the protein system.[2]

---

[2]VIIa did not self-associate under the experimental conditions: 0.05 M Tris buffer, 0.1 M NaCl, and 5 mM $CaCl_2$, pH 7.5, 23° C; it is a monomer. TF, a membrane-bound glycoprotein that serves as the essential cofactor for VIIa, consists of an extracellular domain (residues 1-219), a single trans-membrane domain (residues 220-242), and a cytoplasmic domain (243-263) [for review see Bach (1988)]. The soluble truncation mutant (sTF) consists of residues 1-218 of the extracellular domain. Its 1:1 complex with VIIa has a < 1 nM dissociation constant, and the complex does not form higher order oligomers (Waxman et al., 1992).

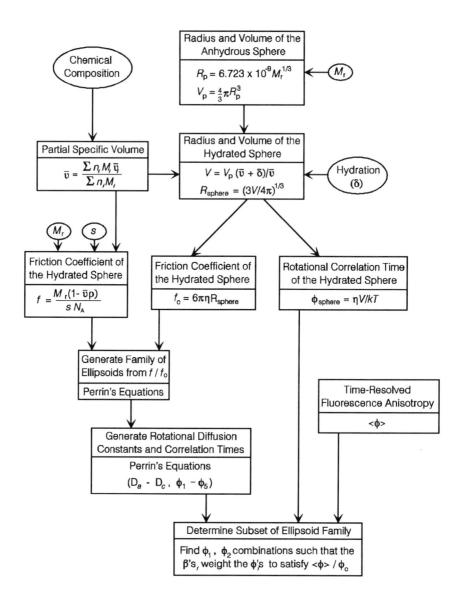

Figure 1. Flow diagram for evaluation of hydrodynamic parameters from analytical ultracentrifugation and time-resolved fluorescence anisotropy experiments, showing inputs, assumptions, procedures, and equations. Adapted after Waxman et al., (1993).

Calculation of R_{sphere}, the Radius of the Hydrated Stoke's Sphere. The initial evaluation of the hydrodynamic observables is calculation of the values expected if the molecule were treated as a rigid sphere. The first step shown in Figure 1 is the calculation of R_p, the radius of an anhydrous sphere with M_r equal to that of the molecule of interest. We used the empirical equation

$$R_p = 6.723 \times 10^{-9} M_r^{1/3} \tag{4}$$

(Teller, 1976). This relationship is based on packing volumes observed in X-ray crystallographic structures of protein-protein complexes. The anhydrous molecular volume, V_p, can be calculated from R_p. It has been shown that the classical textbook formulation,

$$R_o = \sqrt[3]{\frac{3 M_r \bar{\upsilon}}{4 \pi N_A}} \tag{5}$$

underestimates the radii and leads to overestimates of the axial ratios in subsequent calculations (Laue et al., 1992).

The hydration, δ, is used in the next step to generate V, the volume of the *hydrated* sphere.

$$V = V_p \left(\frac{\bar{\upsilon} + \delta}{\bar{\upsilon}} \right) \tag{6}$$

The method of Cohn and Edsall (1943; Laue et al., 1992) is used to estimate $\bar{\upsilon}$

$$\bar{\upsilon} = \frac{\Sigma n_i M_i \bar{\upsilon}_i}{\Sigma n_i M_i} \tag{7}$$

where n_i, M_i, and $\bar{\upsilon}_i$ are the number of moles, molecular weight, and partial specific volumes, respectively, for the i components -- amino acids and carbohydrates -- which make up the protein. For the purposes of this development, δ was estimated at 0.280 g H_2O/g protein. This value is an average of data from sedimentation velocity and small-angle X-ray scattering experiments for a group of proteins with molecular

weights less than 100,000 (Pessen and Kumosinski, 1985). R_{sphere}, the
radius of the *hydrated* Stokes sphere, is easily calculated from V.

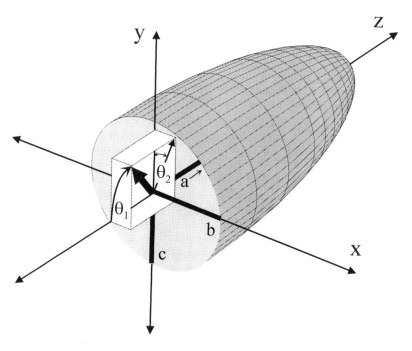

Figure 2: The semiaxes of the general ellisoid are $a \geq b \geq c$. It is assumed that the
absorbance and emission transition dipoles of the fluorophore, represented by the bold
arrow, are coincident. θ_1 is the angle to the major semiaxis a, and θ_2 is the projection
onto the plane defined by the b and c semiaxes. These angles define the orientation of
the transition dipole moments with respect to the molecular axes. Adapted from Waxman
et al., (1993).

Evaluation of f and Calculation of the Friction Ratio. An
asymmetric particle will have a frictional coefficient greater than that
of a spherical particle of the same volume. Thus the friction ratio, f/f_o,
is often used to calculate the axial ratios for prolate and oblate
ellipsoids that are consistent with the behavior of the molecule in the
centrifuge. As discussed above, however, there is no reason to believe
that a protein is well represented by either an oblate or prolate ellipsoid.
In fact, for a given friction ratio the oblate and prolate models provide
the most extreme cases of deviation from a sphere. If we take the

semiaxes of an ellipsoidal structure to be a, b, and c with $a > b > c$ (Figure 2), then for a rigid particle the friction ratio defines a family of rigid ellipsoids ranging in flatness (b/c) and elongation (a/b) from the oblate case to the prolate. The family of ellipsoids consistent with a given friction ratio can be generated by numerical evaluation of Perrin's equation (Perrin, 1934) as outlined by Small and Isenberg (1977). Initially, semiaxis a is set equal to semiaxis b. Semiaxis c is numerically evaluated to obtain the axial ratio for the oblate case. Semiaxis c is then numerically evaluated for increasing values of a/b until the prolate case, $b = c$, is reached. The result of this procedure is a table of b/c versus a/b for ellipsoids with the experimentally-determined friction ratio.

If an asymmetric molecule is not rigid, segmental flexibility will lower the friction ratio compared to that of the fully extended molecule (Laue et al., 1992). Since sedimentation velocity experiments measure the time-averaged shape of the molecule, the friction ratio defines a family of ellipsoids that represent the time-averaged shape of the molecule. Thus, it is impossible, using sedimentation methods alone, to distinguish between a highly asymmetric flexible molecule and a rigid molecule of lower asymmetry.

Fluorescence Anisotropy: Calculation of the rotational correlation time of a sphere. The rate at which fluorophores bound to a macromolecule rotate will be affected by the size, shape, and flexibility of the macromolecule. The depolarization of light emitted from the fluorophore will be affected by these factors and the orientation of the fluorophore with respect to the macromolecular axes. For a fluorophore rigidly attached to a rigid spherical molecule, the rotational correlation time, ϕ_o, can be calculated from the Stokes-Einstein equation

$$\phi_o = \frac{\eta V}{kT} \tag{8}$$

Evaluation of $<\phi>$ and the Rotation Ratio. Two types of experiments are typically performed to assess the rotational correlation time, ϕ, of a fluorophore. In a steady-state fluorescence anisotropy experiment, excitation is continuous and the fluorescence is detected through alternately vertically and horizontally oriented polarizers. The amount of depolarization is expressed as the steady-state anisotropy

$$\langle r \rangle = \frac{I_v - I_h}{I_v + 2I_h} \tag{9}$$

where I_v and I_h are the emission intensities when vertically polarized light is used to excite the sample. The measured steady-state anisotropy of a rigid sphere will be

$$\langle r \rangle_{sphere} = \frac{\langle r \rangle_0}{1 + \tau/\phi_0} \tag{10}$$

where $\langle r \rangle_0$ is the zero-point or 'frozen' anisotropy extrapolated back to infinite solvent viscosity, τ is the intensity lifetime of the fluorophore, and ϕ_0 is the rotational correlation time of the sphere. The degree to which a macromolecule deviates from a rigid sphere may be assessed by comparing the measured $\langle r \rangle$ to $\langle r \rangle_{sphere}$. Asymmetry will cause the measured $\langle r \rangle$ to shift toward $\langle r \rangle_0$ and to exceed $\langle r \rangle_{sphere}$. If the macromolecule is not rigid or the fluorophore has depolarizing motions independent of rotation of the macromolecule, these segmental motions will cause the measured $\langle r \rangle$ to shift toward 0 and to be less than $\langle r \rangle_{sphere}$. In addition, the orientation of the probe with respect to the macromolecular axes is critical in determining the degree to which asymmetry and flexibility cause deviation from $\langle r \rangle_{sphere}$ (see below).

In a time-resolved fluorescence anisotropy experiment the sample is repetitively pulsed with polarized light. By analogy with Equation 9, the time-dependent anisotropy due to rotation is

$$r(t) = \frac{I_v(t) - I_h(t)}{I_v(t) + 2I_h(t)} \tag{11}$$

For a fluorophore rigidly attached to a rigid spherical molecule, the anisotropy decay law is given by

$$r(t) = r_0 e^{-t/\phi_0} \tag{12}$$

where now r_0 is the anisotropy extrapolated back to time zero.

The anisotropy decay law for a rigid ellipsoid is considerably more complicated, and is described by a sum of five exponentials (Tao, 1969;

Belford et al., 1972),

$$r(t) = \sum_{i=1}^{5} \beta_i e^{-t/\phi_i} \tag{13}$$

Calculation of the relative values of the correlation times of these exponentials for ellipsoids of varying axial ratios show that even under ideal circumstances only three values of ϕ_i are sufficiently distinct to be resolvable (Small & Isenberg, 1977). In protein studies, however, often only a single exponential decay with a ϕ equal to the mean correlation time

$$<\phi> = \left[\frac{\sum \beta_i/\phi_i}{\sum \beta_i} \right]^{-1} \tag{14}$$

is observed. The preexponential weighting factors, β_i, and therefore $<\phi>$, are complicated functions of both the asymmetry of the protein and the orientation of the fluorophore's absorbance and emission dipoles with respect to the axes of the protein. As an example (see Figure 2), for a fluorophore that has coincident absorbance and emission dipoles and is attached to a prolate ellipsoid shaped protein, $<\phi>$ will be maximal if the dipoles are aligned parallel to the long axis ($\theta_1 = 0°$) and minimal when the dipoles are aligned parallel to the short axis ($\theta_1 = 90°$).

Because $<\phi>$ depends on orientation of the probe as well as the size, shape, and flexibility of the molecule, the rotation ratio, $<\phi>/\phi_0$, places fewer constraints on the range of ellipsoids that can represent the behavior of the molecule than does f/f_0. However, because $<\phi>/\phi_0$ depends on asymmetry differently than f/f_0, $<\phi>/\phi_0$ can be used to place limits on the family of ellipsoids generated using f/f_0. First, $<\phi>$ is measured and $<\phi>/\phi_0$ is calculated. Then, Perrin's equations are used to calculate the three rotational diffusion coefficients (D_a, D_b, D_c) and five correlation times for each of the members of the friction ratio-consistent table of ellipsoids. Finally, the range of ellipsoids consistent with both f/f_0 and $<\phi>/\phi_0$ is determined. This is accomplished by evaluating, for each ellipsoid, whether a probe orientation can be found such that the β_i's (Equation 14) weight the five exponentials so that the experimentally-determined $<\phi>/\phi_0$ is recovered. The equations for determining the β_i's for different probe angles are outlined by Belford et al. (1972).

In addition to limiting the range of ellipsoids which can represent the hydrodynamic behavior of a protein, time-resolved fluorescence anisotropy decay data can provide important information about molecular flexibility or segmental motion which cannot be inferred from the sedimentation data alone. This is due to the short time frame (ps to ns) over which rotation is measured by fluorescence. In time-resolved fluorescence anisotropy measurements, segmental motion of a region of the protein containing the probe will produce a component in the anisotropy decay with a correlation time less than ϕ_0. The appropriate equation for the anisotropy decay in this case is

$$r(t) \ = \ r_0[\gamma e^{-t/\phi_{segmental}} + (1-\gamma)]e^{-t/\phi_{global}} \qquad (15)$$

where γ is a scaling factor between 0 and 1 (see Lakowicz, 1983). Here, $\phi_{segmental}$ is the correlation time for depolarizing motions of a domain containing the probe and ϕ_{global} is the correlation time for the rotational motion of the protein as a whole. Anisotropy decay data are typically analyzed as a *sum* rather than a *product* of exponentials. For this reason the decay law in the presence of segmental flexibility is observed to be

$$r(t) \ = \ r_0[\ p_{short}e^{-t/\phi_{short}} + p_{long}e^{-t/\phi_{long}}] \qquad (16)$$

where $p_{short} = \gamma$ and $p_{long} = 1 - \gamma$. Thus,

$$\phi_{short}^{-1} = \phi_{segmental}^{-1} + \phi_{global}^{-1} \ ; \quad \phi_{long}^{-1} = \phi_{global}^{-1} \qquad (17)$$

An anisotropy decay which analyzes as a biexponential with one correlation time greater than or equal to ϕ_0 and the other less than ϕ_0 would be diagnostic of segmental motion of the protein domain that contains the probe.

An Illustrative Example. In our experiments on VIIa and the sTF:VIIa complex (Waxman et al., 1993), we specifically fluorescent labelled the active site histidine residue of VIIa by covalent reaction with dansyl-D-Phe-L-Phe-Arg chloromethyl ketone (Waxman et al., 1992). Recognizing that the dansyl probe is specifically placed, we made the assumption that each of the lifetime components of the fluorescence intensity decay was associated with the same molecular motions of the protein (often the intensity decay of a single probe is multi-exponential). In this special case, the mean lifetime, $<\tau>$,[3] conveys information equivalent to that of the individual decay components. Thus, $<\tau>$ may be used to generate apparent rotational correlation times from steady-state anisotropy values. This cannot be done if multiple sites are labelled since labels at different sites are likely to have different intensity decays as well as anisotropy decays. In the latter situation, the intensity and anisotropy decays of each probe are associated in a complicated way.

Apparent mean rotational correlation times of 30 and 43 ns for VIIa and sTF:VIIa, respectively (20°C) were calculated using Equation 10 and an $<r>_0$ of 0.32 for the dansyl probe. (The value for $<r>_0$ was obtained from a Perrin plot ($1/r$ versus T/η) of steady-state anisotropy data for labelled VIIa.) From Equation 8, the rotational correlation times of hydrated spheres (0.28 g water/g protein) with the same molecular weights as these species would be 22 and 32 ns, respectively. These data indicate that neither the shape of VIIa nor that of the sTF:VIIa complex is well represented as a hydrated sphere.

It should be noted that because of the assumptions involved in the above calculations, it is impossible to determine from the fluorescence anisotropy decay data alone whether the large correlation times are due to asymmetry or to higher order association. This question can be resolved by sedimentation equilibrium measurements.

Table I shows the results obtained at each step of the procedure outlined in Figure 1. The M_r values obtained from equilibrium ultracentrifugation of VIIa, sTF, and an equimolar mixture of sTF and VIIa were 49,000, 27,700, and 73,700, respectively; labelling did not result in formation of higher order oligomers. Sedimentation velocity experiments performed using VIIa and the equimolar mixture of sTF and VIIa both showed a single boundary. The translational frictional coefficients shown in Table I are calculated according to Equation 7. f_0 values for VIIa and sTF:VIIa are shown for comparison and respective values of f/f_0 are calculated. Each friction ratio is consistent with a family of ellipsoidal shapes ranging from prolate to oblate. For

---

[3]The intensity weighted mean lifetime $<\tau> = \Sigma\alpha_i \tau_i^2 / \Sigma\alpha_i \tau_i$, where α_i is the amplitude and τ_i is the lifetime of the ith decay component.

TABLE I: Calculation of Hydrodynamic Parameters[a]

Parameter	VIIa	sTF:VIIa
M_r (g/mole)[b]	49000	73700
R_p (cm)	2.50e-07	2.80e-07
V_p (mL)	6.20e-20	9.40e-20
$\bar{v}$ (mL/g)	0.703	0.711
δ	0.28	0.28
V (mL)	8.70e-20	1.30e-19
R_{sphere} (cm)	2.80e-07	3.20e-07
f_0 (g /s)	5.20e-08	5.90e-08
$s^0_{20,w}$ (s)[c]	3.4e-13	3.9e-13
f (g /s)	7.20e-08	9.10e-08
f/f_0[d]	1.39	1.52
ϕ_0 (ns)[e]	22	32
$<\phi>$ (ns)[e]	94	123
$<\phi>/\phi_0$	4.27	3.73

[a]Values, which are taken from Waxman et al. (1993), are rounded off for the purpose of presentation. [b]The estimated error in M_r is $\pm$ 2%. [c]The estimated error is $\pm$ 0.1e-13 s. [d]Propagation of error indicates a $\pm$ 3% error in f/f_0. Intermediate values were not rounded for the actual calculations. [e]Correlation times presented are for 20 °C.

VIIa, the axial ratios for oblate and prolate consistent with the friction ratio are 8.4 and 7.2, respectively. For sTF:VIIa, the respective axial ratios are 11.8 and 9.6.

TABLE II: Time-Resolved Fluorescence Anisotropy Parameters[a]

Sample	T (°C)	η (cp)	β_1	ϕ_1 (ns)	β_2	ϕ_2 (ns)
VIIa	5	1.56	0.36	145	0.04	22.0
	10	1.35	0.36	126	0.04	18.6
	15	1.17	0.36	108	0.04	16.1
	20	1.04	0.36	97	0.04	14.1
	25	0.91	0.36	85	0.04	9.6
sTF:VIIa	5	1.56	0.40	194		
	10	1.35	0.39	168		
	15	1.17	0.40	141		
	20	1.04	0.40	122		
	25	0.91	0.40	112		

[a]The results presented for VIIa were obtained by fixing each β_i to the average value obtained from unrestricted analyses as described by Waxman et al., (1993). The results presented for sTF:VIIa are values obtained from completely unrestricted analyses.

The results from the time-resolved fluorescence anisotropy experiments with VIIa and sTF:VIIa are shown in Table II. As noted above, the ϕ_o for VIIa at 20°C is 22 ns. When this value is compared to the correlation times obtained at this temperature it can be seen that the 97 ns correlation time observed is considerably larger than ϕ_o, and the 14 ns correlation time is considerably smaller than ϕ_o. These results suggest that the anisotropy decay of active site-labelled VIIa is composed of decays due to global and segmental motions. From Equations 15-17 it can be seen that the larger correlation time is equal to global motion correlation time, ϕ_o. A simple calculation (Equation 17) shows that the segmental motion correlation time is 12 ns. Equation 8 can be used to make an order of magnitude approximation of the size of the protein domain responsible for the fast component of the anisotropy decay. This analysis yields $M_r \approx 30,000$ and suggests that a significant fraction of the protein structure is involved.

In contrast to the anisotropy decay obtained for VIIa alone, which was double exponential, the anisotropy decay of the sTF:VIIa complex was single exponential. At 20°C, ϕ_0 for sTF:VIIa is 32 ns. However, the rotational correlation time obtained from analysis of the anisotropy decay data is 123 ns, indicating, in agreement with the sedimentation data, that the sTF:VIIa complex is highly asymmetric. In addition, the lack of a short correlation time indicates a loss of segmental motion in the domain containing the active site of VIIa, where the dansyl probe is located.

In the final step of the analysis outlined in the Experimental Approach we used Perrin's equations (1934) as outlined by Small and Isenberg (1977) to generate the three principal rotational diffusion constants and five rotational correlation times for the family of ellipsoids consistent with the friction ratios obtained for VIIa and sTF:VIIa (Table I). We then determined the subset of each family that was consistent with both the rotation and friction ratios. For VIIa alone, the ellipsoids consistent with both the friction and rotation ratios range from the prolate to an ellipsoid with axial ratios of a/b (elongation) and b/c (flatness) of 1.3 and 7.2, respectively. The corresponding molecular axes are 204×28×28 Å for the limiting prolate shape and 126×98×14 Å for the limiting oblate shape. For sTF:VIIa, the ellipsoids consistent with both the friction and rotation ratios range from the prolate to an ellipsoid shape with axial ratios of a/b and b/c of 2.37 and 6.41, respectively. The corresponding axes are 254×26×26 Å for the prolate shape and 186×78×12 Å for the oblate shape. Thus, for both VIIa and sTF:VIIa, *a subset of flattened shapes is eliminated when the rotation ratio is taken into account.* In particular, this analysis demonstrates that an oblate model cannot account for the hydrodynamic behavior of either VIIa alone or its complex with sTF.

The anisotropy decay data reveals differences in the conformational dynamics of VIIa alone and bound to sTF. Specifically, a biexponential model is necessary to adequately describe the decay of the fluorescence anisotropy of VIIa. In this regard, it should be noted that multiexponential anisotropy decays have been observed for proteins which exhibit segmental flexibility. For example, Yguerabide et al. (1970) observed a biexponential decay with correlation times of 33 and 168 ns for dansyl-lysine bound to IgG. The longer correlation time, which exceeds ϕ_0 for the 150 kD protein, was attributed to the global rotation of the protein while the shorter correlation time, which is less than ϕ_0, was attributed to segmental motions of the Fab fragments. That result is essentially identical with what we have observed for VIIa in the absence of sTF. The shorter correlation time (14 ns at 20°C) is significantly less than ϕ_0 but substantially longer than that expected for side chain motions (< 1 ns). It is likely that this short correlation time represents the jointed motion of a structural domain of VIIa that contains the labeled active site.

Clearly, the centrifugation and fluorescence results each show that both VIIa and sTF:VIIa are highly asymmetric. But as discussed above, the translational friction coefficients alone are not sufficient to distinguish among the family of ellipsoids nor are they capable of distinguishing between a highly asymmetric flexible molecule and a rigid molecule with lower asymmetry. Fluorescence anisotropy decay measurements employing labelling with a site specific probe, however, narrow the range of ellipsoids which are consistent with the hydrodynamic behavior of VIIa and sTF:VIIa and make it possible to detect the presence of segmental motions.

The two physical techniques examined in this chapter are by no means the only combination that could be successfullly employed. Inclusion of additional data for hydrodynamic parameters from other techniques, such as small angle x-ray scattering or electrophoretic mobility, should help to refine shapes even further.

ACKNOWLEDGMENTS

Supported in part by National Institutes of Health grants GM-39750 and HL-29109, and in part by National Science Foundation grant DIR-9002027. We appreciate the technical assistance of Theresa M. Ridgeway, Daryl Lyons, and Steven Eaton. We also wish to acknowledge helpful discussions with Dr. David C. Teller, Dr. Donald F. Senear, Dr. Yale Nemerson, and Dr. Carol A. Hasselbacher.

GLOSSARY OF SYMBOLS

D	translational diffusion coefficient
f	translational friction coefficient
f_0	translational friction coefficient of a hydrated sphere
f/f_0	friction ratio
I_h	horizontally polarized steady-state fluorescence
I_v	vertically polarized steady-state fluorescence
$I_h(t)$	horizontally polarized time-dependent fluorescence
$I_v(t)$	vertically polarized time-dependent fluorescence
k	Boltzmann's constant
M_i	molecular weight of species i
M_r	molecular weight
N_A	Avogadro's number
n_i	number of moles of species i
r_0	time zero anisotropy
$\langle r \rangle$	steady-state anisotropy
$\langle r \rangle_0$	'frozen' steady-state anisotropy
$\langle r \rangle_{sphere}$	steady-state anisotropy of a sphere
$r(t)$	time-dependent anisotropy
R_p	radius of an anhydrous sphere
R_{sphere}	radius of a hydrated sphere
t	time
T	absolute temperature
V	volume of a hydrated sphere
V_p	volume of an anhydrous sphere
α_i	pre-exponential weighting factor of τ_i
β_i	pre-exponential weighting factor of ϕ_i
δ	hydration
η	solvent viscosity
η_0	standard state solvent viscosity
$\langle \phi \rangle$	mean rotational correlation time
ϕ_i	correlation time constant i of the anisotropy decay
ϕ_0	rotational correlation time of a sphere
$\langle \phi \rangle / \phi_0$	rotation ratio
ρ	solvent density
τ	fluorescence lifetime
τ_i	time constant i of the fluorescence intensity decay
$\bar{\upsilon}$	partial specific volume
$\bar{\upsilon}_i$	partial specific volume of species i

REFERENCES

Bach, R. (1988) *CRC Crit. Rev. Biochem. 23*, 339-368.

Belford, G. G., Belford, R. L., & Weber, G. (1972) *Proc. Natl. Acad. Sci. U.S.A. 69*, 1392-1393.

Cohn, E. J., & Edsall, J. T. (1943) Proteins, Amino Acids and Peptides as Ions and Dipolar Ions, p. 157, Rheinhold, New York, NY.

Lakowicz, J. R. (1983) *Principles of Fluorescence Spectroscopy*, pp. 161-162, Plenum Press, New York, NY.

Laue, T. M., Shah, B. D., Ridgeway, T. M., and Pelletier, S. M. (1992) *Analytical Ultracentrifugation in Biochemistry and Polymer Science,* pp. 90-125, Eds. S. Harding & A. Rowe, Royal Society of Chemistry, London.

Perrin, F. (1934) *J. Phys. Rad., Ser. VII,* **V**, 497-511.

Pessen, H., & Kumosinski, T. F. (1985) *Methods Enzymol. 117*, 219-255.

Small, E. W., & Isenberg, I. (1977) *Biopolymers 16*, 1907-1928.

Stafford, W. F. III, & Szent-Györgyi, A.G. (1978) *Biochemistry 17*, 607-614.

Tao, T. (1969) *Biopolymers 8*, 609-632.

Teller, D. C. (1976) *Nature 260*, 729-731.

Teller, D. C., Swanson, E., & de Haën, C. (1979) *Methods Enzymol. 61*, 103-124.

Waxman, E., Ross, J. B. A., Laue, T. M., Guha, A., Thiruvikraman, S. V., Lin, T. C., Konigsberg, W. H., & Nemerson, Y. (1992) *Biochemistry 31*, 3998-4003.

Waxman, E., Laws, W. R. Laue, T. M., Nemerson, Y., & Ross, J. B. A., (1993) *Biochemistry 32*, 3005-3012.

Yguerabide, J., Epstein, H.F., & Stryer, L. (1970) *J. Mol. Bio. 51*, 573-590.

Part III
ACQUISITION AND DATA REDUCTION

ON LINE DATA ACQUISITION FOR THE RAYLEIGH INTERFERENCE OPTICAL SYSTEM OF THE ANALYTICAL ULTRACENTRIFUGE

David A. Yphantis, Jeffrey W. Lary, Walter F. Stafford, Sen Liu, Philip H. Olsen, David B. Hayes, Thomas P. Moody, Theresa M. Ridgeway, Daryl A. Lyons and Thomas M. Laue

INTRODUCTION

The fundamental measurement in analytical ultracentrifugation is the concentration as a function of radial position. The Rayleigh interferometer of the analytical ultracentrifuge produces a cell image in which the concentration at each radial position is presented as the vertical displacement of a set of equally-spaced horizontal fringes (Richards and Schachman, 1959). Manual acquisition of data from interferograms is tedious and automated photographic plate readers still require that photographs be taken, processed, aligned and read before data analysis can be performed. Fortunately, a Rayleigh interference image is well suited for television-camera-based data acquisition. Described here are two types of automated Rayleigh interferometers for the Beckman Model E analytical ultracentrifuge. One type of system relays and magnifies the Rayleigh interference image from the usual photographic plane to a television camera located behind this plane. The other system uses a redesigned camera-cylinder lens combination to create a radially demagnified Rayleigh interference image of the cell on the television camera sensor located on the original cylinder lens mount.

The two optical systems have many features in common. Each system uses a laser-diode light source, a solid-state television camera and a commercially available computer interface to acquire the interference image. The image is aligned so that the columns of pixels on the camera sensor are perpendicular to the radial direction and, therefore, each column corresponds to a radial position. A single-frequency, discrete Fourier analysis at each column of pixels provides the fractional fringe displacement in a manner somewhat similar to that first described by DeRosier and co-workers (DeRosier, *et al.*, 1971). Keeping track of the integral number of fringes traversed enables the image to be converted to fringe displacement versus radial position.

The two optical systems differ principally in their radial resolution. The first, "high-magnification" system, allows the radial resolution to be chosen so that even the steepest concentration gradients can be observed. However, the camera sensor is too small to accommodate the entire cell image, so the camera is mounted on a precision motorized linear stage which moves along the radial axis. This adds to the cost, requires more complex operating software and requires the determination of two calibration constants to calculate the radial position. The second, "low magnification," optical system is simpler and somewhat less expensive, but its radial resolution is lower and fixed by the television camera sensor characteristics.

Both optical systems provide excellent performance. Image acquisition requires 1/30th of a second and image processing less than 10 seconds. The precision of fringe displacement measurements is comparable to that of automated photographic plate readers (DeRosier, *et al.*, 1971; Richards and Richards, 1974; Laue, 1981). By integrating the data acquisition, editing and analysis in a rapid and simple to use package, analytical ultracentrifugation acquires an immediacy not previously associated with it. More importantly, new and more complex experimental protocols become feasible and more information may be obtained from a preparation of unstable molecules. The elimination of the tedious and time consuming procedures associated with processing and measuring photographic images frees experimenters and enables them to concentrate on the results.

DESIGN & DESCRIPTION

Overview

Two different imaging schemes were tested (figure 1). The two schemes illustrate the trade-off that must be made with regard to image resolution versus image size using current camera sensorsHx In the high magnification scheme (Figure 1a), the basic optics of the Model E were retained. The light source consists of a laser, pinhole (optional), objective lens and right-angle prisms arranged essentially as described by Williams (Williams, 1972). The cell image is brought into focus at the normal photographic plane. The television camera is mounted about 25 cm behind the photographic plane. The television camera lens is detached from the camera body so that a focused, magnified image of the photographic plane is observed. By shifting the camera and lens positions along the track, various magnifications can be obtained. The camera and its lens are mounted on a motor-driven stage aligned at right-angles to the optical track so that various portions of the cell image can be observed. The magnification is chosen so that the radial increment (i.e. the distance in cell coordinates crossed by each increment in the column pixel number) is small enough (4-10 μm/column) to follow fringe displacements in regions with concentration gradients near or above

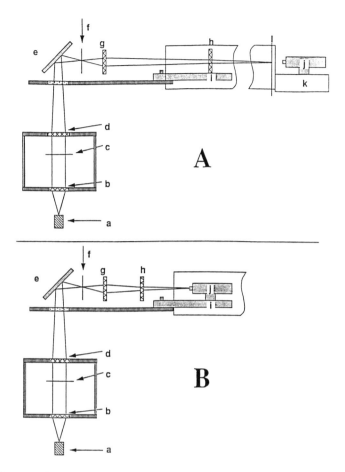

FIGURE 1. Diagrams of the two optical systems described here. The diagrams show the relative position of the optical components. Panel A presents the high magnification optical system. The fundamental optics are identical to those on the Model E (Richards and Schachman, 1959): a) light source, b) collimating lens, c) cell image plane, d) condensing lens, e) 45° mirror, f) focal plane of the condensing lens, g) camera lens, h) cylinder lens, i) tilt table for the cylinder lens, j) television camera on a rotatable mount (k), with a relay lens imaging the focal plane (l) of the Model E optics. Panel B presents the low-magnification optics. Components a–h are essentially as for panel A, except that the focal lengths of the camera lens and cylinder lens have been changed (see text) to bring the image of the cell into focus directly on the camera (j) sensor. The camera is positioned on the cylinder lens tilt table (i) for rotational adjustment. See text for details.

the Wiener skewing limit (Yphantis, 1964). The requirement for a motorized stage adds both hardware and software complexity to this system. Software must be included to move the stage, as well as to determine the radial position based on both the motorized stage position and the pixel column number (i.e. the position on the sensor). It would be far simpler to shrink the cell and reference image so that it fit entirely on the sensor, thus eliminating the need to move the camera. This is the motivation for developing the low magnification imaging scheme.

In the low magnification imaging scheme (figure 1b), the television camera is located approximately at the position of the original cylinder lens. A new camera lens and cylinder lens are placed on the optical track near the location of the original camera lens. The lenses were chosen so that the entire image of the cell and reference holes are brought to focus on the television camera sensor. This eliminates the need for a motorized camera mount. However, using currently available television camera sensors, each column of pixels corresponds to about a 22 μm radial increment, which limits the concentration gradients that can be followed.

Light sources

Laser light sources should be used for on-line acquisition. Pulsed argon-ion lasers (Paul and Yphantis, 1972; Laue, 1981), CW He-Ne lasers (Williams,1972) and laser diodes (Lary *et al.*, 1990; Laue, 1992) have been used successfully. Pulsed argon-ion lasers now are expensive and are no longer used. He-Ne lasers from 2-25 mwatts could be used in a CW mode, restricting them to a single cell, or modulated acousto-optically (Williams, 1972). Modulatable laser diodes with outputs greater than 5 mwatts are available at wavelengths in the red (670 and 730 nm) and near infra-red (780 and 830 nm). Diodes with shorter wavelength output are preferred since, for a given refractive index difference, the fringe displacement is inversely proportional to the wavelength and because the refractive increment is larger at shorter wavelengths (Perlman and Longsworth, 1948). Temperature controlled diode mounts are unnecessary so long as the laser diode case is kept within the specified temperature range. Because of the low duty cycle for pulsed lasers, this is not a problem. Suitable modulatable laser diode systems are available commercially from a number of sources or can be constructed (Laue, 1992; Laue, 1993).

The laser beam can be expanded and collimated as described previously (Williams, 1972; Paul and Yphantis, 1972). The spatial filter described by Williams was removed after it was determined that it affected essentially only the aesthetic qualities of the image. For most lasers, a 10x objective lens provides sufficient beam expansion to fill the collimating lens. Elliptically-shaped beams (axial ratio 4:1) are generated by laser diodes. The long axis of the expanded beam should fill the collimating lens and should be aligned radially. Alternative beam expansion-collimation schemes have been tested for use with laser diodes (Laue,1993). These require a much shorter optical path, can be focused at a lab bench and can be mounted directly underneath the rotor chamber of the Model E.

Cameras

Black and white solid state television cameras are well suited to the acquisition of interference images by virtue of the close and precise spacing of the light sensitive cells (pixels), adequate light sensitivity, insensitivity to blooming or damage from bursts of light and their generally rugged nature. The criteria for choosing an appropriate television camera include: 1) highest possible resolution, accomplished by having narrow pixel spacing and, hence, small pixel size, 2) largest possible array size, accomplished by having the largest number of pixels per row and per column, 3) highest light sensitivity with the lowest noise and 4) broadest range of light intensity providing a linear response. The first criterion arises from the desire to have a high radial resolution. The second criterion results from the desire to image as much of the cell as possible. The widely available high resolution cameras have 500-775 columns of pixels space approximately 10-15 μm apart. The narrow apertures in the Rayleigh optics lead to the need for high light sensitivity, although the recent development of inexpensive, high intensity laser diodes makes this criterion less restricting. The commercial arrays we have tested exhibit shot noise less than 2-bits, with occasional individual pixels exhibiting greater noise levels. In order to minimize the need to adjust the light source intensity, it is convenient to have a sensor that can reproduce fringes over a wide range of intensities without saturating. The filtering performed by the Fourier analysis minimizes the need for accurate reporting of the fringe intensities and reduces the requirement for high-quality images (below).

The spectral response is determined by the properties of the silicon sensors, and is, therefore, fairly consistent for solid state cameras. This parameter is important only if lasers are chosen with wavelengths greater than 930 nm or if an especially low-power laser is chosen (< 5 mwatt) or if low throughput optical components are chosen (e.g. particularly narrow slits or a small-aperture pinhole is used to refine the beam pattern). Finally, for the second imaging scheme, the camera must be small enough to fit in the optical tube. Many commercially available solid state television cameras are suitable for this application. These provide standard video signals appropriate for digital acquisition (e.g. RS-170, CCIR and PAL). The essential characteristics of three cameras used in these studies are presented in Table 1.

The high coherence of the laser light can cause interference artifacts (Williams, 1972). In particular, Newton's rings were visible and their source was traced to the faceplate over the array. Accordingly, the faceplate was removed and the rings were eliminated. This resulted in a more aesthetically pleasing image, but provided no detectable improvement in the measurement of fringe displacements. Similar problems arise from infrared filters attached to the sensor. Removal of the filter is straight forward and results in both higher sensitivity and improved image quality.

Computer interface

Data Translation computer interfaces for the MicroVax II (model DT2651), IBM-PC (model DT2851) and Apple MacIntosh II (model DT2255) have been used to digitize the video image. Software was written in FORTRAN or C and linked to the Data Translation DT-IRIS subroutine library. These programs are available from the authors. Other video image digitizers providing 8 bits of intensity resolution and storage for at least 480 rows by 512 columns of pixels would suffice.

Camera	Pixel size HxV μm	Pixel spacing μm	#rows x #columns	Sensitivity nW/cm² [a]	Saturation μW/cm² [b]	Sensor area HxV mm
Amperex NXA1031	9.9x18.6	9.9	492x610	8	4	6.0x4.5
Cohu 4810	11.5x27.0	11.7	488x754	12	5	8.8x6.6
Dage-MTI CCD 72	11.0x11.0	11.6	581x756	12	5	8.8x6.6

TABLE 1. Camera specifications for the on-line interferometers. [a]Minimum light intensity needed to produce a measureable response. [b]Illumination necessary to fully saturate the array.

Computer system

On line data acquisition systems have been developed for three types of computer systems: 1) Q-bus-based MicroVax II workstations (Digital Equipment Corporation, Maynard, MA) with at least 13 Mbytes of memory, and 210 Mbytes (or more) of hard disk storage running under the VMS operating system and the workstation software; 2) MacIntosh II (Apple Computer Corp., Cupertino, CA) with math coprocessor, 8 Mbyte of memory, 120 Mbytes of hard disk storage; and 3) IBM-PC compatibles employing 80386 (with math co-processor) or 80486 CPUs (Zenith Data Systems, Benton Harbor, MI or Northgate Computer, Prairie View, MN) with at least 4 Mbytes of memory and over 80 Mbytes of hard disk storage running under the MS-DOS operating system. A hard copy graphics device is useful (plotter or printer) and an archival storage device (e.g. tape or removeable cartridge hard disk) is strongly recommended.

High-Magnification optics

For the imaging scheme in Figure 1A, the lenses of the upper optical track of the Model E are retained. The camera and its lens are mounted on a moveable stage which is attached to an optical track that, in turn, is fastened to an adapter-plate bolted onto the film-advance mechanism of the Model E (Figure 1A, k). The camera mount is positioned to allow for the existing gears, thus permitting continued operation of the photographic system. Vertical adjustment is provided by slotted bolt holes in the plate holding the optical track as described previously (Laue, 1992). Rotational adjustment is provided by pivoting the optical track at the site of its attachment to the adapter plate or by attaching the moveable stage to a tilt-table. Two different moveable stages have been tested and their characteristics are presented in Table 2. Each stage is equipped with a computer-accessible position sensor.

The Model E optics bring a 2.1-fold magnified cell image into focus at the photographic plane (Figure 1A). In order to achieve greater magnification, a relay lens was positioned, either on a separate optical carrier or on an extension tube, between the photographic plane and the television camera. Several standard television camera lenses were tested for use as the relay lens and either a 75 mm, f/1.8 or a 50 mm, f/1.5 lens were found to be suitable. The lens and television camera were positioned in such a way that an object at the photographic plane was brought into focus on the camera's sensor. Magnification of the photographic plane could be varied according to the simple lens equation.

Low-magnification optics

In order to fit the entire cell image on the television camera sensor, the standard Model E camera lens was replaced by a 44.5 mm diameter, AR-coated, achromat doublet, but with a focal length of 195 mm (Rolyn Optics, Covina, CA) and the standard Model E cylinder lens was replaced by one 30x60 mm in size, with an 80 mm focal length (Rolyn Optics, Covina, CA). The cylinder lens was attached to the slip ring of a standard Model E UV camera lens mount. This permits the rotational alignment of the cylinder lens with the radial axis of the rotor. The lens mount was modified to accept a set screw to fasten the slip ring. This combination of lenses provides a focused, roughly 0.5-fold demagnified image of the cell approximately at the position originally occupied by the cylinder lens. The camera mount for the second imaging scheme (Figure 1B) uses the Model E cylinder lens mount to hold the camera. This mount provides both the vertical and rotational adjustments.

High-magnification system focusing, alignment and calibration

The centrifuge optics were aligned and focused as described previously (Richards, *et al.* 1971), with two modifications. First, the relay lens was positioned to focus a portion of a ruled scale mounted at the Model E photographic plane. This allows the camera to be used in place of film for the remaining steps of the focusing procedure. Second, for steps that required diffuse light, either an incadescent lamp with an appropriate

Stage	Type[a]	Resolution μm	Run out μm[b]	Speed μm/sec
Oriel 16328	DC	0.1	2	200
Daedal 105021P	Stepper	1.0	1.3	2000

TABLE 2. Motorized stage specifications. The high-magnification systems require that the television camera be moved. [a]The stages can be driven either by a DC motor or by a stepping motor. [b]Maximum cumulative error for a stage translation over the full-range of stage

bandpass filter (e.g. a model P10-633-S <Corion Corp., Holliston MA> for the 633 nm HeNe laser) or the usual diode laser viewed through a rotating translucent plastic disk was used for illumination.

Once the camera is aligned, each column of pixels on the sensor corresponds to a unique radial position. The apparent radial position is given by the column index, which ranges from 1 to 512. Both the motorized stage position and the column index must be considered in the calculation of the radial position. That is, increments of the stage position and of the column index are related by independent magnification factors to increments in radial position in the cell coordinate system. Both magnification factors are determined using the ruled disk available for focusing the Rayleigh optics.

The ruled disk is inserted into the holder, ruled side down, along with the appropriate spacers to bring it to the 2/3 height and arranged in the optics as though the camera lens were being focused. The incandescent light and bandpass filter are used for illumination (above) and the rotor aligned so that a clear image of the 1 mm-spaced lines appears on the television monitor. At the magnification used here, three lines should be in the image at any given time. Programs were written that permit the superposition of a vertical white line on the image at the monitor. This vertical reference line should be exactly parallel to the lines on the ruled disk if everything is aligned properly. To start, the reference line is placed at column 256 (the center of the sensor) and the motorized stage moved until the left edge of one of the ruled-disk lines is adjacent to the reference line. The distance (in mm) needed to move the stage to align each of the lines on the ruled disk with this reference line is noted. The slope from a linear regression of distance as a function of ruling number yields the magnification factor for the motorized stage (M_m). M_m is a unitless ratio, since the ruled lines are precisely 1 millimeter apart, and should equal the magnification factor determined using the standard photographic method (Richards, *et al.*, 1971) if the television camera is focused at the film plane.

The magnification factor relating the position on the sensor to the radial increment is determined by moving the stage so that one of the rulings is transported across the sensor by about 10% of the total sensor width. The experimenter then moves a vertical cursor to cover the ruled marking and the column number is noted. This procedure is repeated several times, and the slope from a linear regression of the increment in pixel column number as a function of the distance of stage movement (in mm) is designated M_c and has units of pixels/mm. At lower magnifications, where several ruled markings are visible in the image, transport of the camera is unnecessary for determining M_c. In this case, slope from a linear regression of the increment in the column index per ruled line is M_c in units of pixels/mm.

For a system which exhibits no aberrations, M_m and M_c are constant. Both M_m and M_c are sensitive to optical aberrations or nonlinearity in the camera transport mechanism (M_c only if the camera is moved during its determination). Artifacts that affect either M_m or M_c (e.g. spherical aberrations of the lenses) can be detected by examining the calibration curves for linearity. Alternatively, graphs of the apparent calibration factor as a function of position may be used. Distortion of video signal leading to noncompliance of the horizontal sweep and the sensor position will affect M_c more than M_m. If necessary, any systematic variation in these calibration factors can be minimized with appropriate software corrections.

In order for the radial position measurements to be valid, the linear motion of the motorized stage must parallel the radial axis of the cell image. Accurate alignment is possible using cross-correlation analysis of the fringe intensity patterns obtained at the two reference images of a counterweight (Laue, 1992). The cross-correlation coefficient is calculated for the images as they appear on the camera sensor, and at each of 480 positions corresponding to shifting one image ± 240 pixels about the other. The maximum in the cross-correlation corresponds to the number of pixels needed to rotate the image for optimum alignment. This measurement is repeated, changing the tilt of the motorized stage between measurements, until the maximum cross-correlation corresponds to near zero pixels. The rotation is constant and only needs to be checked when the optics are refocused.

The final information needed to determine the radial position is the position of the counterweight reference edges. These are determined by moving the motorized stage until the image of the reference edge is in the field of view, then positioning a vertical white line until it is aligned with the inner edge (i.e. that nearest the cell image) of the image. The two reference edge positions are then calculated to be:

$$ref_a = E_a M_m^{-1} + (256 - i_a) \cdot M_c^{-1} \tag{1}$$

$$ref_b = E_b M_m^{-1} + (256 - i_b) \cdot M_c^{-1} \tag{2}$$

where ref_a is the position in microns corresponding to the centripetal reference edge, E_a is the motorized stage position returned by the encoder, M_m^{-1} is the inverse of the magnification factor for the motorized stage, i_a is the column index and M_c^{-1} is the inverse of the camera magnification factor. These same definitions apply to the positions at the outer reference edge, designated as ref_b, E_b and i_b. The midpoint between ref_a and ref_b is 6.5 cm from the center of rotation, so that the radial position can be calculated from any encoder position and column index as:

$$r = 6.5 + \left(E \cdot M_m^{-1} + (256 - i) \cdot M_c^{-1} - \frac{ref_a + ref_b}{2} \right) \cdot 10^{-4} \qquad (3)$$

The factor of 10^{-4} converts the position from μm to cm.

Low-magnification system focusing, alignment and calibration

The focusing of the low-magnification system (Liu and Stafford, 1992) (Figure 1B) is identical to that described above. However, because of the shorter focal length of the camera lens, its positon becomes critical. Therefore, its position is determined with a micrometer mount. Experience indicates that the position of the cylinder lens is less critical and can be place at any position between those found to be optimal for the 2/3 focal positions to 12 mm and 30 mm cell pathlengths. Cylinder lens rotational adjustment is performed as for the normal Model E optics.

The calibration for this system is much simpler since there is no motorized stage. In this case, the image of the ruled disk (at the appropriate vertical position above the mid-cell plane) is focused on the sensor and the slope from a linear regression of pixel number as a function of ruling number yields the only magnification factor M_c. Alternatively, M_c can be determined from the positions of the counterweight reference edges as $M_c = (i_A - i_B)/1.62$, where i_A is the column pixel index for the inner reference edge and i_B is that for the outer reference edge. This determination can be automated by simply monitoring the log of the magnitude of the Fourier analysis as a function of i, and setting i_A and i_B as the positions where the log of the magnitudes fall to 1/2 their value within the reference edge images. Since these images are at either end of the television sensor, there is no difficulty in distinguishing where these positions fall.

The rotational alignment of the camera with the radial direction is made by tipping the camera until the slope of the chamber mask fringes (i.e. with no rotor) is zero across the sensor. The slope is determined from the linear regression of the fringe displacement (below) made at each of several positions across the sensor. The rotational adjustment provided on the standard Model E cylinder lens mount serves to tip the camera. Since the slope can be calculated in a few tenths of a second, this adjustment can be made very rapidly, and it only needs to be made when the optics are re-focused.

The radial position is calculated as:

$$r = 6.5 + \left(\frac{i_A + i_B}{2} - i \right) \cdot M_c^{-1} \qquad\qquad (4)$$

where i_A and i_B are the pixel column numbers corresponding to the counterweight reference edges, i is the pixel column number where the radius is to be calculated and M_c^{-1} is the inverse of the camera magnification factor (above).

Fringe Displacement Measurement

The Rayleigh interference image is well suited for automated analysis using Fourier methods (DeRosier, 1972). Here, however, a single-frequency, discrete Fourier analysis conducted at the spatial frequency of the fringes is used. This method of analysis retains the precision of measurement and insensitivity to noise described by De Rosier *et al.*, but requires far fewer calculations and, therefore, is considerably faster. As used here, the displacement of the maxima in the intensity profile relative to the maxima in the sin function equals the phase, **P**, of the Fourier analysis. Moreover, **P** varies linearly with the fringe shift, so that at any radial position the fractional fringe displacement equals **P**/2π. For the most accurate fractional fringe displacement measurements, certain conditions must be met: 1) the column of intensities must contain an integral number of fringes, 2) the frequency used in the analysis must match the spatial frequency of the fringes, 3) the greatest number of fringes should be included in the analysis and, if possible, the entire envelop of fringes should be included, 4) the single-slit diffraction envelop should be centered on the camera sensor and 5) the finge image should be of as high a quality as possible. The first two criteria relate to the matching of the Fourier analysis to the image. The latter three criteria are determined by the physical set up (e.g. slit width, magnification, cleanliness of the optics, etc.), and these three are arranged in the order of importance.

Two parameters must be set prior to using the single-frequency, discrete Fourier analysis. The first is *ppf*, the number of rows of pixels corresponding to the center-to-center spacing of the fringes. The second is *nf*, the integral number of fringes used in the analysis. Selection of the optimum value for *ppf* is described below. The product of the *nf* and *ppf* must be an integral number corresponding to the total number of rows of pixels, *tnr*, used in the analysis. The values for *nf* is determined as the truncated-integer ratio of the vertical size of the array (e.g. 512 pixels) to *ppf* (e.g. 24.67 pixels/fringe).

The value of *tnr* is determined by multiplying this result by *ppf* (e.g. trunc(512/24.67)=20, and 20 x 24.67 pixels/fringe= 494 pixels= *tnr*).

To speed up the calculation of **P**, two look-up tables are calculated as $\sin(i \cdot 2\pi/ppf)$ and $\cos(i \cdot 2\pi/ppf)$ where i ranges between 1 and 512. Using the values in these look up tables and the intensities from a column of intensities, two sums are calculated, one corresponding to the product of the pixel intensities and the sin function, the other to the product of the intensities and the cosine function:

$$SSUM = -\sum_{j=bottom}^{j=top} \sin(j) * I(j) \tag{5}$$

$$CSUM = \sum_{j=bottom}^{j=top} \cos(j) * I(j) \tag{6}$$

where $I(j)$ is the intensity and the row index of the pixel, j, serves as the pointer to the sin and cosine tables. Values of j range from 256-*tnr*/2 (the top of the portion of the fringe envelope being analyzed) to 256+*tnr*/2 (the bottom of the fringe envelop), thus using an integral number of fringes in the central portion of the television image. Normalization of the sums is achieved by dividing the sums by *tnr*/2.

Two values are calculated from these sums. The first is the Fourier amplitude **A**, calculated by adding the squares of the sin and cosine sums. The second is the Fourier phase **P**, calculated as the arctangent of the ratio of the sin sum and the cosine sum:

$$A = \sqrt{SSUM^2 + CSUM^2} \tag{7}$$

$$P = ARCTAN\left(\frac{SSUM}{CSUM}\right) \tag{8}$$

It is important to note that a full 4-quadrant arctangent function is required. The fractional fringe displacement, Y(**j**) at position **j** is $P/2\pi$.

Testing has shown that the optimum value for *ppf* corresponds to the maximum in the power spectrum and equals the fringe spacing (Laue, 1981; Laue, 1992). It is necessary to determine to optimum *ppf* for each cell and it is best to determine this value every run. To determine the optimum, a reasonable range of values for *ppf* is selected, and **A** calculated for values of *ppf* at intervals of 1/*nf*. The value of *ppf* corresponding to the maximum in **A** is used for the Fourier analysis.

Determining the total fringe displacement

Only the fractional fringe displacement is calculated at each radial position using the Fourier method described. In order to reconstruct the image, then, it is necessary to convert the fractional fringe displacements to total cumulative fringe displacements. This is done by determining the fringe displacement increment between adjacent data points, i.e. $\delta Y = Y(j) - Y(j-1)$, and summing δY going from the top to the bottom of the cell image. The first data point has $\delta Y = 0$, regardless of the true fringe displacement, reflecting the fact that only relative concentrations are available from the Rayleigh system. Calculating the sign of the δY must be done carefully when the fractional fringe displacement is near the crossover between zero and one, especially in regions where there is little gradient in the image. An algorithm must be included to determine whether δY needs to be adjusted by one. This is done by first testing whether the absolute value of δY is greater than 0.5 fringe. If not, then the **j**th fringe displacement, $Y(j)$, is set equal to $Y(j-1) - \delta Y$. If the absolute value of δY is greater than 0.5, and δY is positive, then $Y(j) = Y(j-1) + 1 - \delta Y$, else $Y(j) = Y(j-1) - 1 - \delta Y$. Notice that this test on δY limits the maximum gradient that can be tracked by the on-line interferometer to 0.5 fringe per column of pixels. For the high-magnification system, where the radial increment (in cell coordinates) is about 6.7 μm per column, the maximum traceable gradient of 75 fringes/mm. For the low-magnification system, this limits the maximum traceable gradient to 25 fringes/mm.

PROSPECTS AND CONCLUSIONS

The capabilities and limitations of the Rayleigh interference optics have been described (Richards and Shachman, 1959; Yphantis, 1964; Laue, 1992) and a discussion of the inherent limitations to on-line determination of the fringe displacement and radial position determination has been presented previously (Laue, 1992; Laue, 1993). The optical systems described here have been in use for up to five years in the authors' laboratories. Our experience has been that the precision of the fringe displacement determination is ≈0.003 fringe and the accuracy of blank corrected measurements is about 0.01 fringe. This is sufficient precision for most studies.

In order for analytical centrifugation to be conducted rapidly and with certainty, the automation of the data acquisition is essential. The on-line Rayleigh interferometers presented here fulfill this requirement. However, the extreme rapidity with which precise data can be acquired extends analytical ultracentrifugation two other ways. First, it makes possible the accomplishment of far more complicated experimental protocols than could be considered if photographs were used. This means that studies involving rapid surveys of buffer conditions, ligand titrations and the analysis of interactions can be completed in a timely fashion. On-line systems are useful for examining unstable compounds, where it is often possible to monitor sample stability during the course of an experiment.

The second way an on-line interferometer extends analytical ultracentrifugation is by permitting the development of entirely new methods. For example, the ability to acquire fringe displacement measurements in a few seconds allows, for the first time, the rapid determination of $dc(r)/dt$. By transforming the radial axis to the apparent sedimentation coefficient, it is possible to calculate the differential sedimentation coefficient distribution function on-line (Stafford, 1992). It is quite likely that other methods will be developed that rely on the capabilities of the on-line interferometers.

The instruments presented here were developed for the Model E. However, the high-magnification design has been adapted to the newer XLA analytical ultracentrifuge (Laue, 1993), and a prototype low-magnification design is being developed. Given the relative ease of use of the XLA, its smoother, precession-free operation and its ability to be completed operated from a computer, the future of on-line interferometry seems bright.

ACKNOWLEDGEMENTS

We thank Arthur Anderson, John Levasseur and Raymond Kikas for their useful suggestions in designing the equipment, as well as their careful machining of the parts. Supported in part by Grants PCM 76-21847, BBS 86-12159, BBS 86-15815 and DIR 90-02027 from the National Science Foundation and by the University of New Hampshire Research Office.

Glossary of Symbols

Symbol	Meaning
ref_a	Relative position of the counterweight inner reference hole (in μm)

ref$_b$	Relative position of the counterweight outer reference hole (in μm)
E$_a$	Motorized stage encoder value at the inner reference hole (in μm)
E$_a$	Motorized stage encoder value at the outer reference hole (in μm)
M$_m$	Magnification factor relating the motorized stage motion to position in the cell (unitless)
M$_c$	Magnification factor relating the number of pixels to the position in the cell (pixels/μm)
E	Motorized stage encoder value for any given image
r	Distance from the center of rotation (in cm)
i$_A$,i$_B$	Pixel column numbers corresponding to the edges of the counterweight reference holes
A	Fourier analysis amplitude
P	Fourier analysis phase
ppf	Number of pixels per fringe going from center to center of the fringes

nf	Integral number of fringes over which the Fourier analysis is to be performed
tnr	Total number of rows of pixels encompassing nf fringes
I(j)	8-bit light intensity (0-255) at the jth pixel
δY	Difference in the fractional fringe displacement between the jth and j-1 radial positions
Y(j)	Total fringe displacement at the jth radial position

REFERENCES

DeRosier DJ, Munk P and Cox DJ (1971): Automatic Measurement of Interefernce Photographs for the Ultracentrifuge. *Anal. Biochemistry* 50: 139.

Lary JW, Wu T-W and Yphantis DA (1990): Laser Diode Light Source for Real-Time Interferometry in the Ultracentrifuge. *Biophysical J.* 57: 377a <ABSTRACT>.

Laue TM (1981): Rapid Precision Interferometry for the Analytical Ultracentrifuge. Ph.D. Dissertation, U. of Connecticut, Storrs, CT.

Laue, T.M. (1992) On-line Data Acquisition and Analysis from the Rayleigh Interferometer. In: *Analytical Ultracentrifugation in Biochemistry and Polymer Science*, S.E. Harding and A.J. Rowe, eds. Royal Society of Chemistry, Cambridge, U.K.

Laue, TM (1993) Real-time Interferometry in the Ultracentrifuge. In: SPIE Proceedings, Vol. 1895, Ultrasensitive Laboratory Diagnostics, Cohn, J. ed., Bellingham, WA: SPIE.

Liu S and Stafford WF (1992): A Real-Time Video-Based Optical System for an Analytical Ultracentrifuge Allowing Imaging of the Entire Centrifuge Cell. *Biophys J* 61, A476, #2745.

Paul, CH and Yphantis, DA (1972): Pulsed Laser Interferometry in the Ultracentrifuge. I. System Design. *Analytical Biochem.* 48: 588.

Perlman GE and Longsworth LG (1948): The Specific Refractive Increment of Some Purified Proteins. *J. Am. Chem. Soc.* 70: 2719.

Richards EG and Schachman HK (1959): Ultracentrifuge Studies with Rayleigh Interference Optics I. General Applications *J. Phys. Chem.* 63: 1578.

Richards EG, Teller DC and Schachman HK (1971): Alignment of Schlieren and Rayleigh Optical Systems in the Analytical Ultracentrifuge I. Focusing the Camera and Cylinder Lenses. *Anal. Biochem.* 41: 180.

Richards JH and Richards EG (1974): Light Difference Detector for Reading Rayleigh Fringe Patterns from the Analytical Ultracentrifuge. *Anal. Biochemistry* 62: 523.

Stafford WF (1992): Boundary Analysis in Sedimentation Transport Experiments: A Procedure for Obtaining Sedimentation Coefficient Distributions Using the Time Derivative of the Concentration Profile. *Anal Biochem* 203:295-301.

Williams RC Jr. (1972): A Laser Light Source for the Analytical Ultracentrifuge. *Anal. Biochem.* 48: 164.

Yphantis DA (1964): Equilibrium Ultracentrifugation of Dilute Solutions *Biochemistry*, 3: 297.

EXTENSIONS TO COMMERCIAL GRAPHICS PACKAGES FOR CUSTOMIZATION OF ANALYSIS OF ANALYTICAL ULTRACENTRIFUGE DATA

Joe Hedges, Shokoh Sarrafzadeh, James D. Lear and
Donald K. McRorie

With the advent of computer-compatible data from the analytical ultracentrifuge and powerful, inexpensive computers, the need arises for graphing and numerical analysis packages that also include specific applications for the analytical ultracentrifuge.

General-purpose graphing and numerical analysis packages are available; however, the specific needs of the analytical ultracentrifuge are not addressed in these packages. To meet these specific requirements, we have chosen a graphing program called Origin™ (MicroCal Software, Inc.). This product allows flexibility in handling data analysis from the ultracentrifuge. Further needs can be met through extensions to the product using a software mechanism called a dynamic link library (DLL). This chapter discusses the applications needs for the extensions to Origin and shows how the extensions are implemented.

The Origin software runs under the Microsoft® Windows™ operating system on an IBM PC or compatible computer and takes advantage of Windows' dynamic link capability. Dynamic linking allows two independent programs to communicate via a dynamic (*i.e.*, during the run) interface. This linking permits customization of Origin.

In order to create an extension that is compatible with Origin, one needs a C compiler that supports creation and linking of DLL's. Both Microsoft C++™ and Borland C++® have been successfully tested by us with Origin. Each of these compilers has its own description and instructions for interfacing a DLL.

The interface of Origin to an extension is fairly simple. The data, and any other necessary information, are passed from Origin to the extension via pointers to double-precision floating point arrays. The name(s) of the columnar data are passed from Origin to the extension in an Origin script file. One simply types in the name of the extension to be called and the name(s) of the columns of data to be passed into the script file, and then invokes the script file.

The DLL is invoked at run time by giving the name of the DLL and the name of the main function followed by the names of the columnar data that need to be passed.

Shown below are two examples. The first is a short test example similar to one provided by Origin. The second example shows the actual interface to our extension that allows simultaneous fitting of multiple data files. Redundant details have been left out for clarity.

In general, from within Origin, you input the following from a script window:

> DLL dll_name function_name column1_name column2_name

EXAMPLE 1

In the DLL module (called test), which has been compiled and linked with a C compiler, there is a function called "test1." It takes data from column a, adds the value 5 to it and returns it in the column data_b. Note: the script interface, like the C language, is case-sensitive, *i.e.*, upper- and lower-case letters are different.

> DLL test test1 data1_a 5 data1_b

The DLL code for the module "test" follows.

```
#include <windows.h>
#include <stdio.h>
#include "orgdll.h"
#include "labstr.h"
/*********************************************************
```

LIBMAIN is required Windows entry code

```
******************************************************/

int FAR PASCAL LibMain(hModule, wDataSeg, cbHeapSize,
lpszCmdLine)
HANDLE      hModule;
WORD        wDataSeg;
WORD        cbHeapSize;
LPSTR       lpszCmdLine;
{   return 1;           /* Nothing is done, but it is necessary */
}

/*****************************************************
 WEP is required Windows exit code
******************************************************/

int FAR PASCAL WEP (bSystemExit)
int  bSystemExit;
{   return(1);          / * Nothing is done, but it is necessary */
}

/************ Your DLL code would start here ***************/

/*****************************************************
TEST1 takes three arguments: source, add_amount and destination.
Add_amount is added to the source data and the result is put into the
destination data
******************************************************/

int FAR PASCAL _export Test1(hWnd,lpString)
        HWND              hWnd;
        LPSTR             lpString;
{
double FAR    *lpSource;
double FAR    *lpDestination;
DWORD           dw;
short         i,i1,i2;
double        add_amount;
```

/* these constants are defined in orgdll.h */
char   temp[MAXLINE];
char   source_name[NAME_SIZE];

/* the first token is copied into source_name so source_name now contains "data1_a",
* and the remaining string "5 data1_b" is returned.   */
 lpString = LABUTIL_next_token(source_name,lpString);
 if((dw = SendMessage (hWnd, WM_DLL_OPEN_DATA, 0,
(LONG)(LPSTR) source_name)) == 0L)
   return(1);

/* WM_DLL_OPEN_DATA will find "data1_a" and lock the data handle
and return the far pointer to the data. Origin's worksheet by default holds
double numerics */
 lpSource = (double FAR *)dw;

/* lpSource is now a pointer to the beginning of the given data */

/* next we need to find out the range, which maybe reflects the
DataSelector markers, or the selected worksheet range in Origin. You
must call this immediately after OPEN_DATA since OPEN_DATA finds
the range and stores it in a static location in Origin, which may change
by the next OPEN_DATA*/
 dw = SendMessage(hWnd,WM_DLL_GET_RANGE,0,0L);

 i1=LOWORD(dw);
 i2=HIWORD(dw);

/* now we are ready to read the next token, which should be the number 5 */
 lpString = LABUTIL_next_token(temp,lpString);
 add_amount = LABUTIL_str2real(temp);

/* lpString now points to "data1_b". Before we open the data, we want
to set the active range of the destination data to be the same as the
source, in case they are not. To do that, we send a message to Origin

with a LabTalk command string. */
```
 wsprintf(temp,"set %s -end %d",lpString,i2);
```

/* Origin will process the lParam as a command string if WM_USER
message is sent */
```
 SendMessage(hWnd,WM_USER,0,(LONG)(LPSTR)temp);
```

```
if((dw =SendMessage(hWnd,WM_DLL_OPEN_DATA,0,
(LONG)lpString)) == 0L)
 return(1);
```

/* Define the destination to be a double */
```
lpDestination = (double FAR *)dw;
```

/*************** END OF INTERFACE *******************/

/*************** MAIN PROGRAM **********************/
/* for this demo, we just add add_amount to every point */
```
 for(i = i1; i < i2; i++)
  lpDestination[i] = lpSource[i] + add_amount;
```

/************* END OF YOUR MAIN PROGRAM***********/

/* We are all done, close the data, which unlocks the the handle for the
data set   */
```
 SendMessage(hWnd,WM_DLL_CLOSE_DATA,0,
(LONG)(LPSTR)source_name);
 SendMessage(hWnd,WM_DLL_CLOSE_DATA,1, (LONG)lpString);
```

/*0 = no error, return 1 if errors */
```
return(0);
 }
```

EXAMPLE 2

The syntax for the call from an Origin script window is:

> DLL multi nonLin radius abs std_dev residual id floats answers
> negci posci

We pass in the radius, absorbance, standard deviation, data set identifier and the identifier for the values that are to be floated. Returned from the DLL are the residuals, answers, negative confidence intervals and positive confidence intervals.

```
#include <windows.h>
#include <math.h>
#include <stdlib.h>
#include "orgdll.h"
#include "labstr.h"


char    x_name[NAME_SIZE];
char    y_name[NAME_SIZE];
char    dy_name[NAME_SIZE];
char    itr_name[NAME_SIZE];
char    ans_name[NAME_SIZE];
char    res_name[NAME_SIZE];
char    neg_name[NAME_SIZE];
char    pos_name[NAME_SIZE];
char    file_id_name[NAME_SIZE];


/*********************************************************
                        NONLIN
*********************************************************/
int FAR PASCAL _export nonLin(HWND hWnd,LPSTR lpString)
{
// passed in from the worksheet; in order of appearance
  double FAR *x;
  double FAR *y;
  double FAR *dy;
  double FAR *fxs;
  double FAR *x2;
```

```
    double FAR *itrctl;
    double FAR *answers;
    double FAR *negci;
    double FAR *posci;

DWORD      dw;
short    data_start,\
         data_end;
char    tempstr[MAXLINE];

/* parse the first column name from the string passed in */
  lpString = LABUTIL_next_token(x_name,lpString);

/* find x_name, lock the data handle and return the far pointer */
if((dw = SendMessage (hWnd,WM_DLL_OPEN_DATA,
0,(LONG)(LPSTR) x_name)) == 0L)
  return(1);
/*  radial values */
x = (double FAR *)dw;

/* get the range, this must be done immediately after OPEN_DATA */
  dw = SendMessage(hWnd,WM_DLL_GET_RANGE,0,0L);
data_start=LOWORD(dw);
data_end=HIWORD(dw);

/* parse the next column name from the string passed in  */
lpString = LABUTIL_next_token(y_name,lpString);
if((dw =SendMessage(hWnd,WM_DLL_OPEN_DATA,0,
(LONG)(LPSTR)y_name)) == 0L)
  return(1);
/* absorbance values */
y = (double FAR *)dw;

/* parse the next column name from the string passed in  */
lpString = LABUTIL_next_token(dy_name,lpString);
if((dw =SendMessage(hWnd,WM_DLL_OPEN_DATA,0,
(LONG)(LPSTR)dy_name)) == 0L)
  return(1);
/* std deviation*/
```

```
dy = (double FAR *)dw;

/* parse the next column name from the string passed in  */
lpString = LABUTIL_next_token(res_name,lpString);
if((dw =SendMessage(hWnd,WM_DLL_OPEN_DATA,0,
(LONG)(LPSTR)res_name)) == 0L)
  return(1);
/* residuals */
fxs = (double FAR *)dw;

/* parse the next column name from the string passed in  */
lpString = LABUTIL_next_token(file_id_name,lpString);
if((dw =SendMessage(hWnd,WM_DLL_OPEN_DATA,0,
(LONG)(LPSTR)file_id_name)) == 0L)
  return(1);
/* set identifier */
x2 = (double FAR *)dw;

/* parse the next column name from the string passed in */
lpString = LABUTIL_next_token(itr_name,lpString);
if((dw =SendMessage(hWnd,WM_DLL_OPEN_DATA,0,
(LONG)(LPSTR)itr_name)) == 0L)
  return(1);
/* to float or not to float */
itrctl= (double FAR *)dw;

/* parse the next column name from the string passed in  */
lpString = LABUTIL_next_token(ans_name,lpString);
if((dw =SendMessage(hWnd,WM_DLL_OPEN_DATA,0,
(LONG)(LPSTR)ans_name)) == 0L)
  return(1);
/* best fit answers */
answers = (double FAR *)dw;

/* parse the next column name from the string passed in */
lpString = LABUTIL_next_token(neg_name,lpString);
if((dw =SendMessage(hWnd,WM_DLL_OPEN_DATA,0,
(LONG)(LPSTR)neg_name)) == 0L)
  return(1);
```

```
/* negative confidence intervals */
negci = (double FAR *)dw;

/* parse the next column name from the string passed in  */
lpString = LABUTIL_next_token(pos_name,lpString);
if((dw =SendMessage(hWnd,WM_DLL_OPEN_DATA,0,
(LONG)(LPSTR)pos_name)) == 0L)
  return(1);

/* positive confidence intervals */
posci = (double FAR *)dw;

wsprintf(tempstr,"set %s -end %d",lpString,data_end);
SendMessage(hWnd,WM_USER,0,(LONG)(LPSTR)tempstr);

/* end of origin  interface */

/************* Our nonlinear fitting routine  ****************/

/* Return to Origin *
 SendMessage(hWnd,WM_DLL_CLOSE_DATA,0,
(LONG)(LPSTR)x_name);
 SendMessage(hWnd,WM_DLL_CLOSE_DATA,0,
(LONG)(LPSTR)y_name);
 SendMessage(hWnd,WM_DLL_CLOSE_DATA,0,
(LONG)(LPSTR)dy_name);
 SendMessage(hWnd,WM_DLL_CLOSE_DATA,1,
(LONG)(LPSTR)res_name);
 SendMessage(hWnd,WM_DLL_CLOSE_DATA,0,
(LONG)(LPSTR)file_id_name);
 SendMessage(hWnd,WM_DLL_CLOSE_DATA,0,
(LONG)(LPSTR)itr_name);
 SendMessage(hWnd,WM_DLL_CLOSE_DATA,1,
(LONG)(LPSTR)ans_name);
 SendMessage(hWnd,WM_DLL_CLOSE_DATA,1,
(LONG)(LPSTR)neg_name);
 SendMessage(hWnd,WM_DLL_CLOSE_DATA,1,
(LONG)(LPSTR)pos_name);
}
```

Origin-DLL Communication

To communicate from Origin to an external DLL, a main worksheet is required to hold the parameters for transfer to and from the DLL. This worksheet, data1, is not part of the user interface. Another worksheet, data2, contains the spreadsheet for parameter values and appropriate functions for interface with the user. The command

DLL test data1_a 5 data1_b

instructs the DLL where to output the calculated fit results into the data1 worksheet: in this case, columns a and b. The parameter values corresponding to those in data2 are passed to that worksheet automatically. Further information from data1 can then be accessed by the user through a script window using the command

type -a data1_b[1]

which displays cell1 of column b from the data1 worksheet.

To output parameters (entered by the user) from the data2 worksheet into the data1 worksheet (to send to the DLL), the following command is required:

win -o data1 (dobutton execute);

"win -o" accesses the data1 worksheet from whatever window is currently active and selects the execute button in data1 to transfer parameters from data2 to data1.

To plot results from the DLL, the following commands are required:

Set data1_x x data1_y
Layer -i data1_y
;win -t plot multifit data_plot

The first line sets the x-axis to be column x and the y-axis to be column y. The reason for doing it in this manner is that if the x-axis is changed to a different column in another plot, it can be reset to column x for this initial plot. The second line brings column y into the graphic environment of Origin after column x has been brought in automatically. The third line then calls the appropriate graphic template, called multifit, and plots the data from the columns specified in the previous two command lines.

Data Analysis

To demonstrate the utility of the Nonlin (Johnson *et al.*, 1981) algorithm linked to Origin graphics, the structure of a peptide designed to form a parallel α-helical coiled-coil was examined in the analytical ultracentrifuge. Parallel α-helical coiled-coils are a structural motif in proteins where there is a heptapeptide repeat sequence with hydrophobic residues in the a and d positions of the heptad. The motif is common to many structural proteins such as tropomyosin, myosin, paramyosin and intermediate filament proteins (Cohen and Parry, 1990). Crick (1953) originally proposed that these proteins would fold into two- or three-stranded parallel α-helical coiled-coils with the hydrophobic residues in the a and d positions interlocking to stabilize the structure. The positioning of the hydrophobic residues causes them to fall on the same face of the resulting α-helix to allow appropriate interactions. Analytical ultracentrifugation showed a dimer structure for an analog peptide modeled from tropomyosin (Hodges *et al.*, 1981, Lau *et al.*, 1984). The dimer model was also demonstrated by the crystal structure of a peptide from the transcriptional activator GCN4 (O'Shea *et al.*, 1991). More recently, a trimer structure was confirmed by both crystallography and analytical centrifugation for a peptide designed to evaluate the helix-forming capabilities of various amino acids (Lovejoy *et al.*, 1993). This peptide, coil-Ser, was designed after the original poly-heptapeptide sequence of Hodges *et al.* (1981) to form a two-stranded coiled-coil (O'Neil and DeGrado, 1990). All of these bundled α-helical structures have the characteristic hydrophobic residues in the a and d positions of sequential heptamers (Figure 1).

POSITION	g	a b c d e f g	a b c d e f g	a b c d e f g	a b c d e f g	a b c d e f
Hodges	K	L E A L E G K	L E A L E G K	L E A L E G K	L E A L E G K	L E A L E G
GCN4-p1	Ac-R	M K Q L E D K	V E E L L S K	N Y H L E N E	V A R L K K L	V G E R
coil-Ser	Ac-E	W E A L E K K	L A A L E S K	L Q A L E K K	L E A L E H G-CONH$_2$	
coil-Trp	Ac-E	W E A L E K K	L A A L E W K	L Q A L E K K	L E A L E H G-CONH$_2$	

Figure 1. Amino acid sequences of coiled-coil peptides with the characteristic heptad repeat designated a-g according to the notation of McLachlan and Stewart (1975). Hodges refers to a portion of the peptide (Hodges *et al.*, 1981) based on the structure of tropomyosin. GCN4-p1 is a peptide of the transcriptional activator GCN4 (O'Shea *et al.*, 1991). Coil-Ser and coil-Trp are model peptides designed by O'Neil and DeGrado (1990) to form coiled-coils. Ac- is an acetyl moiety.

The peptide in this study, coil-Trp, has a tryptophan residue substituted for serine in the O'Neil and DeGrado (1990) model. In the presence of 5 M urea, modeling of CD data (not shown) suggested self-association and not just a single species. Analytical ultracentrifugation and data analysis using the Nonlin algorithm linked to Origin were used to characterize the association properties of coil-Trp.

Preliminary analysis was performed to determine weight-average molecular weights for individual samples of coil-Trp with varied conditions of concentration and rotor speed. Examining the trends in these results can provide qualitative information about both the presence of an association and the reversibility of the reaction (Laue, 1992). Table 1 shows results of individual data sets fit, with a weighted fit, as single ideal species using the Origin-Nonlin linked software. Absorbance at 280 nm was measured as a function of radial distance for samples run in 5 M urea, 10 mM MOPS, pH 7.5 at 25°C using two rotor speeds and three sample concentrations.

Table 1. Weight-Average Molecular Weight of Coil-Trp at Different Concentrations and Run Speeds

Concentration (mM)	Speed	
	36,000 rpm	**50,000 rpm**
0.050	8341	8303
0.036	8296	8300
0.030	7907	7439

From the sequence molecular weight of 3424, all molecular weight calculations suggest an aggregation state with a stoichiometry greater than two. The tendencies of the calculated molecular weights to increase with concentration and stay the same with increased rotor speed are consistent with the CD data indicating a reversible mass action association.

Using the Nonlin algorithm linked to Origin, attempts were made to analyze the data in Table 1 by curve fitting the six data files to the same model simultaneously. Models chosen for fitting were varied in stoichiometry to test several possibilities of a self-associating system. Analysis indicated the possibility of two equilibrium systems with maximum stoichiometries of three and four peptides of coil-Trp. Statistically, these two systems were indistinguishable. Therefore, experiments were needed to determine the maximum association state of coil-Trp. Figure 2 shows the curve fit for a sample of the peptide run at 0.2 mM. The best-fit model at this higher concentration, using a weighted fit, was that of a single ideal species of 10,230-dalton molecular weight.

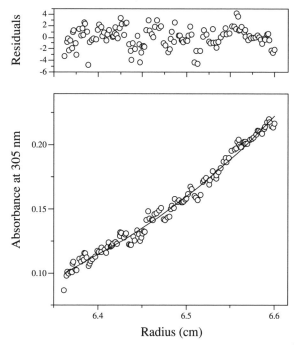

Figure 2. Sedimentation equilibrium analysis of 0.2 mM coil-Trp. The analysis was performed at 25°C in 10 mM 3-[N-morpholino]propanesulfonic acid (MOPS), 6 M urea, pH 6.8. The lower graph shows the concentration profile at equilibrium at 25,000 rpm measured as absorbance at 305 nm as a function of radial distance (open circles) along with the best-fit curve using a single ideal species as a model. The upper graph shows the residuals for this fit. The residuals are in terms of number of standard deviations for each data point because of the weighting used in fitting.

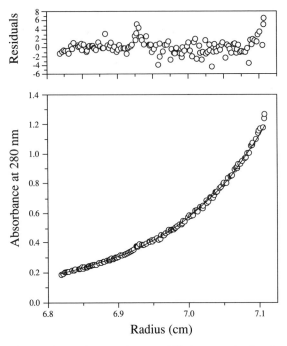

Figure 3. Sedimentation equilibrium analysis of coil-Trp in the absence of urea. The analysis was performed at 25°C in 10 mM MOPS, pH 6.8. The lower graph shows the concentration profile at equilibrium at 30,000 rpm measured as absorbance at 280 nm as a function of radial distance (open circles) along with the best-fit curve for a single ideal species model. The upper graph shows the residuals in terms of standard deviations from the weighted fit.

To confirm the trimer stoichiometry, coil-Trp was run in the absence of urea. As with the higher sample concentration, a weighted fit revealed a single ideal species of trimer molecular weight (Figure 3). In this case, the best-fit molecular weight was 10,800 daltons. Both results are in close agreement with the calculated molecular weight for a trimer of 10,266 daltons. These results support the monomer-dimer-trimer model, which distinguishes between the models with trimer and tetramer as maximum stoichiometries. Also, the results are consistent with the coil-Ser crystallography data indicating a trimer structure for a peptide with a single amino acid substitution (Lovejoy *et al.*, 1993). This consistency could be expected since the replaced amino acid is not involved with the peptide interactions.

With the maximum stoichiometry limited to trimer, the six data files in Table 1 were then analyzed assuming a monomer-dimer-trimer equilibrium. Table 2 shows results for the Origin-Nonlin software using both a weighted and nonweighted fit. The weighted fit uses the standard deviation of an absorbance reading stored in the data file as a weighting factor to force less importance to be placed on data with large standard deviations in the final fit analysis. Table 2 also shows the results of fitting the same data with another software package using the Nonlin algorithm linked to a PC interface (provided by the National Analytical Ultracentrifugation Facility, Storrs, CT, at the course on Analytical Ultracentrifugation: Theory and Practice). This comparison allowed for a confirmation that the communication through the DLL was operating correctly. Confidence intervals could not be compared since the PC Nonlin calculates 67% intervals instead of the 95% in Origin.

Association constants are in terms of absorbance units at this point. To convert to concentration, the extinction coefficient for tryptophan was used, assuming the extinction coefficient for any peptide species is additive relative to the number of Trp residues (Becerra *et al.*, 1991; Ross *et al.*, 1991). So, for $K_{a,1-2}$ the conversion factor is $\varepsilon l/2$ and for $K_{a,1-3}$ the factor is $\varepsilon^2 l^2/3$. Using these conversion factors, the association constants are $1.6 \times 10^5 \text{ M}^{-1}$ and $4.7 \times 10^{10} \text{ M}^{-2}$ for $K_{a,1-2}$ and $K_{a,1-3}$, respectively. From these values, $K_{a,2-3}$ is $2.95 \times 10^5 \text{ M}^{-1}$. This model and K_a values provided a satisfactory fit for the CD data.

Table 2. Fitted Parameters Modeling Coil-Trp as a Monomer-Dimer-Trimer Equilibrium

	Origin Nonlin Weighted Fit	Origin Nonlin Nonweighted Fit	PC Nonlin Nonweighted Fit
Baseline Offset			
File 1	4.18×10^{-3}	1.16×10^{-3}	1.16×10^{-3}
File 2	-8.86×10^{-3}	-1.29×10^{-2}	-1.29×10^{-2}
File 3	-2.20×10^{-4}	-3.14×10^{-3}	-3.14×10^{-3}
File 4	-1.87×10^{-3}	-2.84×10^{-3}	-2.84×10^{-3}
File 5	-6.20×10^{-3}	-1.16×10^{-2}	-1.16×10^{-2}
File 6	1.16×10^{-2}	8.53×10^{-3}	8.53×10^{-3}
A_0 (Absorbance at r_0)			
File 1	0.053	0.064	0.064
File 2	0.045	0.054	0.054
File 3	0.039	0.047	0.047
File 4	0.037	0.044	0.044
File 5	0.029	0.035	0.035
File 6	0.024	0.028	0.028
$\ln(K_{a,1-2})$	3.177	2.602	2.602
$\ln(K_{a,1-3})$	6.638	6.115	6.115
Variance of fit	–	3.44×10^{-5}	3.44×10^{-5}

Note: $\ln(A_0)$ was converted to A_0 in PC Nonlin data; K_a was converted to $\ln(K_a)$ in Origin-Nonlin data.

Glossary of Symbols

K_a association constant

ε molar extinction coefficient

l pathlength

r_0 reference radius

A_0 absorbance at reference radius

References

Becerra, S. P., Kumar, A., Lewis, M. S., Widen, S. G., Abbotts, J., Karawya, E. M., Hughes, S. H., Shiloach, J., Wilson, S. H. Protein-protein interactions of HIV-1 reverse transcriptase: implication of central and C-terminal regions in subunit binding. Appendix: Ultracentrifugal analysis of a mixed association, by M. W. Lewis. *Biochemistry 30*, 11707-11719 (1991)

Cohen, C., Parry, D. A. D. α-Helical coiled coils and bundles: how to design an α-helical protein. *Proteins: Struct. Funct. Genet. 7*, 1-15 (1990)

Crick, F. H. C. The packing of α-helices: simple coiled-coils. *Acta Crystallogr. 6*, 689-697 (1953)

Hodges, R. S., Saund, A. K., Chong, P. C. S., St.-Pierre, S. A., Reid, R. E. Synthetic model for two-stranded α-helical coiled-coils. Design, synthesis, and characterization of an 86-residue analog of tropomyosin. *J. Biol. Chem. 256*, 1214-1224 (1981)

Johnson, M. I., Correia, J. J., Yphantis, D. A., Halvorson, H. R. Analysis of data from the analytical ultracentrifuge by nonlinear least-squares technique. *Biophys. J. 36*, 575-588 (1981)

Lau, S. Y. M., Taneja, A. K., Hodges, R. S. Synthesis of a model protein of defined secondary and quaternary structure. Effect of chain length on the stabilization and formation of two-stranded α-helical coiled-coils. *J. Biol. Chem. 259*, 13253-13261 (1984)

Laue, T. M. Short column sedimentation equilibrium analysis for rapid characterization of macromolecules in solution. *Technical Information DS-835*. Palo Alto, Calif., Spinco Business Unit, 1992.

Lovejoy, B., Choe, S., Cascio, D., McRorie, D. K., DeGrado, W. F., Eisenberg, D. Crystal structure of a synthetic triple-stranded α-helical bundle. *Science 259*, 1288-1293 (1993)

McLachlan, A. D., Stewart, M. Tropomyosin coiled-coil interactions: evidence for an unstaggered structure. *J. Mol. Biol. 98*, 293-304 (1975)

O'Neil, K. T., DeGrado, W. F. A thermodynamic scale for the helix-forming tendencies of the commonly occurring amino acids. *Science 250*, 646-651 (1990)

O'Shea, E. K., Klemm, J. D., Kim, P. S., Alber, T. X-ray structure of GCN4 leucine zipper, a two-stranded, parallel coiled coil. *Science 254*, 539-544 (1991)

Ross, P. D., Howard, F. B., Lewis, M. S. Thermodynamics of antiparallel hairpin-double helix equilibria in DNA oligonucleotides from equilibrium ultracentrifugation. *Biochemistry 30*, 6269-6275 (1991)

A GRAPHICAL METHOD FOR DETERMINING THE IDEALITY OF A SEDIMENTING BOUNDARY

David B. Hayes and Thomas M. Laue

INTRODUCTION

Analytical ultracentrifugation can provide useful thermodynamic and hydrodynamic information for a wide variety of chemical systems. It is apparent that through the use of modern electronics and computers, analytical ultracentrifugation is undergoing a renaissance. A modernization of ultracentrifugation is underway. The automation of acquisition, reduction and analysis of data has been quite successful. However, what has been more difficult to automate is the *art* of determining how much information is accessible from an experiment. This is a problem that becomes apparent to anyone starting a project that involves analytical sedimentation.

Often, it seems that the questions being addressed by a researcher can be stated quite succinctly, but that the available results yield only equivocal conclusions. Even worse, there are times when a researcher is uncertain that the data are sufficient to answer a particular question. One such situation arises when analyzing a sedimenting boundary. In such cases, the question often asked is whether or not a boundary is sedimenting ideally. If so, then the sedimentation coefficient may be interpreted using straight-forward hydrodynamic analyses, a meaningful diffusion coefficient can be obtained and a correct buoyant molecular weight can be calculated from s/D. What is desired is some way of knowing that the data are worth analyzing. This problem is particularly vexing if the purity of the sedimenting material is not known, or if it is suspected that convection has perturbed the boundary shape. Under these circumstances, what appears to be a reasonable boundary may yield inaccurate results.

There are several means of testing the shape of a sedimenting boundary, including the direct fitting of the concentration profiles to various approximations of the Lamm equation (Philo, this volume; Hayes, 1991). Presented here is a graphical method of determining the ideality of a sedimenting boundary. While a diffusion coefficient can be attained from this graph, the real purpose of this analysis is to convey to an experimenter the quality of a boundary.

The goal is to provide a transformation of the data to indicate whether or not any time-dependent flow is occurring other than from ideal sedimentation and diffusion of a single moving boundary. The presence of a single boundary will result in a graph of lines that overlap with time. Either the presence of two or more components each with a distinct s value, or the presence of non-ideal sedimentation will result in a graph of lines that systematically deviate from overlapping.

THEORY

If the variable y is defined as:

$$
y = \left(\frac{1 - \dfrac{r}{r_o} e^{-s\omega^2 t}}{[\varepsilon(1 - e^{-\tau})]^{\frac{1}{2}}} \right)
\tag{1}
$$

where r is a position in the cell, r_o is the meniscus position, $\tau = 2s\omega^2 t$ and $\epsilon = 2D/(s\omega^2 r_o^2)$, an asymptotic approximation of Faxen's exact solution can be expressed as described by Fujita (Fujita, 1975, pp. 64-71):

$$
C(r,t) = \frac{C_o e^{-\tau}}{2} [1 - \phi(y)]
\tag{2}
$$

where C(r,t) is the time dependent concentration distribution, C_o is the initial concentration and Φ is the error function. This equation is valid under the assumptions of Faxen's solution: 1) the boundary is freely sedimenting not restricted by a top or bottom boundary, 2) that $\epsilon \ll 1$ and 3) that $\tau \ll 1$. These assumptions are not terribly restrictive, and are met when a boundary has pulled away completely from the meniscus. Noting that:

$$
\delta r = \frac{r_o \sqrt{\varepsilon(1 - e^{-\tau})}^{\frac{1}{2}}}{e^{\frac{\tau}{2}}} \delta y
\tag{3}
$$

and holding time constant, it is possible to take the derivative of this equation to remove the integration, leaving:

$$
\left(\frac{\delta C}{\delta r} \right)_t = \frac{C_o e^{-\tau} e^{-s\omega^2 t}}{\sqrt{\pi} r_o \sqrt{\varepsilon(1 - e^{-\tau})}^{\frac{1}{2}}} e^{-y^2}
\tag{4}
$$

Expanding the term $(1-e^{-\tau})$ and rearranging leaves:

$$2\left(\frac{\delta c}{\delta r}\right)_t \sqrt{\pi t}\, e^\tau \left(1 + \frac{\tau}{2} + \frac{\tau^2}{6} + \frac{\tau^3}{24} \cdots\right)^{\frac{1}{2}} = \frac{C_o}{\sqrt{D}}\, e^{-y^2} \qquad (5)$$

This equation provides the basis for a diagnostic transformation of the concentration data from sedimentation velocity experiments.

The procedure for the transformation is as follows. The raw data are fit by a parametric spline and $\delta C/\delta r$ is computed and evaluated at each radial position. By using the approximations:

$$e^\tau \approx \left(\frac{r}{r_o}\right)^2 \qquad (6)$$

and:

$$\tau \approx -2 s^* \omega^2 t \qquad (7)$$

the left-hand side of equation 5 is evaluated at each radial position.

The purpose of this transformation becomes evident when examining the right hand side of equation 2. In the kernel of the error function integral:

$$-y^2 = 0 \quad when \frac{r}{r_o}\, e^{-s\omega^2 t} = 1 \ or\ s^* = s \qquad (8)$$

The maximum of any transformed sedimentation distribution at any time during the experiment will have a radial position corresponding to the sedimentation coefficient and a height proportional to the original concentration and to the inverse of the square root of D:

$$maximum\ peak\ height\ = \frac{C_o}{\sqrt{D}}$$

Thus, the height and position of the peak of this transformed boundary will remain constant over time for any single boundary. However, when two or more boundaries of differing s are mixed, the height and position of the apparent peaks will change with time as the distance between the two boundaries increases. Likewise, if a single boundary is spreading at a rate that is not strictly proportional to the square root of time, the peak height will decrease. Accordingly, this graph is sensitive to the unmixing of a solution and to the ideality of the rate of change of the boundary shape.

Due to the effects of diffusion, this transformed $\delta C/\delta r$ value of the y-axis will remain constant with time only at the peak of a single component. At other radial positions, the height of the graph depends in a complicated way on s, D and time. When plotted against the radial position, diffusion will cause the boundaries to become wider with time. When graphed against the s^* axis, sedimentation will cause the boundaries to become sharper with time (since diffusion proceeds as the square root of time while sedimentation proceeds relative to time).

However, it has been noticed (Hayes, 1991) that when this transformation is graphed against an x-axis corresponding to a corrected concentration:

$$C_{cor}(r) = \int_{r_o}^{r} \left(\frac{r}{r_o} \right)^2 dC \tag{10}$$

the plot for a single boundary remains constant with time at all positions of this x axis.

The term correcting for radial dilution used in computing the x-axis assumes tht the sedimentation coefficient is constant as the boundary moves down the cell. Therefore, any concentration dependence of s will appear on this graph as a narrowing of a peak along the x axis.

This diagnostic plot can provide the following information: 1) if the x-axis differs in length over time, the assumption that radial dilution is corrected by the factor $(r/r_o)^2$ is incorrect, 2) if the transformations of boundaries observed at different times are not constant, the boundary is not a single ideal Faxen boundary, due either to non-ideality or to the presence of multiple components, and 3) for an ideal boundary, the apparent diffusion coefficient can be computed from the maximum ordinate as described above.

METHODS

Data simulation

Sets of simulated data were generated using the approximation to the Faxen solution (Fujita, 1975):

$$C = \frac{C_o e^{-\tau}}{2} \left[1 = \phi \left(\frac{1 - \left(\frac{r}{r_o} e^{-\tau} \right)^{\frac{1}{2}}}{[\varepsilon (1 - e^{-\tau})]^{\frac{1}{2}}} \right) \right] \tag{11}$$

where Φ is the error function. Data were generated using rpm= 56,000 and r_o=5.95 cm in all cases. For the single component, s = 5.00 S, D = 7.00 F and C_o= 1.0 were used. These values closely approximate those expected for bovine serum albumin. A mixture also was simulated, containing equal quantities of s = 4.00, 4.40 and 6.00 S, and D = 6.00, 5.60 and 4.00 F, respectively. Though any amount of random noise could be added to these data, the level of the noise was kept small (rms = 0.0001) so that the error introduced by the method could be observed.

Protein preparation

One ml of a 100 mg/ml sample of bovine serum albumin (BSA) was passed over an 80x0.75 cm G-150 Sephadex column equilibrated with 0.05 M NaAcetate (pH 4.40), 0.15 M NaCl (Kegeles and Gutter, 1951). One ml fractions were collected. Two peaks were observed, one corresponding to BSA monomer and the other to a higher molecular weight contaminant. The peak fractions were pooled as described (Hayes, 1991).

Sedimentation

Samples were centrifuged at 56,000 rpm at 20 °C and in an AN-H rotor. A Beckman Model E ultracentrifuge equipped with an on-line interferometer (Yphantis, et al., this volume) was used to collect data. For the sample containing BSA monomer, sixteen scans were used in the analysis. These were taken as eight pairs of scans separated by one minute and with the pairs acquired at intervals of 600-780 seconds apart. These scans covered the portion of the experiment spanning from 2160 to 8220 seconds following the start of sedimentation (Hayes, 1991). For the contaminant, twenty-four scans taken in a similar fashion during the period spanning from 2040 to 9900 seconds from the start of sedimentation. The integral of $\omega^2 t$ was recorded throughout the experiment, including during acceleration (Laue, 1981; Hayes, 1991).

Data manipulations

The raw data were converted from an ordered set of points to a smooth function using a parametric cubic spline to fit the data. A high-resolution graphical interface (MVAG20, Hayes, 1991) for the DEC Laboratory Graphics Package was written to link into the FITPACK routines (Dierckx, 1987). All details concerning these programs and their use are available (Hayes, 1991).

RESULTS

Simulated data

Data simulating a single component was computed for ten different times, transformed and graphed as described. It is apparent from figure 1 that the x-intercept (the corrected concentration) is 1.0. Using this value, and the maximum of the ordinate (which is 1195), a diffusion coefficient of 7 F (1 $F = 1x10^{-7}$ cm^2/sec) can be calculated.

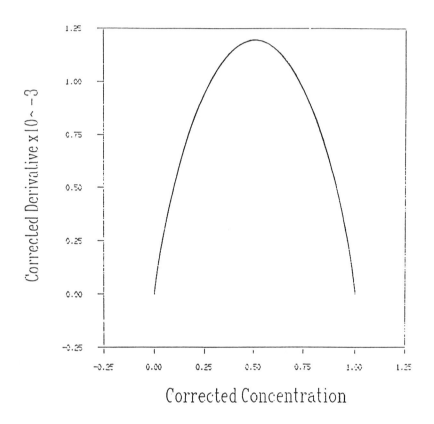

FIGURE 1. A graph of the corrected radial derivative (equation 5) for ten sets of data simulating a single component (see Methods). The superposition of the curves is expected, as explained in the text. The maximum in the peak yields a diffusion coefficient of 7.00 F, in accordance with the theory.

A rather different picture emerges, though, when data simulating a mixture of components is analyzed (figure 2). For these data, two poorly-separated boundaries are visible (Hayes, 1991). The experiment does not have sufficient resolution to distinguish the 4.0 and 4.4 s components. Without knowledge of the generating parameters, two components would be assumed. Accordingly, the remainder of the analysis will assume only two components are present. In this way, it is possible to observe the consequences of missing a component.

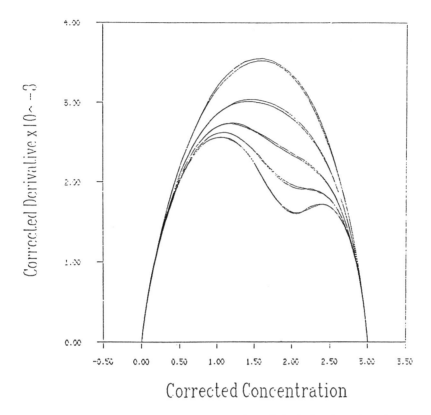

Corrected Concentration

FIGURE 2.   A graph of the corrected radial derivative for data sets simulating three components (see Methods). The time order of the data sets is apparent: the earlier times result in higher, less well-defined peaks, while the later times result in two peaks being resolved. The non-superposition of the curves is anticipated (see text).

The graph of the transformed data clearly shows that there is more than one component, but only two maxima appear. It is difficult to extrapolate the curves back to the x-axis. Thus, it is not readily possible to estimate the relative concentrations of the components. Using crude estimates of the two components being in a 2:1 ratio (small to large), the apparent diffusion constants are 5.82 F and 3.06 F for the slower and faster components respectively. Clearly these are in error, for the slower component is a mixture of materials with diffusion constants of 6.00 and 5.40 F, and the faster component is 4.00 F. No credence, then, should be given to diffusion coefficients calculated when the curves are not superimposed. Under these circumstances, the plot only yields a warning that the boundary shape should not be analyzed to determine a diffusion coefficient.

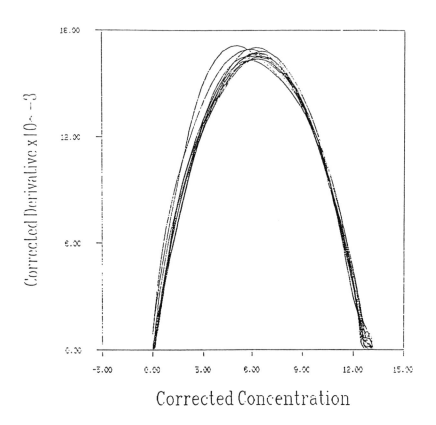

Corrected Concentration

FIGURE 3. A graph of the corrected radial derivative data sets acquired with BSA monomer. The time order of the data sets is not distinguishable from the graph itself. The two aberrant lines start above 0.0 fringes and correspond to data acquired before the boundary had completely cleared the meniscus (Hayes, 1991). The superposition of the curves indicates that the boundary can be analyzed as a single ideal component. The diffusion coefficient determined from the maximum is close to that expected for BSA (see text).

Other analyses in which only the 4.0 and 4.4 s component were modeled (data not shown) yielded a single peak, but the peak height systematically decreased with time in a fashion similar to that for the left hand peak in figure 2. This is what is expected since the apparent diffusion coefficient for the boundary cannot be described with a single constant, but, instead, appears to increase with time as the boundary spreads due to sedimentation.

Experimental data

A graph of data for purified, monomeric BSA is presented in figure 3. The protein concentration was about 4 mg/ml, yielding 13 fringes of data. It is apparent that most of the curves are superimposing. The two aberrant lines, which start above 0.0 fringes, are the earliest data sets, and apparently show the effects of not clearing the meniscus (i.e. violating one of the criteia for the Faxen approximation). The maximum peak height yields an estimate of 6.40 F for the diffusion constant, which differs from the expected value of 6.81 F (Hayes, 1991) by about 7%. Based on this diffusion coefficient and the measured sedimentation coefficient, an apparent molecular weight of 70,200 was calculated (Hayes, 1991).

However, the main purpose of this plot is to reassure an experimenter that further analysis of the boundary shape is justified. It is difficult to determine the maximum from this graph. Therefore, other means of determining s and D are preferred. The most promising method is to use a direct fit to data (Philo, this volume; Hayes, 1991). From this sort of analysis values of s (5.29 S) and D (6.78 F) were obtained that are in good agreement with the anticipated values, and which yield a molecular weight estimate of 68,800, which is in excellent agreement with the expected value.

The leading chromatographic peak also was subjected to sedimentation velocity analysis. In the oral history of biochemistry, this larger material has been ascribed variously to BSA dimer and to IgG. When subjected to sedimentation velocity analysis, two components are apparent in $g(s^*)$ plots (Hayes, 1991), the smaller one both in quantity and size at 5.2 S, and the dominant peak at 7.5 S. The faster moving boundary seems to be rather symmetrical, suggesting that it may result from a single component (figure 4). However, a graph of the transformed data presented in figure 5 suggests that neither boundary should be analyzed as a single component. Two peaks appear on this graph, but it is apparent that the lines are not superimposing.

Indeed, if the last time points are used to determine D (these would correspond to the best separation of components) a value

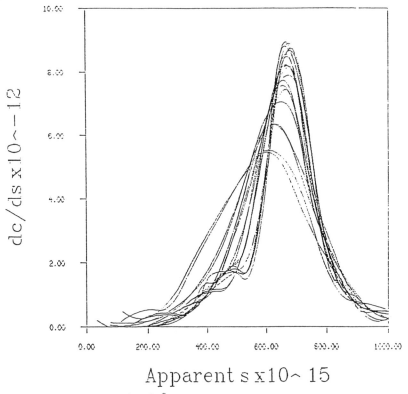

Apparent s x10⌃ 15

FIGURE 4. A graph of $\delta C/\delta s^*$ for the leading chromatographic peak reveals two boundaries. The earlier times are the broader, shorter distribution functions and the later times yield the narrower distributions (Hayes, 1991; Stafford, 1992). At first glance it may seem that the faster boundary corresponds to a single component. But the transformed data in figure 5 suggest otherwise.

of 8.0 F is obtained (Hayes, 1991), nearly double that expected for either BSA or IgG. Clearly the faster moving boundary should not be analyzed as a single component.

CONCLUSIONS

It is important that certain and simple methods are developed that will help guide experimenters as they analyze sedimentation data. Many of the present methods rely on curve fitting (Philo, this volume; Hayes, 1991; Johnson , et al. , this volume), which is the preferred

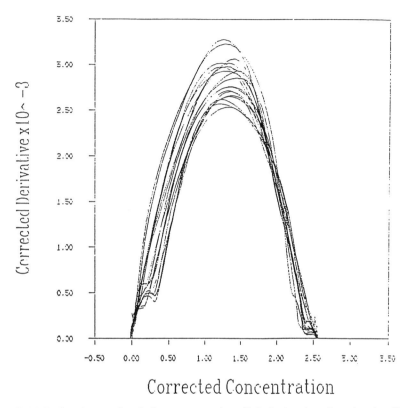

Corrected Concentration

FIGURE 5. A graph of the corrected radial derivative for the leading chromatographic peak. The non-superposition of the curves suggest that neither boundary represent a single component. The time order of the curves can be discerned here, with the later data yielding lower curves. This systematic shift in the curves corresponds to the boundary spreading faster than can be accounted for by any single component, and thus suggests that this boundary is composed of more than one material. The apparent diffusion coefficient for the later time points (8.0 F), is ridiculously high for a 7.5 S component.

method for determining parameters. One problem with curve fitting, though, is that a proper model must be chosen, and it is not always clear what sort of model is appropriate. The result, especially for a novice, is a tremendous amount of struggling with trying to find an appropriate model to fit their data. While experience tends to make this process somewhat simpler, there is still a need for rapid and sure ways of *discarding* some models. The graphical method presented here is an attempt to provide such a method.

Have we succeeded? This question can be divided into three? First, is the method certain? Second, is the method rapid? Third, is the method simple?

Is the method certain? The greatest advantage of this method is that there are no a priori assumptions made about the samples' nature beyond those of the Faxen approximation. Thus, for a narrow set of conditions (a single ideal sedimenting boundary) the method presented here works: superposition of the lines means that the boundary can be considered a single ideal component. Non-superposition of the lines eliminates the simplest model, but is only provides modest guidance as to what models would be more appropriate. There is still an element of judgement needed as to whether or not the lines can be considered superimposing. Noise will result in non-superimposing curves, and it is important to observe whether the curves systematically shift as data from later times are added to the graph. Systematic deviations are a clear sign that the data are not conforming to that expected for a single ideal boundary. Non-systematic deviations are expected, particularly since a derivative of the data ($\delta C/\delta r)_t$ is being taken. Thus, from the standpoint of certainty, the method is a qualified success.

Is the method rapid? The computations needed to construct these graphs and present them can be done in a few seconds on a PC. From this standpoint, the method is successful.

Is the method simple? In principle, all of the calculations could be automated. In fact, though, it is necessary for the experimenter to point out to the program the meniscus position, and to help with smoothing the data. The first problem can be overcome by having the program examine the data in the vicinity of cell ordinarily including the meniscus (e.g. for the Fourier-based Rayleigh systems, the amplitude of the transform increases dramatically as the meniscus is crossed). The latter problem, that of smoothing the data so that clean derivatives can be taken, is more difficult to resolve. Here we have used a parametric cubic spline. However, we have found that the knots of the spline must be positioned carefully to prevent distortion of the data. In the end, user intervention was required (Hayes, 1991). This added considerable time and complexity to the data handling, and seriously detracts from the general application of this method. It should be possible, though, to replace the radial derivative with the time derivative, thus significantly reducing the noise (Stafford, 1992).

So, have we succeeded? Our conclusion would have to be more negative than positive. Should the search for such methods be abandoned? No. It is still important to find ways to guide experimenters towards an honest appraisal of their data. In this way, researchers will be certain that the maximum amount of information has been extracted from their data, while at the same time having confidence that they are not over-interpreting the results.

ACKNOWLEDGEMENTS

Supported in part by Grants BBS 86-15815 and DIR 90-02027 from the National Science Foundation , and by the University of New Hampshire Research Office.

Glossary of Symbols

Symbol	Meaning
r	Distance from the center of rotation (cm)
C	Concentration in arbitrary units
r_o	Meniscus position (cm)
t	Time (sec)
s	Sedimentation coefficient (sec)
s^*	Apparent sedimentation coefficient (sec)
D	Diffusion coefficient (cm^2/sec)
R	Gas constant (erg/mole °K)
ω	Angular velocity (sec^{-1})

T	Absolute temperature (K)
χ	$(r/r_o)^2$ (dimensionless)
τ	$2s\omega^2 t$ (dimensionless)
ϵ	$2D/(s\omega^2 r_o^2)$ (dimensionless)
Φ	Error function
y	Dimensionless quantity defined in terms of s, ω, τ and ϵ

REFERENCES

Dierckx P (1987): FITPACK User Guide, Department of Computer Science, K.U. Leuven, Leuven, Belgium.

Fujita H (1975): Foundations of Ultracentrifugal Analysis, J. Wiley and Sons: New York.

Hayes DB (1991): Automated Analysis of Sedimentation Velocity Distribution Functions. Masters thesis, University of New Hampshire, Durham.

Kegeles G and Gutter FJ (1951): The Determination of Sedimentation Constants from Fresnel Diffraction Patterns. *J. Am. Chem. Soc.* 73: 3770.

Stafford, WF ,III. (1992): Boundary Analysis in Sedimentation Transport Experiments: A Procedure for Obtaining Sedimentation Coefficient Distributions Using the Time Derivative of the Concentration Profile. *Anal. Biochem.* 203: 1.

Part IV

SOME SPECIFIC EXAMPLES

ANALYTICAL ULTRACENTRIFUGATION AND ITS USE IN BIOTECHNOLOGY

Steven J. Shire

INTRODUCTION

Analytical ultracentrifugation has played a critical role in laying the foundations for modern molecular biology. Among its achievements are the demonstration that proteins are macromolecules rather than complexes of smaller units (Svedberg and Fahraeus 1926), and direct support for the semi-conservative replication of DNA as proposed by Watson and Crick (Meselson and Stahl 1958). Unlike techniques such as SDS polyacrylamide gel electrophoresis (SDS PAGE) or gel permeation chromatography (GPC), analytical ultracentrifugation can be used to determine absolute molecular weights without the use of molecular weight standards or interference from the sieving matrix used for separation. Mass spectrometry technology has improved tremendously over the last 5 years and routinely enables researchers to determine molecular weights of macromolecules far more accurately than by analytical ultracentrifugation. However, molecular weights of associating macromolecules in solution are still best determined by centrifugation. Quantitation of these interactions by determining association constants is most easily done by sedimenting solutions to equilibrium and fitting the resulting concentration gradient to a specific association model. The interactions between molecules in oligomeric proteins, self-associating proteins, and interacting systems such as receptor-ligand complexes can be investigated by analytical ultracentrifugation.

With the advent of recombinant DNA technology many of the problems tackled in the new field of molecular biology were of a qualitative nature. Essentially, researchers were interested in whether the proteins expressed from genetically engineered microorganisms had a measurable activity and if the proteins were aggregated during expression and purification. As recombinant DNA technology evolved and became more commercialized, techniques were developed to express and purify large quantities of human proteins from mammalian cell lines. As more of the expressed proteins have entered the stage for development into human pharmaceuticals there has been a greater need to use physico-chemical techniques to determine the best conditions for purification and ultimately formulation of the protein drug. Thus, because of the explosion of protein drugs being developed, there has been a renewed interest in analytical ultracentrifugation.

This chapter summarizes some of our efforts at Genentech to use analytical ultracentrifugation to address a variety of problems encountered in the biotechnology industry. The work summarized includes determinations of the molecular weight of glycoproteins in solution, analysis of a protein ligand-receptor association, analysis of protein self association and finally analysis of the interaction of a protein with a surfactant that is used as a formulation component.

DETERMINATION OF SOLUTION MOLECULAR WEIGHT OF GLYCOPROTEINS

Many of the proteins being developed as pharmaceuticals exist as glycosylated proteins. It is unclear how the composition or content of carbohydrate affects behavior on typical gel permeation chromatography media. Often the standards that are used to calibrate the chromatography are globular proteins and are inadequate to obtain an accurate molecular weight of the glycosylated protein in solution. The apparent molecular weights determined may be sufficiently in error to result in the erroneous conclusion that the protein exists in an associated state in solution. The experiments presented here show three examples using glycoproteins produced at Genentech, Inc: rgp120, the recombinant DNA derived envelope glycoprotein of human immunodeficiency virus type 1, sTNF-R1, the extracellular domain of human tumor necrosis factor (TNF-α) type 1 receptor, and apolipoprotein(a), a large plasma glycoprotein. Gel permeation chromatography suggested that these molecules exist as multimers in solution. In order to determine if these conclusions are correct, molecular weights were determined by analytical sedimentation equilibrium experiments.

sTNF-R1, THE EXTRACELLULAR DOMAIN OF HUMAN TUMOR NECROSIS FACTOR (TNF-α) TYPE 1 RECEPTOR

The extracellular domain of human tumor necrosis factor (TNF-α) type 1 receptor (sTNF-R1) was expressed in 293S human embryonic kidney cells (Pennica, Kohr et al. 1992). The secreted soluble receptor was purified by chromatography on a TNF-α affinity column and reversed phase HPLC. The molecular mass of this protein is ~19 k based on the theoretical amino acid composition derived from the protein sequence encoded by the cDNA. This protein has three potential N-linked glycosylation sites and this may contribute to the observed heterogeneity as detected by SDS PAGE. The purified protein migrates on SDS PAGE as two major bands with apparent molecular weights of 28 k and 33 k, and a faint minor band at ~ 24 k. The presence of these bands is due to small differences in N-terminal and C-terminal processing as well as heterogeneity in the carbohydrate composition. Analysis by gel permeation chromatography yields an apparent molecular weight of 55-60 k, and suggests that this protein exists as a dimer in solution.

rgp120 IIIb, THE RECOMBINANT DNA DERIVED ENVELOPE GLYCOPROTEIN OF HUMAN IMMUNODEFICIENCY VIRUS TYPE 1, STRAIN IIIB

rgp120 IIIb, (rgp120) was expressed as a fusion protein in Chinese hamster ovary cells and purified by immunoaffinity chromatography (Leonard, Spellman et al. 1990). The fusion protein has a molecular weight of ~53 k and consists of a short segment of the herpes gD1 protein fused to the gp120 protein at the amino terminus. However, SDS PAGE analysis reveals a broad band centered at approximately 120 k because it is glycosylated at 24 Asn residues. The broadness of the electrophoretic band is typical of glycosylated proteins and may reflect the heterogeneity of the carbohydrate moieties. Gel permeation chromatography yields apparent molecular weights which range from 200 k to 280 k suggesting that the rgp120 self associates.

rApo(a), RECOMBINANT DNA DERIVED APOLIPO-PROTEIN(a), A HIGH MOLECULAR WEIGHT PLASMA GLYCOPROTEIN

Apo(a) is normally associated with low density lipoprotein (LDL) in plasma as lipoprotein(a) (Lp(a)). High plasma concentrations of Lp(a) have been implicated as an important risk factor in the development of atherosclerosis (Lackner, Boerwinkle et al. 1991). Apo(a) isolated from Lp(a) exists as high molecular weight isoforms. The apo(a) is also attached to LDL through disulfide linkages and reduction results in a denatured and aggregated protein. Recombinant DNA derived apo(a) (rApo(a)) was expressed and secreted in 293s human embryonic kidney cells. The purified rApo(a) has an apparent mass of 500 k by SDS PAGE and 1200-1400 k by gel permeation chromatography suggesting that this recombinant protein also exists as an aggregated protein (Koschinsky, Tomlinson et al. 1991).

Ultracentrifugation of the Glycoproteins

sTNF-R1 , rgp120 IIIB isolate and rApo(a) were sedimented to equilibrium in either a Beckman Model E or Optima XLA analytical ultracentrifuge. Buoyant molecular weights, $M(1-\bar{v}\rho)$, were determined by analyzing the absorbance gradient as described below. Molecular weights were computed from the buoyant molecular weights using estimated values for the partial specific volume of the glycosylated protein, and densities of the solvent buffer measured with a Metler-Parr model DMA 35 density meter. The partial specific volume for sTNF-R1 was calculated assuming that the mass of sTNF-R1 is between 10 and 30% carbohydrate and that the contribution from a carbohydrate chain is 0.63. The partial specific volumes for rgp120 and rApo(a), were estimated from the amino acid and carbohydrate composition.

Sedimentation Equilibrium Data Analysis

The sedimentation equilibrium data for sTNF-R1 and rgp120 were analyzed as a single ideal species, whereas for rApo(a) a two ideal species model was used for the analysis. The absorbance, A_r at any radial position, r, is related to the molecular weights, M_1 and M_2 of species 1 and 2, by

$$(1) \quad A_r = \sum_{i=1}^{2} A_{i,r_o} \exp\{(\omega^2/2RT) M_i (1-\bar{v}_i\rho)(r^2-r_o^2)\}$$

where A_{r_o} is the absorbance at a radial reference distance, r_o, ω is the angular velocity, R the gas constant, T, the Kelvin temperature, $\bar{v}_i$, the partial specific volume of species i and ρ the solution density. The reference radial distance, r_o, for the analysis was set to a radial position 2/3 of the column height. The data were fitted to this equation with the general curve fitting routines in the commercially available graphics software package, KaleidaGraph™.

Results and Discussion

Analysis as a single ideal species of the sedimentation equilibrium concentration gradient in the cell yields buoyant molecular weights. In order to convert these values to absolute molecular weights it is necessary to determine the partial specific volume for the glycoprotein. Experimentally this is done with high precision density measurements as a function of protein concentration. Standard physico-chemical methods such as pycnometry require an inordinately large amount of material. It is possible to obtain $\bar{v}$ values using density meters which make use of the mechanical oscillator technique (Kratky, Leopold et al. 1973), but this technique also requires a fair amount of protein which may not always be available. Alteration of solvent density using $D_2{}^{16}O$ or $D_2{}^{18}O$ in sedimentation velocity or sedimentation equilibrium experiments also can be used to determine $\bar{v}$ (Edelstein and Schachman 1967). An alternative method is to compute the partial specific volume from a weight average of the partial specific volumes of the component amino acid residues. These calculated and observed partial specific volumes of many proteins are in good agreement (Charlwood 1957; McMeekin and Marshall 1952). In the case of a glycoprotein, a similar calculation can be used to estimate partial specific volume if the carbohydrate content is known, and again good agreement between experimental and calculated values has been shown for glycoproteins (Gibbons 1972). Obviously if the carbohydrate content is unknown, it is difficult to make estimates for partial specific volume for a glycoprotein. However, the types of carbohydrate structures found in glycoproteins are generally understood, and it is possible to estimate the contribution from these oligosaccharide chains. Table 1 shows the typical type of N-linked carbohydrates found in glycoproteins along with an estimate of partial specific volume for each type of structure. The estimated values range from 0.62 to 0.64, and thus it would appear that an estimate for the average contribution of 0.63 is not unreasonable. What remains then is to compute the partial specific volume for a glycoprotein with unknown carbohydrate composition and this is discussed further below.

The concentration gradients for sTNF-R1 and rgp120 glycoproteins after attaining sedimentation equilibrium are shown in Figures 1-2 along with the fit to a single ideal species model. The residuals are also given in the figures and aside from some obvious variations of

Table 1. *Estimated partial specific volume for typical N-linked oligosaccharides. All three classes have identical core stucture to the right of the dashed lines. Variations may occur due to increased branching or the presence of other sugars. The values for partial specific volume are calculated on a weight basis using calculated partial specific volumes for individual sugar residues (Gibbons, 1972). The range computed for the complex form is based on a range of typical structures found in proteins.*

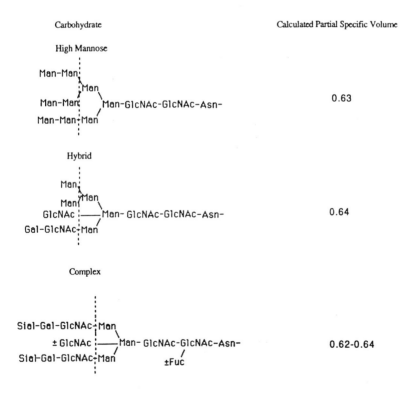

Carbohydrate	Calculated Partial Specific Volume
High Mannose	0.63
Hybrid	0.64
Complex	0.62-0.64

absorbance, (probably due to dirt or imperfections in the quartz windows of the cells) these data suggest that a single ideal species model is adequate in both analyses. The concentration gradient for one of the rApo(a) samples is shown in Figure 3 analyzed with either a single or two ideal species model. The residuals generated from the curve fitting are also shown in the bottom panel of Figure 3. Table 2 summarizes the buoyant molecular weight and resulting absolute molecular weights determined from the estimated partial specific volumes. It is clear from these data that the sTNF-R1 and rgp120 glycoproteins are adequately described by a single ideal species model and are essentially monomeric in solution rather than dimers as suggested by the gel permeation chromatography. On the other hand, a single ideal species model does not account for the rApo(a) concentration gradient at sedimentation equilibrium. Analysis using a two single ideal species model yields an excellent fit along with a random distribution of the residuals (Figure 3).

The determined molecular weight for rgp120 is ~104 k which is in very good agreement with the expected molecular weight of 102 k based on amino acid and average carbohydrate composition. This agreement is quite good considering that the partial specific volume was estimated from amino acid and average carbohydrate composition. In the case of sTNF-R1 the carbohydrate composition is unknown. Calculated molecular weights assuming 10, 20 and 30% carbohydrate with an average partial specific volume of 0.63 are shown in Table 2. As can be seen, the final values are relatively insensitive to the carbohydrate content and show that this protein is a monomer rather than a dimer as suggested by the gel permeation chromatography experiments. The determined molecular weights for the rApo(a) evaluated as a single ideal species gave an average value of 359 ± 29 k, significantly different from the value of 323 k expected for a glycosylated apo(a) monomer. The two ideal component analysis yielded an average molecular weight (as a result of 8 determinations) of 325 ±27 k for the major component and 1300±500 k for the minor component. The absorbance of each component before centrifugation was estimated from the model fitting parameters using (Marque 1992)

$$(2) \quad A_{0,i} = (A_{i,r_0}/\alpha_i)\{\exp(\alpha_i(r_b^2 - r_0^2)) - \exp(\alpha_i(r_m^2 - r_0^2))\} / (r_b^2 - r_m^2)$$

where $A_{0,i}$ is the loading absorbance of component i, A_{i,r_0} is the absorbance of component i at the reference radial position r_0, $\alpha_i = (\omega^2/2RT) M_i (1-\bar{v}_i\rho)$, and r_b and r_m are the radial positions at the base and meniscus of the cell, respectively. Estimates of the amount of multimer present in the different lots of apo(a) (Table 2) are obtained from the computed component absorbance values assuming that the absorptivity of the monomer is not different than that of the multimer. The percentage of monomer present varied depending on the apo(a) preparation but was always greater than 90%. In order to determine if the higher molecular weight species is in equilibrium with

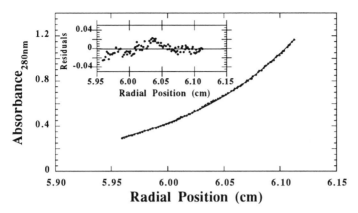

Figure 1. *Sedimentation equilibrium of sTNF-R1 at a loading concentration of 75 μg/mL in phosphate buffered saline. The centrifugation was carried out at 15000 rpm in a Beckman Optima XLA analytical ultracentrifuge over a period of 18 to 24 hours at 20°C. The solid line is the result of a fit to a single ideal species model. The inset shows the residuals to this fit expressed as the difference between experimental and fitted values divided by experimental values.*

monomer or is a separate non-equilibrating component requires additional experiments at higher centrifugal fields since analysis of data obtained at one rotor speed using a monomer-nmer model, or two ideal species model is indistinguishable from each other (Figure 3). Sedimentation velocity experiments further verified the presence of a higher molecular weight component (Phillips, Lembertas et al. 1993). When this high molecular weight contaminant was removed by density gradient centrifugation, the remaining protein was found to be monomeric. Thus, the experiments with the purified rApo(a) clearly show that the appropriate model for the sedimentation equilibrium analysis is a two component model.

In conclusion, the sedimentation equilibrium technique can be used to obtain molecular weights of glycoproteins in solution. Estimation of partial specific volumes using amino acid and carbohydrate composition may be sufficient for these analysis. Moreover, it may even be possible to estimate the partial specific volume for a glycoprotein with unknown carbohydrate composition. In contrast, the molecular weights obtained by gel permeation chromatography of the three glycoproteins discussed are unreliable. The sedimentation equilibrium technique, which does not rely on calibration with standard proteins, is an appropriate method to determine molecular weight of glycoproteins in solution.

Table 2. *The molecular weight of three glycoproteins determined by sedimentation equilibrium.*

glycoprotein	M(1-vρ)	v	M
rgp120 IIIB	33000	0.68	104000
rapo(a) (n=8)			
single ideal species:	111700±9300	0.686	359000±29000
two ideal species			
species 1 (95±2%):	101200±8400	0.686	325000±27000
species 2 (5±2%):	404600±156000	0.686	1300000±500000
sTNF-R1 (n=3)	7970±870	0.68-0.70	26000±2800

Effect of Assumed % Carbohydrate on Molecular Weight of sTNF-R1

% carbohydrate of sTNF-R1:	10	20	30
estimated $v_{sTNF-R1}$:	0.70	0.69	0.68
$M_{sTNF-R1}$:	26900±2800	26000±2700	25200±2700

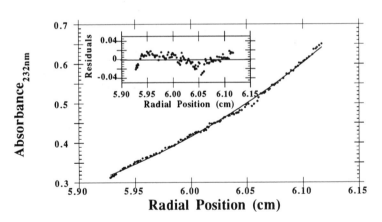

Figure 2. *Sedimentation equilibrium of rgp120 IIIB isolate at a loading concentration of 0.4 mg/mL in 20 mM sodium phosphate, 117 mM NaCl, pH 6.5. The centrifugation was carried out in a Beckman Optima XLA analytical ultracentrifuge at 10000 rpm for ~48 hrs at 20°C. The solid line is the result of a fit to a single ideal species model. The inset shows the residuals to this fit expressed as the difference between experimental and fitted values divided by the experimental values.*

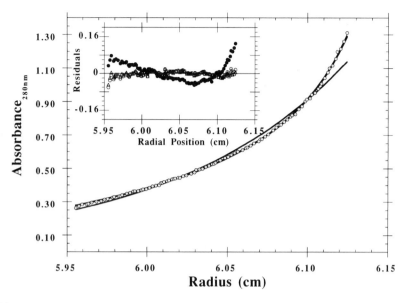

Figure 3. *Sedimentation equilibrium of rApo(a) at a loading concentration of 0.2 mg/mL in phosphate buffered saline at pH 7.2. This protein was sedimented at 4°C to equilibrium in a Beckman Optima XLA analytical ultracentrifuge at 5000 rpm. The bold solid line is the result of fitting the data to a single ideal species model. The curves which go through the data points, are indistinguishable from each other and are the result of a fit to either a two ideal species model or an association monomer-n mer model. The association monomer- n mer model yields a value of 5.4±0.1 for n and -0.68 for ln $k_{1,n}$. The inset shows the residuals (as the difference between experimental and fitted values divided by the experimental values) for the single ideal species model (closed circles), two ideal species model (open circles) and the monomer-n mer association model (open triangles).*

ANALYSIS OF RECEPTOR-LIGAND ASSOCIATION: EVIDENCE FOR TUMOR NECROSIS FACTOR (TNF-α) INDUCED RECEPTOR AGGREGATION

TNF-α is a cytokine with molecular weight of 17.5 k produced by activated macrophages and lymphocytes. This cytokine, which induces necrosis of tumors by interfering with the tumor blood supply, binds to specific receptors on cell surfaces. It appears that the binding of TNF-α to the specific receptors triggers biochemical events in cells leading to the observed necrotic activity. The extracellular domain of human TNF type 1 receptor was cloned as described previously and shown to exist as a monomer in solution. TNF-α exists as a trimer (Arakawa and Yphantis 1987) and there is much controversy regarding the active form of the molecule, i.e.

whether TNF-α binds to receptor as a monomer, dimer or trimer
(Smith and Baglioni 1987) (Peterson, Nykjaer et al. 1989) (Heller,
Song et al. 1990). Many protein ligands such as growth hormone
bind to more than one receptor at the cell surface and it is believed
that such an event triggers the biological response (Cunningham,
Ultsch et al. 1991). Analytical ultracentrifugation was used to
determine if TNF-α formed a complex with more than one receptor.

*Analytical Ultracentrifugation of rhTNF-α and sTNF-R1:rhTNF-α
complex*

rhTNF-α, sTNF-R1 and sTNF-R1:TNF-α complex in phosphate
buffered saline at pH 7.3 were   analyzed by sedimentation
equilibrium. Analysis of TNF-α alone gave a molecular weight of
48.8 k (Table   3). Although the determined molecular weight is
slightly lower than that expected for a trimer of TNF-α (~51 k) it is
consistent with previous observations of dissociation of trimers at low
protein concentration at pH 6 and 7 (Narhi and Arakawa 1987). As
previously discussed, the value of the molecular weight determined for
sTNF-R1, 26 ± 2.8 k, shows that under the conditions of these
experiments sTNF-R1 exists predominantly as a monomer in solution.

Table 3. *Molecular Weight Analysis by Sedimentation Equilibrium.
The buoyant molecular weight [M(1-v̄r)], partial specific volume (v̄)
estimated as described in the text and the molecular weight for each
component is shown for n samples ± one standard (for n=2,average)
deviation. The partial specific volume for sTNF-R1 was estimated by
assuming 20% cabohydrate content and molecular weight computed
from the buoyant molecular weights (Table 2). The partial specific
volume for sTNF-R1:TNF-α trimer complex was estimated for a 2:1
model.*

sample	M(1-vρ)	v	M
TNF-α (n=3)	12550±22	0.739	48770±90
complex (n=2)			
species 1:	33150±1600	0.71	115700±5600
species 2:	8470±1600	0.68	26700±5000

The sTNF-R1:TNF-α complex was prepared by mixing sTNF-
R1 and TNF-α trimer at a 3:1 molar ratio before analysis by
sedimentation equilibrium. The partial specific volumes for the sTNF-
R1:TNF-α complex were estimated for both 3:1 and 2:1 ratios and for
20% glycosylation of sTNF-R1. A single ideal species model could
not fit the data adequately. Since the mixture for the centrifuge
analysis has two components as determined by gel permeation
chromatography, this was to be expected (Pennica, Kohr et al. 1992)
Accordingly, the centrifuge data for the mixture were analyzed as a
two component ideal species model (Figure 4).   The smaller

molecular weight species has a similar buoyant molecular weight

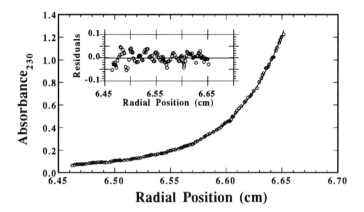

Figure 4. *Sedimentation equilibrium of an sTNF-R1: TNF-α mixture in phosphate buffered saline at pH 7.2 at a total loading concentration of ~ 0.1 mg/ml (3:1 receptor to TNF-α trimer molar ratio). The centrifugation was carried out at 20°C in a Beckman Optima XLA analytical ultracentrifuge at 15000 rpm. The absorbance gradient was analyzed with a two ideal species model. The inset shows the residuals to the fit expressed as the difference between experimental and fitted values divided by the experimental values.*

found for STNF-R1, and is presumably uncomplexed soluble receptor (Table 3). The models at 2:1 and 3:1 stoichiometries resulted in stoichiometric ratios of 2.5±0.5 and 2.4±0.6 sTNF-R1 molecules per TNF-α trimer, respectively. Calculations with the same models, but with the assumption of either 10% or 30% glycosylation for sTNF-R1 yield values for the stoichiometric ratios within 10% of the values computed assuming 20% glycosylation for sTNF-R1. These results are consistent with the STNF-R1:TNF-α complex containing at least two and possibly three molecules of sTNF-R1 per TNF-α trimer. It has been shown that a variety of TNF biological activities can be initiated by the clustering of TNF receptors on cell surfaces by specific antibodies (Engelmann, Holtmann et al. 1990; Espevik, Brockhaus et al. 1990; Tartaglia, Weber et al. 1991) Since TNF-α associates to trimers and these trimers are capable of binding more than one receptor it is plausible that TNF-α mediated receptor oligomerization is required for triggering the biological activities induced by TNF-α.

ANALYSIS OF PROTEIN SELF-ASSOCIATION: PORCINE

RELAXIN (pRlx), AND RECOMBINANT DNA DERIVED HUMAN (rhRlx)

Relaxin is a protein hormone which plays a major role in the reproductive biology of various species (Sherwood 1988) (Bryant-Greenwood 1982). This protein appears to modulate the restructuring of connective tissues in target organs in order to generate the required changes in organ structure during pregnancy and parturition. In particular, relaxin regulates a number of biological responses of reproductive tissues in pregnant animals including softening of cervical tissue and inhibition of uterine myometrial contractions (Steinetz, Beach et al. 1959). Relaxin's ability to induce cervical ripening at parturition has been investigated in a clinical setting using relaxin isolated from porcine corpus lutea and ovaries (Evans, Dougan et al. 1983; MacLennan, Green et al. 1981; MacLennan, Green et al. 1986).

Relaxin consists of two polypeptide chains which are linked by inter and intrachain disulfide bonds. Although the amino acid sequence homology is low between relaxin and insulin there are a number of striking homologies between these protein hormones. The positions of the disulfide links are similar in both proteins and there is a conservation of the gly residues immediately adjacent to the cysteine residues in the B chain (Sherwood 1988). These residues are crucial to maintaining the proper structural folding for insulin and therefore it is not surprising that relaxin has a similar tertiary structure to that of the insulins. Although there are these striking similarities between insulin and relaxin, the proteins exhibit very different physiological and biochemical properties. The aggregation properties of insulin are well known (Blundell, Cutfield et al. 1971; Hodgkin and Mercola 1972) and it is thought that these properties are related to the *in vivo* storage and regulation of insulin (Blundell, Dodson et al. 1972). In particular, studies have been done on zinc-free insulin using analytical ultracentrifugation and in each case the weight average molecular weight has been determined as a function of concentration (Goldman and Carpenter 1974; Jeffrey and Coates 1966). Unlike insulin, analytical ultracentrifugation studies of porcine relaxin suggest that this protein does not exhibit self association behavior (Sherwood and O'Byrne 1974). However, it has been suggested that the activity of porcine relaxin administered subcutaneously (SQ) in the mouse interpubic ligament stretch bioassay is related to the aggregation of porcine relaxin induced by the addition of the dye, benzopurpurine 4-B (Schwabe and Harmon 1978). Pharmacokinetic studies (Ferraiolo, Cronin et al. 1989) suggest that relaxin complexed with the the the dye is slowly released from the SQ site , and this slow release results in a biological response in the assay. Circular dichroism spectroscopy has been used to study the benzopurpurine 4B-porcine relaxin complex (Schwabe and Harmon 1978). The far UV CD spectrum of zinc-free bovine insulin is similar to the spectrum of the relaxin dye complex, but differs from the CD spectrum for porcine relaxin without benzopurpurine (Figure 5). Since zinc-free insulin forms dimers it was suggested that the CD spectral differences between porcine relaxin with and without the benzopurpurine were due to formation of relaxin dimer. The association of relaxin was studied by circular dichroism and analytical ultracentrifugation to determine if spectral differences in the far UV CD are related to relaxin dimer formation.

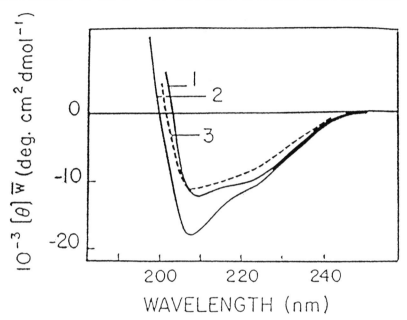

Figure 5. *The effect of benzopurpurine 4-B on the far UV circular dichronism spectra of porcine relaxin. Curve 1: Zinc free insulin (50 μM); Curve 2: Porcine relaxin (50 μM); Curve 3: Porcine relaxin (50 μM) with benzopurpurine 4-B (25 μM). Reproduced with permission from Schwabe and Harmon, 1978.*

Analytical Ultracentrifugation of rhRlx and pRlx.

Preparations of rhRlx and pRlx (Shire, Holladay et al. 1991) at 0.05, 0.1 and 0.2 mg/mL were centrifuged to sedimentation equilibrium. The partial specific volumes for porcine and human relaxin, 0.743 and 0.741 respectively, were computed from the amino acid composition using the additivity rule and values for the individual amino acid residues (Cohn and Edsall 1965). A typical concentration gradient at sedimentation equilibrium for rhRlx at a loading concentration of 0.1 mg/mL is shown in Figure 6. The solid line is a result of fitting the data at all three loading concentrations simultaneously to a non-ideal monomer-dimer self-association model using a simplex algorithm (Shire, Holladay et al. 1991), and the bottom panel shows the residuals to the fit. The rhRlx data were also fitted to a monomer-dimer-trimer, monomer-dimer-tetramer and

isodesmic association models, and the   porcine relaxin data were
analyzed as a non-ideal monomer-dimer association. The resulting
parameters from the fits of two separate experiments are summarized
as the mean values of the association constants and deviation from the
mean in Table 4. The root mean square lack of fit is also presented as

Table 4. *Analysis of rhRlx and pRlx Sedimentation Equilibrium Data
Using Simplex Algorithm. Reproduced with permission from Shire, et
al., 1991.*

model	association constants	rmslf
	rhRlx in 10 mM Citrate, 0.13 M NaCl pH 5.0	
	32000 rpm	
monomer-dimer	K_2=100 ±4 $(g/L)^{-1}$	0.0040±0.0003
monomer-dimer-trimer	K_2=104±9 $(g/L)^{-1}$ K_3=1.09±0.03 $(g/L)^{-2}$	0.0040±0.0002
monomer-dimer-tetramer	K_2=170 ±52 $(g/L)^{-1}$ K_4= 0.1±0.1 $(g/L)^{-3}$	0.0040±0.0001
isodesmic	K= 3.0±0.03 $(g/L)^{-1}$	0.007±0.001
	22000 rpm	
monomer-dimer	K_2=77 ±19 $(g/L)^{-1}$	0.005±0.0003
	porcine relaxin in 10 mM Citrate, 0.13 M NaCl pH 5.0	
	32000 rpm	
monomer-dimer	K_2= 0.7±0.6 $(g/L)^{-1}$	0.0043

a mean value and deviation from the mean. The fitted values for the
second virial coefficients which were essentially zero (from 5×10^{-5} to
2×10^{-7}) demonstrate that non-ideality was not significant during
relaxin sedimentation. The monomer-dimer association constant for
porcine relaxin of 0.27±0.22 $(g/L)^{-1}$ yields a weight average
molecular weight of 6700±500 at a total relaxin concentration of 0.5
mg/mL which is in good agreement with previously reported values of
6300 determined by sedimentation equilibrium measurements

(Sherwood and O'Byrne 1974). Unlike porcine relaxin which is monomeric, human relaxin self associates in solution. Using the mean value for the monomer-dimer association constant, K_2, of 100 $(g/l)^{-1}$ the computed weight average molecular weight at a total human relaxin concentration of 0.5 mg/mL is 11140. A similar value for K_2 is obtained at lower rotor speeds (Table 4) which demonstrates that the self association is a reversible equilibrium dependent on total protein concentration (Teller 1972).

The rhRlx data appear to be fit equally well on the basis of the root mean square lack of fit to discrete monomer-dimer-trimer and monomer-dimer-tetramer models. However, for both higher association models the higher aggregate association constant is at least 100 fold lower in magnitude whereas the monomer-dimer association

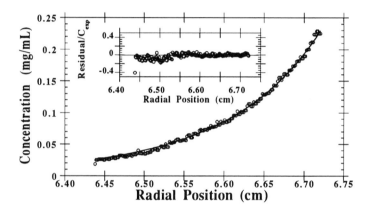

Figure 6. *Concentration distribution of recombinant human relaxin at a loading concentration of 0.1 mg/mL during sedimentation equilibrium at 32000 rpm. The centrifugation was carried out in a Beckman Model E analytical ultracentrifuge in an An-F analytical ultracentrifuge rotor. The solid lines are a result of fitting three solvent-solution pairs of data at 0.05, 0.1 and 0.2 mg/mL simultaneously with a simplex algorithm to a monomer-dimer model. The inset shows the residuals to the fit for the 0.1 mg/mL loading concentration, and the resulting parameters are given in Table 4. Reproduced with permission from Shire et al., 1991.*

constants are similar. The concentrations of trimers or tetramers would be far less than 0.1% of the total protein concentration and therefore the human relaxin association is well characterized as a monomer-dimer equilibrium. Moreover, it is unlikely that human relaxin associates as an indefinite association scheme since the root mean square lack of fit for an isodesmic model is almost twice that of the discrete association models (Table 4). Although analytical ultracentrifugation of rhRlx clearly shows that this protein self-associates to dimers , GPC analysis of final formulated protein only detects monomers because of the large dilution that occurs during the chromatography (data not shown).

Circular Dichroism and Association of rhRlx and pRlx

The Far UV CD spectra of rhRlx and pRlx (Figure 7) differ significantly in the 230-215 nm range, which lead to a lower value for the ratio of the ellipticities at 208 and 222 nm for human relaxin when compared to porcine relaxin. It was suggested that the differences in this wavelength region between zinc-free insulin and porcine relaxin may be attributed to the fact that zinc-free insulin is dimeric whereas porcine relaxin is monomeric under the conditions studied (Schwabe and Harmon 1978). In order to determine whether changes in the far UV CD spectra are related to relaxin dimer formation the CD spectra was determined for rhRlx at 5 mg/mL and 10 μg/mL. The monomer-dimer rhRlx association constant determined by analytical ultracentrifugation shows that dilution of the protein from 5 mg/mL to 10 μg/mL leads to an almost 5 fold increase in the weight fraction of monomer. However, this large increase in monomeric rhRlx does not alter the Far UV CD spectrum of human relaxin (Figure 7) whereas the near UV CD (Figure 8) does show a significant change in the region usually assigned to tyrosine residues. On the other hand, the environment of the tryptophans probably does not change greatly upon dissociation of the dimer since there are no changes in the broad band near 295 nm. (Figure 8). This suggests that upon dissociation to the monomer the lone tyrosine environment is altered sufficiently to allow for greater mobility of the tyrosine residue and hence a decrease in CD signal strength. This observation is supported by the recent determination of the crystal structure of human relaxin (Eigenbrot, Randal et al. 1991). This protein crystallizes as a dimer, and the lone tyrosine residue is centered at the dimer interface. This tyrosine residue is replaced by an arginine in porcine relaxin, and this substitution may affect the self association properties of the protein. In contrast to the near UV CD, the fact that the far UV signal doesn't change upon dilution of the rhRlx suggests that the dissociation of the human relaxin dimer is accompanied by little overall change in secondary structure.

CHARACTERIZATION OF RECOMBINANT HUMAN TISSUE FACTOR (rhTF)-SURFACTANT MIXED MICELLES

Human tissue factor is a blood coagulation protein which exists as a glycosylated integral membrane protein (Bach 1988; Nemerson 1988). This protein functions as a cofactor in conjunction with factor VII or VIIa and calcium ions to increase the rate of conversion of factor X to Xa. The production of factor Xa results in conversion of prothrombin to thrombin which converts fibrinogen to fibrin and generates a fibrin clot. This coagulation pathway, referred to as the extrinsic pathway, does not require factor VIII, and therefore tissue factor may be useful as a therapy for hemophiliac patients with antibodies or inhibitors to a factor VIII (Figure 9). In order for tissue factor to function in this pathway it needs to be incorporated into a membrane structure. Recombinant human tissue factor that lacks the cytoplasmic domain was formulated into phosphate buffered saline

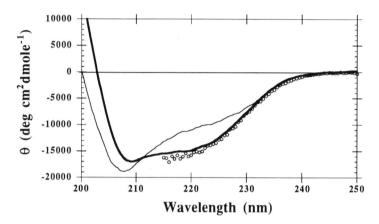

Figure 7. *Effect of relaxin concentration on the far UV circular dichroism spectra of recombinant human relaxin. The Far UV circular dichroism of porcine (thin solid line) and human (bold solid line) relaxin at 0.5 mg/mL were obtained in an Aviv modified Cary 60 spectropolarimeter using a thermostated 0.01 cm cylindrical cuvette maintained at 20°C. The scans are the result of of an average of three scans using an average time for each single data point collection of five seconds. Data were collected at 0.2 nm intervals at a spectral band width of 1.5 nm. The CD data for relaxin at 10 μg/mL (open circles) were obtained in a 1 cm thermostated cylindrical cuvette, and are the result of of an average of ten scans using an average time for each single data point collection of ten seconds. The weight fraction of monomer estimated from the determined association constant of 100 (g/L)$^{-1}$ (Table 4) is 0.13 at 0.5 mg/mL and 0.62 at 10 μg/mL. Reproduced by permisision from Shire et al., 1991.*

with either $C_{12}E_8$ or octylglucoside surfactants. These surfactants are at concentrations greater than the critical micelle concentration. The studies presented here were designed to determine if surfactant micelles are required for activity of tissue factor, the average size of such mixed micelles and the stoichiometry of the complex (Cipolla and Shire 1992).

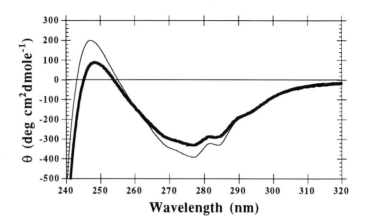

Figure 8. *Effect of relaxin concentration on the near UV circular dichroism of recombinant human relaxin. The near UV circular dichroism spectra were obtained in an Aviv modified Cary 60 spectropolarimeter. Human relaxin at 0.5 mg/mL (thin solid line) was thermostated at 20°C in a 1 cm cell whereas relaxin at 20 µg/mL (bold solid line) was in an unthermostated 10 cm cylindrical cuvette.The temperature in the sample compartment was at ~ 27°C during the data collection process. The CD data were collected at 0.25 nm intervals at a spectral band width of 0.5 nm and are the result of an average of three scans using an average time for each single data point collection of five seconds for the 0.5 mg/mL samples and the result of an average of ten scans using an average time for each single data point collection of ten seconds for the 20 µg/mL sample.The weight fraction of human relaxin monomer estimated from the determined association constant of 100 (g/L)$^{-1}$ (Table 4) is 0.13 at 0.5 mg/mL and 0.50 at 20 µg/mL. Reproduced by permission from Shire et al., 1991.*

Preparation and formulation of rhTF

Recombinant DNA derived human tissue factor was expressed in E. coli as a 243 amino acid non glycosylated peptide that included the extracellular and transmembrane domains (Paborsky et al. 1989; Cipolla and Shire 1992). The expressed protein was purified from E. coli extracts using immunoaffinity chromatography. The rhTF was formulated into PBS (10 mM Sodium Phosphate, isotonic with NaCl, pH 7.3) formulations that contained either $C_{12}E_8$ (0.04%) or octylglucoside (0.8%) surfactants.

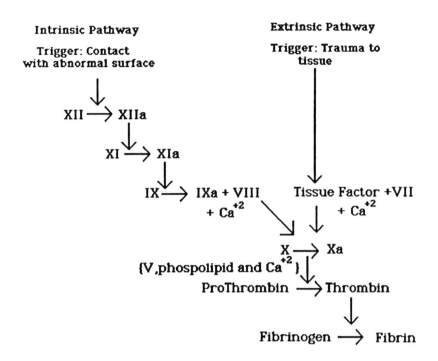

Figure 9. *Schematic showing the intrinsic and factor VIII independent extrinsic blood coagulation pathways.*

Activity of rhTF as a function of surfactant concentration

The activity of rhTF as a function of $C_{12}E_8$ concentration was measured with a clotting assay that assesses the ability of rhTF to clot Factor VIII deficient human plasma. The activated partial thromboplastin time as a function of $C_{12}E_8$ concentration shows that below ~0.005% $C_{12}E_8$ there is an abrupt decrease in the activity of rhTF as reflected in an increase in the clotting time (Figure 10) . The reported critical micelle concentration for $C_{12}E_8$ is ~0.004% (Kawashima et al., 1985), and therefore it appears that the micelle is critical for rhTF activity. It is known that tissue factor requires lipids for activity. Furthermore, tissue factor is a membrane-associated protein and without any phospholipids present it is not unreasonable to assume that the apoprotein is associated with surfactant micelles through the transmembrane domain. The formulated apo protein may be inserted in the micelle and this mixed micelle system might allow for efficient transfer into phospholipid bilayers producing an active tissue factor-lipid complex. Sedimentation velocity and equilibrium analytical ultracentrifugation were used to characterize the hydrodynamic size of such mixed micelles and the number of tissue factor molecules associated with the surfactant micelles.

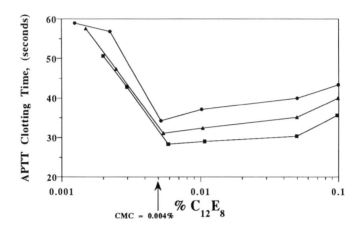

Figure 10. *Clotting Activity (Activated Partial Thromboplastin Time) of rhTF as a Function of % $C_{12}E_8$. Tissue Factor was eluted from an affinity column at 63 µg/mL in 0.033% C12E8, 10 mM Sodium Phosphate, isotonic with NaCl, pH 7.2. The rhTF was diluted into identical buffer containing $C_{12}E_8$ at one of six concentrations: 0.001%, 0.002%, 0.005%, 0.01%, 0.05%, and 0.1%, spanning the reported critical micelle concentration of 0.004% $C_{12}E_8$. Samples were assayed for clotting activity.*

Analytical Ultracentrifugation of rhTF-Surfactant Formulations

Sedimentation velocity and equilibrium experiments were performed using a Model E analytical ultracentrifuge equipped with a photoelectric UV absorption scanner. The equilibrium data were fit to a single ideal species using the non-linear curve fitting routines in commercially available software (Kaleidagraph™). The sedimentation equilibrium data can be analyzed reasonably well using a single ideal species model, (Figures 11 a and b).The buoyant molecular weights for a number of experiments are summarized in Table 5. The residuals (non-linear curve fit- experimental values) are shown in the lower panels of each figure and suggest that a single ideal species is an adequate model. The sedimentation velocity experiments yield sedimentation coefficients corrected to standard conditions of water at 20°C = 2.2 s and 2.8 s for $C_{12}E_8$ and octylglucoside formulated proteins respectively (Figures 12a and b).

The sedimentation coefficient is related to the stokes radius R_S as follows (Tanford, Nozaki et al. 1974):

$$(3) \quad s = M(1-\bar{v}\rho)/(6\pi\eta N R_S)$$

Therefore the sedimentation velocity data can be combined with the buoyant molecular weights determined by sedimentation equilibrium to give the stokes radius of the mixed micelle. rhTF- $C_{12}E_8$ mixed micelles have a stokes radius = 42.6 Å whereas rhTF-octylglucoside micelles have a stokes radius = 33.9 Å

Table 5. *Buoyant molecular weights determined by sedimentation equilibrium of rhTF.*

	C12E8		
	Co (mg/mL)	rotor speed(rpm)	$M(1-\bar{v}\rho)$
Expt. 1	0.1	15000	10930
	0.1	22000	9141
	0.075	22000	10095
Expt. 2	0.1	15000	11294
	0.1	22000	12236
	0.075	15000	11801
	0.075	22000	10258
		mean value ± σ:	10822±1070

	Octylglucoside		
	Co (mg/mL)	rotor speed(rpm)	$M(1-\bar{v}\rho)$
Expt. 1	0.1	15000	11971
	0.075	15000	13420
	0.075	22000	10820
Expt. 2	0.1	15000	11971
	0.1	22000	10653
	0.075	15000	11801
	0.075	22000	10492
		mean value ± σ:	11292±1190

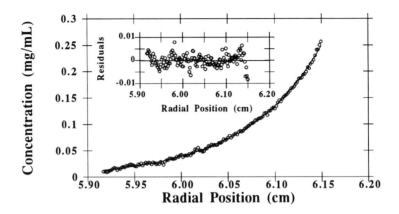

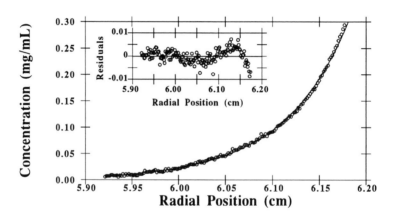

Figure 11. *Sedimentation Equilibrium Analytical Ultracentrifugation of rhTF. Photoelectric scanner output as a function of radial position is shown for rhTF in 0.04% C12E8 (Fig. 11a) and 0.8% octylglucoside (Fig. 11b). The ultracentrifugation was performed in charcoal filled 6 channel equilibrium cells in An-F rotors at either 15000 or 22000 rpm at 20°C. Equilibrium was ascertained to be reached after comparison of subsequent scans taken 12 hours apart. These data were fit to a single ideal species model using a non linear curve fitting procedure. The resulting residuals from the fits are shown in the inset.*

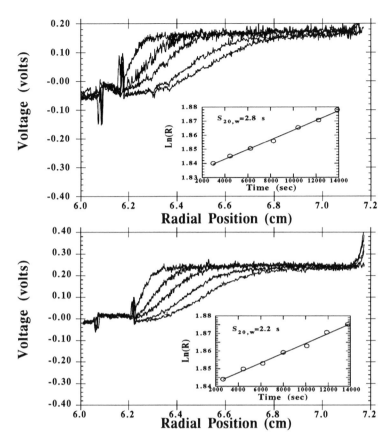

Figure 12. *Sedimentation Velocity Analytical Ultracentrifugation of rhTF. Photoelectric scanner output as a function of radial position is shown for rhTF in 0.04% C12E8 (Fig.12a) and 0.8% octylglucoside (Fig.12b). The ultracentrifugation was performed in charcoal filled double sector cells in An-D rotors at 36000 rpm at 20°C. All scans are not shown for clarity. The inset figures show the sedimentation coefficient analysis. Sedimentation coefficients were corrected to standard conditions of water at 20°C using experimentally determined viscosities and densities of the surfactant buffers.*

In order to compute molecular weight of the protein-surfactant micelles it is necessary to evaluate $\bar{v}$ for the complex from (Tanford, Nozaki et al. 1974):

$$(4) \quad \bar{v} = (\bar{v})_p + \delta_S \bar{v}_S / (1 + \delta_S)$$

where $\bar{v}_p$ and $\bar{v}_S$ are the partial specific volumes of the protein and surfactant and δ_S is the amount of surfactant bound to protein expressed on a gram to gram basis. The partial specific volume for non-glycosylated truncated rhTF computed from the amino   acid composition is 0.736 cm^3/g. What remains is to determine partial specific volumes for the surfactants in the PBS formulation buffer and values for δ_s. We determined the partial specific volume for both surfactants at $19.8 \pm 0.02°C$ in the PBS formulation buffer with an Metler-Parr model DMA 60 high precision density meter. The partial specific volume was determined experimentally by measuring the density of solvent alone (ρ_s, in g/mL) and the density of solutions (ρ_1, ρ_2,..., ρ_i, in g/mL) containing known concentrations of solute (C_1, C_2,..., C_i, in g/mL):

$$(5) \ \bar{v}_i = (1/\rho_s)[1 - (\rho_i - \rho_s)/C_i].$$

$\bar{v}$ is determined by plotting the individual $\bar{v}_i$'s versus C_i and extrapolating to $\bar{v}$ at $C = 0$. When the density of solution is directly proportional to the concentration of solute, that is when

$$(6) \ \rho_i = \rho_s + (C_i)(slope),$$

and

$$(7) \ \bar{v} = (1 - slope)/\rho_s.$$

The density as a function of $C_{12}E_8$ concentration does not show a noticeable change in the slope (Figure 13a) probably because of the low value, 0.00003-0.00005 g/mL for the critical micelle concentration (CMC) of $C_{12}E_8$ (Kawashima et al., 1985). The partial specific volume as a function of $C_{12}E_8$ concentration (Figure 13b) yields a mean value from two determinations of $\bar{v} = 0.964 \pm 0.001$ in the PBS formulation. This compares favorably with a previously determined value of 0.973 for $C_{12}E_8$ at 25°C in 0.1 M NaCl (Tanford, Nozaki et al. 1977). In the case of octylglucoside the CMC is substantially greater than for $C_{12}E_8$ as reflected by the shift in the slope of the plot of density versus octylglucoside concentration (Figure 14a) at 0.006 g/mL which is consistent with the range reported for the CMC for this surfactant.

Below the CMC,

$$(8) \ \rho_i = 1.00579 + 0.17605 \ C_i,$$

and above the CMC,

$$(9) \ \rho_i = 1.00608 + 0.12809 \ C_i.$$

For values below the CMC, determination of $\bar{v}$ is identical to that described above. Substituting the value of the slope from equation (8) into equation (7), yields $\bar{v} = 0.819$ cm^3/g for concentrations of octylglucoside below the CMC. To determine v for concentrations of octylglucoside above the CMC, the expression relating ρ_i to c_i from equation (9) as well as the value of $\rho_s = 1.00579$ from (8) are substituted into equation (5), yielding $\overline{v}$ as a function of the octylglucoside concentration:

$$(10) \quad \overline{v} = 0.867 - 0.00029/c_i.$$

The asymptotic value of $\overline{v}$ for micellar octylglucoside is 0.867 cm^3/g (Figure 14b). This value is identical to the reported value for the partial specific volume of octylglucoside in water at 22°C (Kameyama and Takagi 1990).

Although it is difficult to obtain a value for the parameter δ_s it is possible to calculate the parameter for mixed micelles with different number of rhTF molecules by using the aggregation number and molecular weight for a surfactant monomer. $C_{12}E_8$ micelles have an aggregation number = 120 to 127 and molecular weight of 65 to 69 k (Tanford, Nozaki et al. 1977) and octylglucoside micelles have an aggregation number = 87 and molecular weight of 25 k (Kameyama and Takagi 1990). The computed δ_s values combined with experimentally determined values for v_s can then be used to compute the molecular weight of the mixed micelle from the experimentally determined buoyant molecular weight. Subtraction of the micelle molecular weight and division by the protein monomer molecular weight yields a computed stoichiometry that can be compared to the original assumed stoichiometry for the computation of δ_s. Table 6 summarizes these computed values for the octylglucoside and $C_{12}E_8$ mixed micelles. The determined values for the number of rhTF molecules per octylglucoside micelle, 1.0-1.2 are relatively insensitive to the value chosen for δ_s, whereas there is a wide range, (0.5-1.7) for $C_{12}E_8$. In the case of $C_{12}E_8$ the entire computation is self consistent, i.e. the computed stoichiometry is similar to the assumed stoichiometry at a stoichiometry of 1.2 rhTF molecules per $C_{12}E_8$ micelle. Overall these data are consistent with an association of ~1.2 rhTF molecules with either a $C_{12}E_8$ or octylglucoside surfactant micelle. A slightly greater value of 1.5 was previously obtained for octylglucoside micelles (Cipolla and Shire 1992). However, the determination of this value was based on a molecular weight of 5 k for the octylglucoside micelle which was obtained by gel permeation chromatography (Rosevear, VanAken et al. 1980).The more reliable value of 1.2 rhTF molecules per micelle was based on an octylglucoside micelle molecular weight of 25 k that was determined by low angle laser light scattering and analytical ultracentrifugation (Kameyama and Takagi 1990).

a

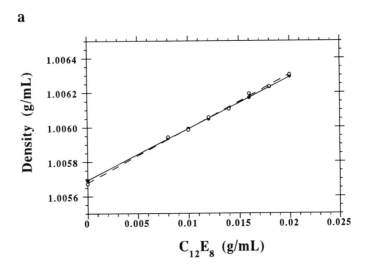

b

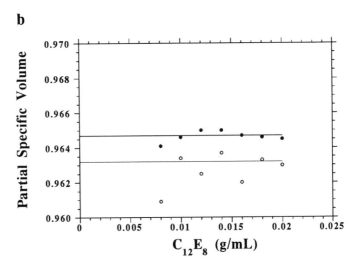

Figure 13. *Density (13a) and apparent partial specific volume (13b) as a function of $C_{12}E_8$ concentration. Phosphate buffered saline solutions containing 0% and 2% $C_{12}E_8$ were appropriately mixed resulting in a series of solutions at different $C_{12}E_8$ concentrations. Density measurements (at 19.80±0.02 °C) were made using a Metler-Parr precision density meter which measured the oscillation period of a quartz tube filled with these solutions.*

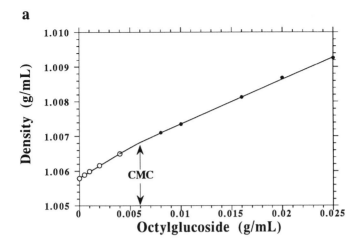

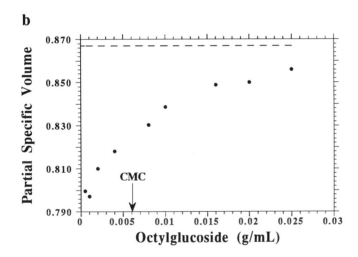

Figure 14. *Density (14a) and apparent partial specific volume (14b) as a Function of octylglucoside concentration. Phosphate buffered saline solutions containing 0% and 2.5% octylglucoside were appropriately mixed resulting in a series of solutions at different octylglucoside concentrations. Density measurements (at 19.80±0.02°C) were made using a Metler-Parr precision density meter which measured the oscillation period of a quartz tube filled with these solutions. The partial specific volume above the CMC was determined from the asymptote and is indicated by a dashed line in Figure 14b.*

Our data are not precise enough to rule out the possibility that some micelles contain more than one tissue factor molecule. Nonetheless although the buoyant molecular weights from 2 determinations have an ~10% error these data are consistent with the majority of the micelles containing one tissue factor molecule. Moreover, it is unlikely that there is a significant population of micelles containing more than one tissue factor molecule since a single ideal species model can adequately describe the sedimentation equilibrium data.

SUMMARY AND PERSPECTIVES OF THE FUTURE USE OF ANALYTICAL ULTRACENTRIFUGATION IN BIOTECHNOLOGY.

Many of the efforts in the broad sector termed biotechnology revolve around the development of protein pharmaceuticals. The determination of molecular weight and presence of protein aggregates is of prime importance in the development of protein drugs. Protein aggregation may affect activity and the biological half-life of a protein which can have an impact on the efficacy of the drug (Shao, Li et al. 1993). The immunogenicity of a protein may also be affected by aggregation which may not only alter the efficacy of the protein but may also pose a safety risk (Moore and Leppert 1980; Pinkard, Weir et al. 1967). Successful use of recently developed humanized antibodies that will be administered chronically will be very dependent on minimizing any immunogenic responses to the antibody. Accordingly, much attention is given to ensuring that the final formulated protein does not contain signigicant amount of aggregate, and that protein aggregation does not occur during storage of the drug. Often the molecular size distribution in solution is obtained by gel permeation chromatography (GPC). The advantages of GPC are that it is relatively cheap and rapid. However, as demonstrated by some of the examples presented in this chapter, molecular weight determinations of glycoproteins are often in error when determined by GPC with protein calibration standards. Since GPC determines hydrodynamic volume, this is not surprising because of the increased effective hydrodynamic diameter due to the additional carbohydrate structures attached to the globular protein structure. This problem can be rectified by interfacing low angle light scattering detection with the chromatography in order to obtain absolute molecule weight assignments of chromatographed peaks (Takagi 1990). However, it is not uncommon for proteins to interact with the gel filtration support media. Often, detergents are added to prevent protein adsorption and the true molecular weight of associating species is no longer being determined by the chromatographic method. Frequently GPC is also used to determine molecular size distribution during stability determination of a large number of formulations for a protein drug. It is impractical to run each GPC column with the formulation as the mobile phase and thus one buffer system is usually chosen for the experiment. This may result in complications, especially if the pH or ionic strength influences the state of association of the protein. Sedimentation equilibrium with 6 channel cells and rotors that accommodate at least 3 cells allow for multiple determinations of up to 9 formulations in one experiment. Despite these drawbacks the GPC method is usually preferred in a quality control environment because of the short time required to run the chromatography. Recent advances with short solution columns (Correia and Yphantis 1992) allow sedimentation equilibrium experiments to be completed within a couple of hours, and thus the time factor may be less of a concern. The sedimentation equilibrium technique also offers the advantage of analyzing reversible protein self association that can go undetected because of

Table 6. *Determination of stoichiometry for rhTF mixed micelles.*

C12E8

assumed stoichiometry	δ_S	$\bar{v}$	$1-\bar{v}\rho$	MW	computed stoichiometry
1.0	2.45	0.898	0.0958	113000	1.7
1.1	2.23	0.893	0.100	107800	1.5
1.2	2.04	0.889	0.105	103300	1.3
1.3	1.88	0.885	0.109	99300	1.2
1.4	1.75	0.881	0.113	96000	1.1
1.5	1.63	0.877	0.117	92800	0.9
1.6	1.53	0.874	0.120	90200	0.8
1.7	1.44	0.871	0.123	87700	0.8
1.8	1.36	0.867	0.127	85500	0.7
1.9	1.29	0.864	0.130	83600	0.6
2.0	1.23	0.862	0.132	81900	0.5

Octylglucoside

assumed stoichiometry	δ_S	$\bar{v}$	$1-\bar{v}\rho$	MW	computed stoichiometry
1.0	0.912	0.798	0.195	57900	1.2
1.1	0.829	0.795	0.198	56900	1.2
1.2	0.760	0.793	0.201	56200	1.1
1.3	0.701	0.790	0.204	55400	1.1
1.4	0.651	0.788	0.206	54800	1.1
1.5	0.608	0.786	0.208	54200	1.1
1.6	0.570	0.784	0.210	53700	1.1
1.7	0.536	0.782	0.212	53300	1.0
1.8	0.506	0.780	0.214	52800	1.0
1.9	0.480	0.778	0.215	52400	1.0
2.0	0.456	0.777	0.217	52100	1.0

the large dilutions that occur during the column chromatography as shown in the case of rhRlx. Although relaxin dimers are in equilibrium with monomers, the ability to undergo association could have an impact on long term storage of the protein since there is an increased chance for formation of a covalent linkage between molecules.

Interactions of proteins with formulation components such as surfactants can be studied by analytical centrifugation as shown by the example of rhTF. The number of rhTF molecules that were associated with the surfactant micelle was computed from the buoyant molecular weight by making estimations of the partial specific volume for the complex. It is also possible to obtain the results more directly by altering the density of the solution so that the contribution from the micelles are essentially blanked out and only the molecular weight of protein associated with the micelles is determined (Tanford, Nozaki et al. 1974). Alternatively, using the Rayleigh interference optical system such as found on the Model E analytical ultracentrifuge allows for the determination in the same experiment of the concentration gradient due to the mixed micelle as well as the protein alone, since the major contributor to the absorbance at 280 nm is the rhTF. The development of real time interferometric optical systems (Laue 1992) will make such measurements more routine. In addition, since most proteins have the same refractive increment per mg/mL, the use of both interference and UV absorption will result in rapid determinations of absorptivity values for routine spectrophotometric concentration determinations. Such an analysis can also be extended to associating systems to determine whether extinction coefficients are altered when the protein self associates.

Heterologous association of protein molecules, such as a protein-receptor interaction as described in the case of TNF-α can be analyzed by analytical ultracentrifugation. The ability to analyze strongly interacting systems such as antibody-antigen complexes will be made easier as improvements in low concentration detection systems such as fluorimetry are incorporated into the analytical ultracentrifuge.

As a final comment, the model E has been the major instrument used for analytical ultracentrifugation. Improvements in data acquisition and analysis have been made by a number of groups that have contributed to a refinement of the technology. However, for most users that do not have the desire to be involved in development of instrumentation, the technique has remained out of reach. The recent introduction of a commercial instrument that has an improved UV absorption detection system (Giebler 1992) and modern PC data acquisition capabilities finally makes the general use of analytical ultracentrifugation in the biotechnology industry a real possibility.

Acknowledgements

The author thanks the following collaborators for supplying

purified recombinant proteins: Dr. Tim Gregory (human tissue factor and gp120), Dr. Diane Pennica and Dr. David Goeddel (TNF-α and sTNFR1), Dr. Ernst Rinderknecht (human relaxin), and Dr. Tom Zioncheck (rApo(a)). Thanks are also extended to Dr. Geoffrey Tregear for supplying purified porcine relaxin. Special thanks to Yvonne Chen for her careful determination of density and partial specific volume of octylglucoside and C12E8 surfactants. The author also thanks Dr. Toshio Takagi for interesting discussions regarding the more recent and reliable determination of octylglucoside micelle molecular weight. Finally I wish to thank Milianne Chin for helping me to get this chapter into final publication format.

GLOSSARY OF SYMBOLS

r	radial position
A_r	absorbance at radial position r
M_i	molecular weight of species i
A_i, r_o	absorbance contributed by species i at a radial reference position
r_o	radial reference position
ω	centrifugal angular velocity
R	gas constant
T	Kelvin temperature
$\bar{v}_i$	partial specific volume of species i
ρ	solution density
$A_{o,i}$	initial absorbance of species i prior to centrifugation
r_b	radial position at the base of the cell
r_m	radial position of the meniscus
R_s	Stokes radius
S	sedimentation coefficient
η	solution viscosity
N	Avagodro's number
$\bar{v}_p$	partial specific volume of protein
$\bar{v}_s$	partial specific volume of surfactant
δ_s	amount of surfactant bound to protein on a gram to gram basis
C_i	weight concentration of species i

REFERENCES

Arakawa, T. and D. A. Yphantis. (1987): Molecular weight of recombinant human tumor necrosis factor-α. *J. Biol. Chem.* 262: 7484-7485.

Bach, R. R. (1988): Initiation of coagulation by tissue factor. *CRC Critical Reviews in Biochemistry.* 23: 339-368.

Blundell, T. L., J. F. Cutfield, S. M. Cutfield, E. Dodson, D. C. Hodgkin, D. A. Mercola and M. Vijayan. (1971): *Nature (London).* 231: 506-511.

Blundell, T. L., G. Dodson, D. Hodgkin and D. A. Mercola. (1972): Insulin: The structure in the crystal and Its reflection in chemistry and biology. *Adv. Protein Chem.* 26: 279-402.

Bryant-Greenwood, G. D. (1982): Relaxin as a new hormone. *Endocrine Rev.* 3(1): 62-90.

Charlwood, P. A. (1957): Partial specific volume of proteins in relation to composition and environment. *J. Am. Chem. Soc.* 79: 776-781.

Cipolla, D. and S. J. Shire. (1992): Characterization of human tissue factor-surfactant mixed micelles. The Protein Society. San Diego : 80, abstract number S198.

Cohn, E. J. and J. T. Edsall. (1965): Proteins, amino Acids and peptides as ions and dipolar ions. New York, Hafner.

Correia, J. J. and D. A. Yphantis. (1992): Equilibrium sedimentation in short solution columns. In: *Analytical Ultracentrifugation in Biochemistry and Polymer Science,* S. E. Harding, A. J. Rowe and J. C. Horton, eds. Cambridge: The Royal Society of Chemistry.

Cunningham, B. C., M. Ultsch, A. M. De Vos, M. G. Mulkerrin, K. R. Cluser and J. A. Wells. (1991): Dimerization of the extracellular domain of the human growth hormone receptor by a single growth hormone molecule. *Science.* 254: 821-825.

Edelstein, S. J. and H. K. Schachman. (1967): The simultaneous determination of partial specific volumes and molecular weights with microgram quantities. *J. Biol. Chem.* 242(306-311):

Eigenbrot, C., M. Randal, C. Quan, J. Burnier, L. O'Connell, E. Rinderknecht and A. A. Kossiakoff. (1991): X-Ray structure of human relaxin at 1.5 Å. *J. Mol. Biol.* 221: 15-21.

Engelmann, H., H. Holtmann, C. Brakebusch, Y. S. Avni, I. Sarov, D. Nophar, E. Hadas, O. Leitner and D. Wallach. (1990): Antibodies to a soluble form of a tumor necrosis factor (TNF) receptor have a TNF-like activity. *J. Biol. Chem.* 265: 14497-14504.

Espevik, T., M. Brockhaus, H. Loetscher, U. Nonstad and R. Shalaby. (1990): Characterization of binding and biological effects of monoclonal antibodies against a human tumor necrosis factor receptor. *J. Exptl. Med.* 171: 415-426.

Evans, M. I., M.-B. Dougan, A. H. Moawad, W. J. Evans, G. D. Bryant-Greenwood and F. C. Greenwood. (1983): Ripening of the human cervix with porcine ovarian relaxin. *Am. J. Obstet. Gynecol.* 147(4): 410-414.

Ferraiolo, B. L., M. Cronin, C. Bakhit, M. Roth, M. Chestnut and R. Lyon. (1989): The pharmacokinetics and pharmacodynamics of a human relaxin in the mouse pubic symphysis bioassay. *Endocrinology.* 125(6): 2922-2926.

Gibbons, R. A. (1972): Physico-chemical methods for the determination of the purity, molecular size and shape of glycoproteins. In: *Glycoproteins, Part A*, A. Gottschalk, eds. Amsterdam: Elsevier.

Giebler, R. (1992): The Optima XL-A: A new analytical ultracentrifuge with a novel precision absorption optical system. In: *Analytical Ultracentrifugation in Biochemistry and Polymer Science*, S. E. Harding, A. J. Rowe and J. C. Horton, eds. Cambridge: The Royal Society of Chemistry.

Goldman, J. and F. H. Carpenter. (1974): Zinc binding, circular dichroism, and equilibrium sedimentation studies of insulin (bovine) and several of its derivatives. *Biochemistry* 13: 4566-4574.

Heller, R. A., K. Song, D. Villaret, R. Margolskee, J. Dunne, H. Hayakawa and G. M. Ringold. (1990): Amplified expression of tumor necrosis factor receptor in cells transfected with Epstein-Barr virus shuttle vector cDNA libraries. *J. Biol. Chem.* 265: 5708-5717.

Hodgkin, D. C. and D. A. Mercola. (1972): In: *Handbook of Physiology I*, D. Steiner, eds. Washington, D. C.: American Physiological Society.

Jeffrey, P. D. and J. H. Coates. (1966): An equilibrium ultracentrifuge study of the effect of ionic strength on the self-association of bovine insulin. *Biochemistry* 5: 3820-3824.

Kameyama, K. and T. Takagi. (1990): Micellar properties of octylglucoside in aqueous solutions. *J. Colloid Interface Sci.* 137: 1-10.

Kawashima, N. Fujimoto, N., and Meguro, K. (1985): Determination of critical micelle concentration of several nonionic surfactants by azo-hydrazone tautomerism of anionic dye. *J. Coll. Int. Sci.* 103: 459.

Koschinsky, M. L., J. E. Tomlinson, T. F. Zioncheck, K. Schwartz, D. L. Eaton and R. M. Lawn. (1991): Apolipoprotein(a): Expression and characterization of a recombinant form of the protein in mammalian cells. *Biochemistry.* 30: 5044-5051.

Kratky, O., H. Leopold and H. Stabinger. (1973): The determination of the partial specific volume of proteins by the mechanical oscillator technique. Methods in Enzymology. New York, Academic Press.

Lackner, C., E. Boerwinkle, C. C. Leffert, T. Rahmig and H. H. Hobbs. (1991): Molecular basis of apolipoprotein(a) isoform size heterogeneity as revealed by pulsed-field gel electrophoresis. *J. Clin. Invest.* 87: 2153-2161.

Laue, T. M. (1992): On-line data acquisition and analysis from the Rayleigh interferometer. In: *Analytical Ultracentrifugation in Biochemistry and Polymer Science*, S. E. Harding, A. J. Rowe and J. C. Horton, eds. Cambridge: The Royal Society of Chemistry.

Leonard, C. K., M. Spellman, L. Riddle, R. J. Harris, J. N. Thomas and T. J. Gregory. (1990): Assignment of intrachain disulfide bonds and characterization of potential glycosylation sites of the type I recombinant human immunodeficiency virus envelope glycoprotein (gp120) expressed in chinese hamster ovary cells. *J. Biol. Chem.* 265: 10373-10382.

MacLennan, A., R. C. Green, G. D. Bryant-Greenwood, F. C. Greenwood and R. F. Seamark. (1981): Cervical ripening with combinations of vaginal prostaglandin $F_{2\alpha}$, estradiol and relaxin. *Obstet Gynecol.* 58(5): 601-604.

MacLennan, A. H., R. C. Green, P. Grant and R. Nicolson. (1986): Ripening of the human cervix and induction of labor with intracervical purified porcine relaxin. *Obstet. Gynec.* 68(5): 598-601.

Marque, J. (1992): Personal communication.

McMeekin, T. L. and K. Marshall. (1952): Specific volumes of proteins and their relationship to their amino acid contents. *Science.* 116: 142-143.

Meselson, M. and F. W. Stahl. (1958): *Proc. Natl. Acad. Sci.* 44: 671.

Moore, W. V. and P. Leppert. (1980): Role of aggregated human growth hormone (hGH) in development of antibodies to hGH. *J. Clin. Endocrin. Metabol.* 51: 691.

Narhi, O., L. and T. Arakawa. (1987): Dissociation of recombinant tumor necrosis factor-α studied by gel permeation chromatography. *Biochem. Biophysical Res. Comm.* 147: 740-746.

Nemerson, Y. (1988): Tissue factor and hemostasis. *Blood.* 71: 1-8.

Paborsky, L. R., Tate, K. M., Harris, R. J., Yansura, D. G., Band, L., McCray, G., Gorman, C. M., O'Brien, D. P., Chang, J. Y., Swartz, J. R., Fung, V.P., Thomas, J. N. and Vehar, G. A. (1989): Purification of recombinant human tissue factor. *Biochemistry* 28:8072-8077.

Pennica, D., W. J. Kohr, B. M. Fendly, S. J. Shire, H. E. Raab, P. E. Borchardt, M. Lewis and D. V. Goeddel. (1992): Characterization of a recombinant extracellular domain of the type I tumor necrosis factor receptor: evidence for tumor necrosis factor-α induced receptor aggregation. *Biochemistry* 31: 1134-1141.

Peterson, C. M., A. Nykjaer, B. S. Christiansen, L. Heickendorff, S. C. Mogensen and B. Moller. (1989): Bioactive human recombinant tumor necrosis factor α: An unstable dimer? *Eur. J. Immunol.* 19: 1887-1894.

Phillips, M., A. V. Lembertas, V. N. Schumaker, R. M. Lawn, S. J. Shire and T. F. Zioncheck. (1993): Physical properties of recombinant apolipoprotein (a) and its association with LDL to form an Lp(a)-like complex. *Biochemistry.* 32: 3722-3728.

Pinkard, R. N., D. M. Weir and W. H. McBride. (1967): Factors influencing immune response: I. Effects of the physical state of the antigen and use of lymphoreticular cell proliferation on the response to intravenous injection of bovine serum albumin in rabbits. *Clin. Exp. Immunol.* 2: 331.

Rosevear, P., T. VanAken, J. Baxter and S. Ferguson-Miller. (1980): Alkyl glucoside detergents: A simpler synthesis and their effects on kinetic and physical properties of cytochrome c oxidase. *Biochemistry.* 19: 4108-4115.

Schwabe, C. and S. J. Harmon. (1978): A comparative circular dichroism study of relaxin and insulin. *BBRC.* 84(2): 374-380.

Shao, Z., Y. Li, R. Krishnamoorthy, T. Chermak and A. K. Mitra.

(1993): Differential effects of anionic, cationic, nonionic, and physiologic surfactants on the dissociation, alpha-chymotryptic degradation and internal absorption of insulin hexamers. *Phar. Res.* 10: 243.

Sherwood, C. D. and E. M. O'Byrne. (1974): Purification and characterization of porcine relaxin. *Arch Biochem Biophys.* 160: 185-196.

Sherwood, O. D. (1988): Relaxin. In: *The Physiology of Reproduction*, E. Knobil and J. Neill, eds. New York: Raven Press.

Shire, S. J., L. A. Holladay and E. Rinderknecht. (1991): Self-Association of human and porcine relaxin as assessed by analytical ultracentrifugation and circular dichroism. *Biochemistry.* 30: 7703-7711.

Smith, R. A. and C. Baglioni. (1987): The active form of tumor necrosis factor is a trimer. *J. Biol. Chem.* 262: 6951-6954.

Steinetz, B. G., V. L. Beach and R. L. Kroc. (1959): The physiology of relaxin in laboratory animals. In: *Recent Progress in the Endricrinology of Reproduction*, C. W. LLoyd, eds. New York: Academic Press.

Svedberg, T. and R. Fahraeus. (1926): A new method for the determination of the molecular weight of the proteins. *J. Am. Chem. Soc.* 48: 430.

Takagi, T. (1990): Application of low-angle laser light scattering detection in the field of biochemistry. *J. Chrom.* 506: 409-416.

Tanford, C., Y. Nozaki, J. A. Reynolds and S. Makino. (1974): Molecular characterization of proteins in detergent solutions. *Biochemistry.* 13: 2369-2376.

Tanford, C., Y. Nozaki and M. F. Rohde. (1977): Size and shape of globular micelles formed in aqueous solution by n-alkyl polyoxyethylene ethers. *J. Phys. Chem.* 81: 1555-1560.

Tartaglia, L. A., R. F. Weber, I. S. Figari, C. Reynolds, M. A. Pallidino and D. V. Goeddel. (1991): *Proc. Natl. Acad. Sci. U. S. A.* 88: 9292-9296.

Teller, D. C. (1972): Characterization of proteins by sedimentation equilibrium in the analytical ultracentrifuge. *Meth. Enzym.* 27: 346-441.

APPLICATIONS OF ANALYTICAL ULTRACENTRIFUGATION IN STRUCTURE-BASED DRUG DESIGN

Thomas F. Holzman and Seth W. Snyder

Drug Design & Delivery, Abbott Laboratories, Abbott Park, IL.

INTRODUCTION

As a technique analytical ultracentrifugation encompasses a family of related hydrodynamic methods which are employed to monitor either transport (sedimentation velocity) or equilibrium (sedimentation equilibrium) processes. Recent development of the Beckman Optima XLA analytical ultracentrifuge, to eventually replace the Model E, makes it possible to routinely apply these methods to biophysical problems associated with the development of effective, targeted, pharmacophores. In order to evaluate the contributions that ultracentrifugation can make to the process of structure-based drug design it is essential to first define the drug design cycle (Fig. 1). The characterizations of

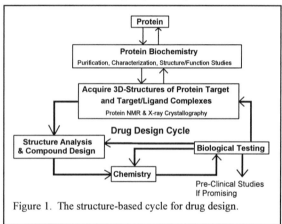

Figure 1. The structure-based cycle for drug design.

structure and function provided by Protein Biochemistry in this process precede and fuel all subsequent three-dimensional structure determinations. The role of computationally-based and structure-based 3D information in the design cycle has recently been elaborated (Propst and Perun, 1989, and references cited therein). Typically, the cycle begins with the identification of a tissue source of the "activity" of interest: i.e. the protein or the DNA encoding the protein of interest. Whether the desired protein is from a natural or recombinant source it is first purified and characterized. Some recent examples from our laboratories include the purification and characterization of two classes of natural and recombinant, highly-expressed, peptidyl-prolyl isomerases, cyclophilin (Holzman et al., 1991, and references cited therein) and FKBP (Edalji et al., 1992). The availability of these proteins permitted determination of the bioactive conformations of the, respective, ligands Cyclosporin A and Ascomycin (Fesik et al., 1990; Fesik et al., 1991; Neri et al., 1991; Petros et al., 1991) and eventually the 3D structures of protein-ligand complexes formed with these compounds (Meadows *et al.*, 1993; Thériault *et al.*, 1993). This type of detailed structural information is then available for a chemistry effort to modify a known drug or to synthesize a drug *de novo*, so as to preserve crucial structural interactions while improving specificity and potency.

From the standpoint of drug design a significant proportion of new pharmaceutical targets are proteins. In most cases little, or nothing, is known about the natural quaternary structural interactions of these protein targets, or potential alterations in these interactions by drug molecule binding. Indeed, it is quite likely that a protein of interest may only have been recently discovered, or may only recently have become associated with an enzymatic activity or important physico-chemical function of interest. Thus, it is in the area of Protein Biochemistry, where molecular form and function are examined, that analytical ultracentrifugation has an opportunity to contribute to structure-based drug design. This chapter focuses on examples of the uses of, and results from, ultracentrifugation in the study of pharmaceutically relevant proteins which are potential targets for drug design.

ACQUISITION AND ANALYSIS METHODS

Simple Analog-to-Digital Data Acquisition Systems for the Beckman Model E or Prep-UV Scanner Analytical Ultracentrifuges. The absorbance signal is sampled and stored using an IBM PC-based Keithley Series 500 Data Acquisition System equipped with an ADM-2 16-bit analog-to-digital converter board and controlled using Asyst software (Keithly Instruments) (Holzman *et al.*, 1990). The ADM-2 board samples the analog signal from either the Model E or Prep-UV scanner at a rate (frequency) chosen by the operator using Asyst software. With the ADM-2 typically set to receive a 0 to 0.5 volt signal (offset binary), the 16-bit converter provides a resolution of $0.5/2^{15}$ volts (15.3 μV). For a 0.4 V signal this corresponds to 26,214 steps along the Y-axis (absorbance). For the Prep-UV scanner the electronic absorbance calibration system delivers a 1.0 absorbance calibration signal at ~80% of the full-scale electronic response. Thus for 0.4 V full-scale the 1.0 absorbance calibration signal corresponds to 20,972 steps and gives a theoretical digital resolution of the optical signal corresponding to 4.77×10^{-5} absorbance units. Because the integers returned by the A/D board is directly proportional to absorbance, determination of the actual voltage output is unnecessary. Absorbance values are calculated using a linear transformation and the zero absorbance baseline and 1.0 absorbance calibration value. Data from the Model E scanner calibration are handled similarly, except that the entire stair-step calibration series (to either 1.0 or 2.0 optical density units) is utilized to automatically convert from A/D increments to optical density.

Cell scans with either the Model E or Prep-UV are under software control from Asyst running under MS-DOS. During a scan across the cell image (~60 seconds), the analog signal is sampled at a rate of ~77 Hz for a period of 65 seconds. This provides 5000 data points total. During acquisition the data are stored in an integer array of 5000 elements. Scans of samples in Yphantis multichannel centerpieces give ~300-500 useful data points across each channel; samples in long-column centerpiece give ~2000-2500 useful data points.

Following acquisition, data are stored as integer arrays using Asyst. For both the Model E and Prep-UV scanner the Asyst software is used to recognize and record the positions of the inner and outer reference hole boundaries and to subsequently compute the radial position of each absorbance increment measured. Measurements of solution/solvent density and protein partial specific volume are accomplished using a Mettler ultrasonic density meter Model DA-310, which is accurate to five decimal places.

   The Beckman Optima-XLA. The control and acquisition software for the XLA is provided by Beckman and has been described in detail (Giebler, 1992).

   Software Packages for Acquisition and Analysis: Microsoft DOS and Microsoft Windows-Based. We use several approaches for analyzing data from the Model E, Prep-UV Scanner, or Optima XLA. The specific software used depends upon the number and type of samples to be analyzed and the number and type of models to fit to the data. In practice, sedimentation equilibrium analysis for a single sample for monodisperse behavior (below) is accomplished using either Asyst software routines developed in-house running under MS-DOS or as an MS-DOS application under Microsoft Windows. Alternatively, the Windows-based program, Origin, from Microcal, Inc is useful for fitting of individual data sets. If the data set consists of more than two or three samples, and the samples exhibit complex behavior (see below), data analyses are substantially faster using Asyst in a batch processing mode with output ported to model-fit files in Lotus *.wk* file format for subsequent input into an MS-Windows-based spreadsheet-presentation program like Micrografx Charisma or Microsoft Excel. Since sedimentation velocity data are always contained in multiple acquisition files our preferred analysis approach for these experiments is also in-house Asyst routines operating in batch process mode. Files are typically processed with and without application of multiple noise-frequency cut-off filters and with and without second-moment analysis; the resulting *.wk* files are also ported to Charisma or Excel software for viewing.

   Sedimentation Equilibrium Data Analysis for Samples which are Monodisperse or Undergo a Discrete Oligomerization Process. At sedimentation equilibrium the behavior of an ideal solute, analyzed at infinite dilution in the ultracentrifuge, will obey an exponential (1) relationship and its linear transform (2) :

$$C_r = C_m e^{\left\{\frac{\overline{M}_w \omega^2 \left(1-\overline{v}\rho\right)\left(r_i^2-r_m^2\right)}{2RT}\right\}} \quad \text{or} \quad C_r = C_m e^{\left\{\overline{M}_w C_0 \left(r_i^2-r_m^2\right)\right\}} \quad (1)$$

$$\ln C_r = \frac{\overline{M}_w \omega^2 \left(1-\overline{v}\rho\right)\left(r_i^2 - r_m^2\right)}{2RT} + \ln C_m \quad (2)$$

   Where the apparent solute weight-average molecular weight is defined by the slope of the plot of $\ln C_r$ versus $(r_i^2-r_m^2)$. The C_0 term is the combined set of equation constants $(\omega^2\{1-v\rho\}/2RT)$. Deviations from linearity are typically due to solute polydispersity and/or nonideality. For most dilute samples of otherwise homogeneous proteins, the usual cause of polydispersity is self-association of solute monomers to form larger discrete oligomers or dissociation to monomers from a normally constituted oligomeric protein. The distribution of each species present in a polydisperse solution during centrifugation is described by an equation of the form (1). Rather than dealing with logarithmic transformations of data (2) from the photoelectric scanner, a preferred analysis involves deriving best-fit sums of exponentials to the observed absorbance values (Edelstein *et al.*, 1970; Crepeau *et al.*, 1974)). For example, at equilibrium the distribution of total protein concentration between monomer, dimer and tetramer, as a function of radial position, is described as follows:

$$P_{r^2} = A_{31}\, e^{\left(C_0 \bullet \overline{M}_w \bullet r^2\right)} + A_{32}\, e^{\left(2 \bullet C_0 \bullet \overline{M}_w \bullet r^2\right)} + A_{33}\, e^{\left(4 \bullet C_0 \bullet \overline{M}_w \bullet r^2\right)} \quad (3)$$

Where r^2 is equal to the square of the radial position relative to the sample meniscus and $P_{(r^2)}$ is equal to the absorbance at that radial position. The monomer molecular weight (M_w) for the three-term fit, along with the pre-exponential constants are either initial values fixed by the user, or are allowed to float during data fitting. Using the initial values for the molecular weight and pre-exponential coefficients, the fitting routine iteratively adjusts the values of these coefficients to minimize the difference between the observed data and values calculated from the analytical expression. The number of iterations the program goes through is determined by a preselected value for the tolerance of fit. Once the preset tolerance is attained the program constructs plots of the actual data overlaid with the calculated data set and a plot of the residual differences between the observed data set and the calculated data set. In order to determine the minimum number of useful terms in the analytical expression for describing the observed data set, the residual differences for each analytical expression of two to five exponential terms are compared. For most data, fit to explicit oligomerization models, from either the Model E scanner, prep UV scanner, or Optima-XLA we find that use of more than three terms do not provide a significant improvement in the distribution of residuals.

   Sedimentation Equilibrium Data Analysis for Samples Exhibiting Indeterminate Poly/Paucidispersity. From an operational view-point, data are first subjected to analysis by (1) and (2) to determine the solute molecular weight based on simple monodisperse behavior. If the data appear to fit these relationships then analysis is complete. If the data, analyzed by (2), are curvilinear then, in order to calculate the molecular weight distribution present, the data must be fitted to an explicit model, such as (3), which accounts for the molecular weight distribution. In practice, assessing the degree-of-fit to various possible oligomerization models is very often a time-consuming task which is complicated by the absence of data which are both *timely* and *high-quality* for "real-world" samples [see also Johnson & Straume, this volume]. Although a detailed model-fit is preferred in the long-run, it is often very helpful employ a model-independent data analysis method (Holzman, 1992a). In such an approach a multi-term (usually three-to-five) analytical expression like (3) can be fit to the data over the range of radial positions and absorbancies under examination. This later process can be performed for any sample without prior knowledge of the mechanisms leading to the particular molecular weight distributions observed. Although this approach does not in-and-of itself define a model for the observed data, it is useful for representing data sets for molecular weight distribution analyses and for rapidly comparing data sets acquired under different experimental conditions. The basis of this method is as follows. An Asyst program was developed to provide non-weighted fits to sums of exponentials of up to five terms for equations of the form:

$$P_{r^2} = A_{51}\, e^{\left(C_0 \bullet \overline{M}_{5w} \bullet r^2\right)} + A_{52}\, e^{\left(2 \bullet C_0 \bullet \overline{M}_{5w} \bullet r^2\right)} + \cdots + A_{55}\, e^{\left(8 \bullet C_0 \bullet \overline{M}_{5w} \bullet r^2\right)} \qquad (4)$$

The fitted pre-exponential "A" values and the user-chosen monomer molecular weight, M_{5w} are designed only to provide a good fit of the analytical expression to the observed data. In this particular expression the exponential terms are incremented in a step-wise fashion, 1-2-4-6-8, to indicate the formation of a dimer followed by dimeric oligomer growth. Because of the constraints used, the individual terms of the fit are unlikely to represent distributions of individual components. Once the best-fit coefficients are calculated for each analytical expression, the expressions which provide appropriate fits are used to evaluate changes in the molecular weight distributions for a set of experimental conditions. Plots are constructed for weight-average molecular weight as a function of the absorbance and for the fractional contribution of each weight-average species to absorbance as a function of the molecular weight distribution across the cell. The former plot is constructed using Asyst, Charisma, Excel, or Origin software from data files constructed in Asyst. The later, more useful plot is usually prepared using Charisma from data files constructed in Asyst; it requires somewhat more complex computations, the basis for which can be summarized as follows. When plotted in logarithmic form (2) the weight-average molecular weight at a particular radial position in the cell is defined by the tangent to the observed distribution at that position. The tangent is simply the first-derivative of the fitted analytical expression (4) evaluated at that point:

$$\frac{d(\ln P_{r^2})}{d(r^2)} = \frac{d}{d(r^2)}\left\{ \ln\left(A_{51}\, e^{\left(C_0 \bullet \overline{M}_{5w} \bullet r^2\right)} + A_{52}\, e^{\left(2 \bullet C_0 \bullet \overline{M}_{5w} \bullet r^2\right)} + \cdots \right.\right.$$
$$\left.\left. \cdots + A_{55}\, e^{\left(8 \bullet C_0 \bullet \overline{M}_{5w} \bullet r^2\right)} \right) \right\} \qquad (5)$$

which becomes:

$$P' = \frac{1}{P_{r^2}}\left\{ A_{51} C_0 \overline{M}_{5w}\, e^{\left(C_0 \bullet \overline{M}_{5w} \bullet r^2\right)} + \cdots + A_{55}\, 8 C_0 \overline{M}_{5w}\, e^{\left(8 \bullet C_0 \bullet \overline{M}_{5w} \bullet r^2\right)} \right\} \qquad (6)$$

Now the derivative, P', evaluated at any individual value of r^2 is equal to the value of $M_{w,r^2\ app.}$ and for a single point, this can be rewritten as:

$$P_{r^2} = A_{r^2}\, e^{\left(C_0 \bullet \overline{M}_{w,r^2} \bullet r^2\right)} \qquad (7)$$

This expression is then applicable to all points in the data set simply through variation in r^2, A_{r^2}, and $M_{w,r^2\ app}$; (7) can be rewritten as:

$$\ln(P_{r^2}) = \ln(A_{r^2}) + C_0 \bullet \overline{M}_{w,r^2} \bullet r^2 \qquad (8)$$

At a particular point, r^2, in the sample distribution (i.e. P_r2), where $M_{w,r2}$ is the tangent to the curve, $\ln(P_r2)$ is calculable across the sedimented sample. The area under the line tangent to the observed distribution at point r^2 may then be determined by integration across the bounds of the data set. This area represents the *apparent* optical density contributed by a *species* of $M_{w,r2}$ to the entire distribution of molecular weights observed in the sedimenting sample. For purposes of clarity let $r^2=x$ and the integral is then:

$$\int_{x_0}^{x_1} \ln(P_x)dx = \ln(A_x)\int_{x_0}^{x_1} dx + C_0 \overline{M}_{w,x} \int_{x_0}^{x_1} xdx$$

$$= \ln(A_x) \cdot (x_1 - x_0) + \frac{C_0 \cdot \overline{M}_{w,x}}{2} \cdot (x_1^2 - x_0^2) \quad (9)$$

For any *particular* value of $M_{w,x}$ at point x, the fraction of solute with that molecular weight given by:

$$\text{Fraction} = \frac{\ln(A_x) \cdot (x_1 - x_0) + \frac{C_0 \cdot \overline{M}_{w,x}}{2} \cdot (x_1^2 - x_0^2)}{\int_{x_0}^{x_1} [\ln(A_x) \cdot (x_1 - x_0) + \frac{C_0 \cdot \overline{M}_{w,x}}{2} \cdot (x_1^2 - x_0^2)]dx} \quad (10)$$

Thus for each value of x the fractional contribution of the ith species to the total solute present is given by:

$$i\text{ th Fraction} = \frac{\ln(A_x)(x_1 - x_0) + \frac{C_0 \overline{M}_{w,x}}{2}(x_1^2 - x_0^2)}{(x_1 - x_0)\{\sum_{i=x_0}^{i=x_1} \ln(A_i)\} + \frac{C_0(x_1^2 - x_0^2)}{2}\{\sum_{i=x_0}^{i=x_1} \overline{M}_{wi,x}\}} \quad (11)$$

From this analysis it is evident that a plot of the fractional contribution of the i th species to M_w *versus* M_w will have a maximum value at the M_w providing the largest mass contribution. A shift in the M_w distribution towards larger species it will be reflected in a shift in the maximum M_w value toward higher molecular weights; the converse is also true. A plot of fractional contribution *versus* M_w is particularly useful in examining the effects of solvent conditions, solute concentrations, and ligand binding on the M_w distributions present in associating systems.

   Sedimentation Velocity Data Analysis. The treatment of data from sedimentation velocity experiments, as based on useful forms of Faxén's two-component solution to the Lamm equation, has been presented in detail (Fujita, 1962; see also Fujita, this volume). The useful expressions are:

$$c = \left((c_0 e^{-\tau})/2\right)\left[1 - \Phi\left(\frac{1-(xe^{-\tau})^{1/2}}{[\varepsilon(1-e^{-\tau})]^{1/2}}\right)\right] \quad (12)$$

Where $x = (r/r_0)^2$, $\tau = 2s\omega^2 t$, $\varepsilon = 2D/s(\omega r_0)^2$, Φ is the error function of the term enclosed in brackets, and s and D are, respectively, the sedimentation and diffusion coefficients. Thus, under suitable conditions (Fujita, 1962) (12) can be used to approximate the profile of a sedimenting boundary. Two other useful expressions (Fujita, 1962) derived from (12) are:

$$\ln(r/r_0) = s\omega^2(t\text{-}t_0) \quad (13) \quad \text{and} \quad (A/H)^2 = D(2\pi/s\omega^2)(\exp(2s\omega^2 t) - 1) \quad (14)$$

Expression (13) gives the sedimetation coefficient, s, while (14) gives the square of the Area/Height ratio (Fujita, 1962) (of the first-derivative of the sedimenting boundary) which is proportional to D (and s). In practice we use the second-moment (Goldberg, 1953) value of s computed as previously described (Holzman et al., 1982).

REPRESENTATIVE EXAMPLES
Sedimentation Equilibrium Analysis

Ligand-Induced Association of CMP-KDO Synthetase in D_2O. The enzyme CMP-KDO Synthetase or CKS (CTP:CMP-3-deoxy-D-manno-octulosonate Cytidylyltransferase) is required for growth in almost all gram negative bacteria. It functions to activate a required 8-carbon sugar, KDO, for incorporation into outer bacterial cell-wall lipopolysaccharide. In addition to providing structural integrity, the outer bacterial membrane influences virulence and host-pathogen interactions. The enzyme is one of several involved in the synthesis and assembly of the outer cell membrane of gram-negative bacteria (Unger, 1981). Once the enzyme was cloned and expressed at sufficient levels (Goldman et al., 1986) it became a target for inhibition and thus a candidate for the application of structure-based drug design. When studied by NMR as purified protein it became evident that, at the concentrations required for typical NMR experiments (~0.1-5 mM protein), samples of CKS exhibited some line-broadening (R. Gampe and S. Fesik, unpublished observations) thought to be uncharacteristic for a protein of ~27kD (H-form) and suggested a certain fraction of the protein might be aggregating. In addition to potential interferences with data analyses it was not clear that the active form of the protein was, in fact, a monomer. Thus, in order to facilitate the study of CKS by NMR, the binding of substrate, CTP, and inhibitor, KDOI, to fully *deuterated* enzyme was first analyzed using titration calorimetry in D_2O. Subsequently, the sedimentation equilibrium behavior of each titrated sample was analyzed (Holzman et al., 1992c).

The binding stoichiometries observed by calorimetry were consistent with a monomer having one site each for CTP and KDOI. However, calorimetry also revealed that binding of the substrate and inhibitor were mutually *cooperative*. That is: the K_d of CKS alone for CTP alone was 23 μM, the K_d for KDOI alone was 70 μM, the K_d of a CKS:CTP mixture for KDOI was 6.3 μM and the K_d of a CKS:KDOI mixture for CTP was 2.0 μM (Holzman et al., 1992c). Thus based on calorimetry, in the absence of hydrodynamic data, it could be suggested that the active enzyme is a monomer and any tendency to aggregate must be unrelated to its active tertiary structure. However, this proved not to be the case. In Figure 2 the data for several samples of the enzyme are analyzed for ith species fractional contribution to optical density. It is evident that, even in the absence of added ligands, the enzyme was not monomeric. Furthermore, the shifts in M_w-distribution widths and positions along the M_w-axis, in response to added ligand, indicate ligand binding

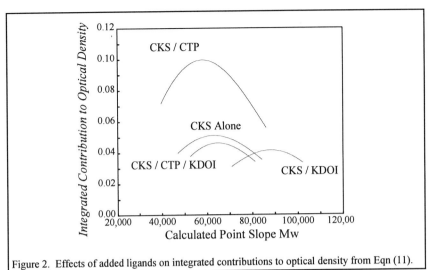

Figure 2. Effects of added ligands on integrated contributions to optical density from Eqn (11).

modulates the tendency to aggregate. These effects are summarized in Figure 3. First, the deuterated enzyme alone is not monomeric (M_w = 29,441), it is in equilibrium with higher order components. At a higher protein concentration the breadth of the Mw distribution for enzyme alone is narrowed and spans the range near dimer (M_w = 58,882). Second, in the presence of the substrate alone, CTP, the effect is repeated and again the breadth of the M_w distribution spans the range of dimer. In comparison, in the presence of the inhibitor alone, KDOI, the distribution is shifted to a much higher molecular weight range, close to that of tetramer (M_w = 117,764). Finally, in the presence of both CTP and KDOI the molecular weight range again approaches that of dimer.

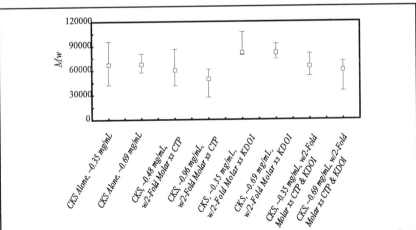

Figure 3. Summary of CKS self-association behavior estimated from model-independent analyses at 8.8°C, 16000 rpm for 24 hours. The weight-average molecular weight distribution ranges are indicated by horizontal hash-marks at the ends of each vertical line. The weight-average molecular weight at which the i th species gives a maximum fractional contribution to optical density is marked with a "□" for each sample.

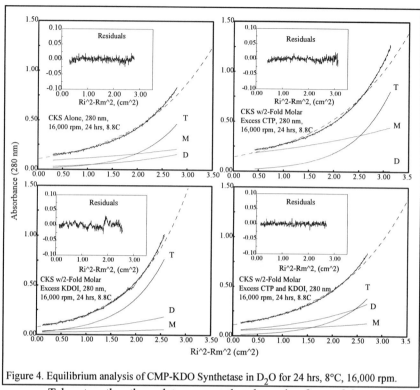

Figure 4. Equilibrium analysis of CMP-KDO Synthetase in D_2O for 24 hrs, 8°C, 16,000 rpm.

Taken together these data suggest that the active form of the enzyme is likely a quaternary complex composed of two individually active monomers. The molecular weight ranges observed indicated that it might be possible to fit the aggregation behavior to a discrete equilibrium model of monomer ⇔ dimer ⇔ tetramer (M⇔D⇔T) as described by Eqn (3) above. Figure 4 depicts four panels containing experimental data plotted as absorbance *versus* r^2 for CKS alone, CKS with CTP, CKS with KDOI, and CKS with both CTP and KDOI. In each figure the solid line passing through the data represents the fit of the M⇔D ⇔T model to the distribution of sedimented protein. For comparison to distributions expected for monodisperse samples, dashed lines represent the (poor) fits of single exponential terms to each set of data. The calculated values of the individual monomer, dimer, and tetramer distributions for each sample are also presented. It is evident that the M⇔D⇔T model was adequate to fit the data sets. The fitted M, D, and T distributions reveal the effects of ligand modulation of enzyme oligomerization. CKS alone contains substantial amounts of monomer, tetramer, and dimer. Addition of CTP to CKS substantially increases the relative amounts of tetramer and monomer at the expense of dimer. Addition of KDOI to CKS substantially increases the level of tetramer at the expense of both monomer and dimer. Addition of both CTP and KDOI to CKS increases the level of dimer at the expense of monomer and tetramer. In the absence of added ligands the M⇔D K_d was estimated to be ~16.4 mM and the D⇔T K_d was estimated to be ~1.3 mM. In the presence of a substrate, CTP, the enzyme exhibited much weaker M⇔D interactions and much stronger D⇔T interactions. In the presence of an analog of the eight carbon sugar substrate, KDOI, M⇔D interactions were strengthened and D⇔T interactions were essentially unaltered. M⇔D interactions were strengthened in the presence of both CTP and KDOI and D⇔T interactions were weakened. The data indicate that the enzyme, whose quaternary structure was previously uncharacterized, exhibits a complex pattern of oligomerization behavior in response to added substrate/inhibitor. The enhanced formation of dimer in the presence of both CTP and KDOI, at the apparent expense of Monomer and Tetramer, suggested the active form of the enzyme is a homodimer of active monomers. The three-dimensional structure of this protein has recently been solved at Abbott (C. Park and C. Abad-Zapatero, unpublished data). The crystal form of the protein has two monomers per asymmetric unit and is consistent with the enzyme comprising a homodimer of active monomers.

Sedimentation Velocity Analysis

β-Amyloid Aggregation. Alzheimer's disease (AD) is one of the most frequent causes of dementia in the elderly. It affects about 10% of individuals over age of 65 in most developed countries (Evans *et al.,* 1989). One of the hallmarks of the disease is the development of extra-cellular deposits of neuritic protein plaques which are composed primarily of the 39-42 residue A4 peptide

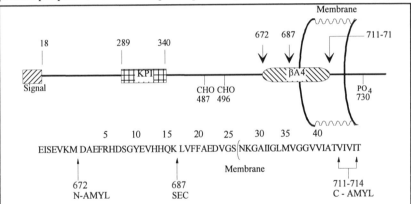

Figure 5. Schematic diagram of human APP containing the ßA4 peptide. N-Amyl=amyloidogenic cleavage of APP to give the N-terminus of ßA4 (DAEF...), C-Amyl = amyloidogenic cleavage of APP to give the C-terminus of ßA4, SEC = secretory cleavage of APP which precludes amyloid formation. The soluble form of amyloid has also been characterized as a protein termed proteinase nexin II and contains a kunitz proteinase inhibitor domain (KPI).

(ßA4 or ß-amyloid) (Glenner and Wong, 1984; Kang *et al.,* 1987). Recent evidence suggests that deposition of the A4 peptide and formation of the neuritic plaques may play a primary pathologic role in the disease (Chartier-Harlin *et al.,* 1991; Kowall *et al.,* 1991; Yanker *et al.,* 1990). The A4 peptide (Fig. 5) is derived from a 110-135 kD integral membrane protein *amyloid precursor protein* (ßAPP) (Goldgaber *et al.,* 1987; Kang *et al.,* 1987). At least six forms of ßAPP are generated by alternative mRNA splicing of the ßAPP gene. Four of them, designated ßAPP770, ßAPP751, ßAPP711 and ßAPP695, contain the 39-42 residue A4 peptide. Amyloid formation occurs by proteolytic processing of one or more forms of βAPP at a site within a membrane spanning region near the carboxyl end of βAPP and at a site N-terminal to the membrane spanning region. Recent evidence links an aggregated form of ßA4 to cellular toxicity in tissue culture (Cotman *et al.*, 1993; Giordano *et al.*, 1994). If a particular aggregated form of ß-amyloid ßA4 mediates toxicity then it becomes of paramount interest to both define the total range of aggregated states amyloid can occupy and the *particular* aggregated form which is toxic.

We have recently begun to characterize the kinetics of aggregation of ßA4 and the hydrodynamic properties of the aggregates (Snyder *et al.*, 1993a,b). At present this information has been used to prepare defined aggregates of amyloid to begin investigation of the mechanisms by which amyloid toxicity is mediated (Giordano *et al.*, 1994). Sedimentation velocity analysis of highly

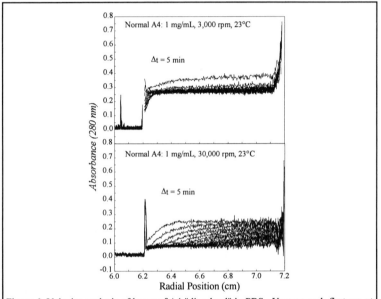

Figure 6. Velocity analysis of human ßA4 "dissolved" in PBS. Upper panel: first run at 3,000 rpm sediments very large species (s~50,000); Lower panel: a jump to 30,000 rpm for the same sample sediments soluble aggregates (s~30). The absorbance not sedimenting at 30,000 rpm was shown to be monomeric.

purified synthetic ßA4 1-40 (Fig. 5) prepared by simple "dissolution" in phosphate-buffered-saline (Snyder *et al.*, 1993b) reveals the presence of three discrete classes of peptide (Fig. 6). First, very large species were observed, s ≅ 50,000, which represent either large undissolved aggregates or large aggregates formed upon attempted dissolution (Fig. 6). Second, species were observed, s ≅ 33, which were *soluble* aggregates (Figs. 6,7). Last, after the larger aggregated forms were centrifuged to the base of the cell, monomeric ßA4 was observed in solution at equilibrium (Fig. 6). In Figure 8, a sample of the "soluble" A4 is analyzed as described by Fujita in Eqn. (12). If the amyloid peptide aggregate sediments as a single solute species it should exhibit behavior characteristic of a single sedimenting boundary. Further, using iterative analysis methods it is

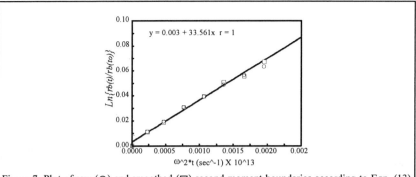

Figure 7. Plot of raw (O) and smoothed (□) second-moment boundaries according to Eqn. (13) from velocity data in Fig. 6, lower panel.

possible to fit both sedimentation and diffusion coefficients to each absorbance profile.

Fitting the data in Figure 8 gives the predicted boundaries presented in the figure. The time-dependent evolution in fitted values of s and D for each fitted boundary are also presented. At early times in the velocity run the single-component fitted-boundaries give a poor representation of the actual data. Later in the run, however, the fitted boundaries more closely approximate the observed data. This "improvement" in fit also is reflected in the behavior of the computed values of s and D (Fig. 8). The increase in these values, at early times after attaining 30,000 rpm, followed by the plateau in each value, indicates the

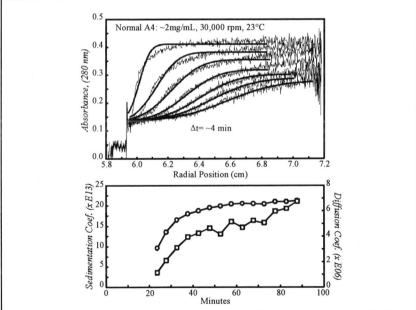

Figure 8. Velocity analysis of human βA4 "dissolved" in PBS. Upper panel: run at 30,000 rpm after ~20 min at 3,000 rpm to sediment *large* aggregates. Solid lines through each sedimenting boundary represent the fit according to Eqn. (12) for a single solute. Lower panel: time evolution of fitted values of s (O) and D (□).

sample of A4 is heterogeneous. That is, the A4 must comprise at least two components one of which sediments faster than the other.

The fit to a two-component sedimentation velocity model is demonstrated for one boundary scan in Figure 9. In this analysis the two-component fit to the sedimenting boundary comprised a linear sum of two terms, each identical to Eqn. (12). The two component fit gave one range of species with an s of 16×10^{-13} sec and a D of 1.8×10^{-6} cm^2/sec and a second with an s of 34×10^{-13} sec and a D of 11×10^{-6} cm^2/sec. Further analyses (not shown) suggest the two component fit is sufficient to describe the sedimentation velocity behavior of the "soluble" aggregated A4 under these particular conditions in PBS.

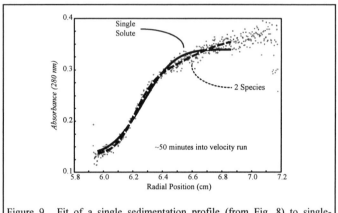

Figure 9. Fit of a single sedimentation profile (from Fig. 8) to single-component and two-component behavior according to Eqn. (12).

In order to study the amyloid aggregation processes we have examined (Snyder et al., 1993b) the common experimental solution conditions employed by researchers in studying the effects of amyloid on cells in tissue culture or as an isolated species. Analysis of the sedimentation behavior of various amyloid forms (Fig. 5) including 1→40, 1→42, 1→28, 25→35, and the reverse sequence 40→1, indicates that only when dissolved in DMSO were the peptides 1→40, 1→42, and 40→1 truly monomeric (Snyder et al., 1993b). In DMSO these peptide were soluble as monomers up to ~30 mg/mL (5-6 mM) (Snyder et al., 1993b). Under other commonly used solution conditions including acidic H$_2$O or 0.1% trifluoroacetic acid with 30% acetonitrile there were substantial amounts of aggregated peptide, even in those samples which were apparently optically clear. Shorter peptides, derived from the N-terminal or C-terminal regions of ß-amyloid were monomeric in all aqueous solvents examined. The aggregation of the longer peptides in aqueous solution is a molecular assembly process where final macromolecular form is driven by peptide sequence. The end products of this process for two different amyloid forms are illustrated in Figures 10 and 11 for 1→40 and 1→42. The simple difference in sequence (Fig. 5), i.e.: the addition of Ile-Ala to the C-terminus of 1→40, gives rise to a

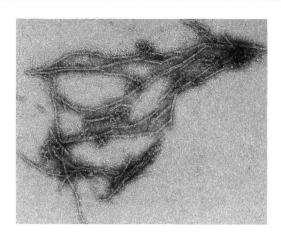

Figure 10. Electron micrograph of aggregated synthetic human ßA4 1→40 at ~35,000x. Analysis suggests three apparent forms: 1) small globules, 2) long fibrils, 3) long fibrils with "curly" ends. The fibrillar form appears to be composed of two strands and may have a long-period twist along the fibril axis length. The diameter of each strand is approximately the length of the ß-amyloid molecule (80-150 °A).

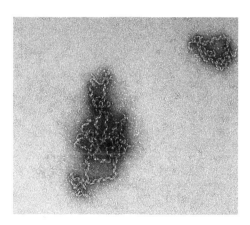

Figure 11. Electron micrograph of aggregated synthetic human ßA4 1→42 at ~35,000x. Analysis suggests two apparent forms: 1) small globules, 2) long "curly" fibrils. The curled fibrillar form has a readily apparent helical twist with a short-period repeat 2-4x the diameter of the fibril coil forming the helix. It is not known if each coiled strand is comprised of several sub-strands.

dramatic change in the self-assembled macromolecular form of the aggregated amyloid. By comparison, the reverse sequence, 40→1, also aggregates, but into small amorphous "globules" rather than discrete fibrillar forms (data not shown). The nature of these self-assembled forms and the kinetics of their assembly is under investigation as a function of sequence and solution conditions.

In summary, an understanding of the ß-amyloid aggregation process may make it possible to alter aggregation behavior. This could be accomplished through the use of small molecules which selectively block the aggregation process or the particular toxic structural feature which amyloid aggregate presents to the cell.

ACKNOWLEDGEMENTS

Individuals contributing to the samples described here include: R. Gampe and S. Fesik from the Protein NMR group, and G. Wang and G. Krafft from the Probe Design group. The electron micrographs of aggregated ß-amyloid were taken by Jennifer Holly and Dr. Robert Josephs of the Department of Molecular Genetics and Cell Biology at The University of Chicago. A special thanks to Thomas J. Perun and Jack Henkin for their support of these techniques at Abbott. S. Snyder was supported by NIH-NIA AG10481-01.

GLOSSARY OF SYMBOLS

A/D = analog-to-digital

A_{31}, A_{32}, ... etc. = pre-exponential coefficients the values of which are determined through fitting

C_r = concentration at radial position r

C_m = concentration at meniscus

C_0 = equation constant defined in Eqn (1)

R = gas constant

T = temperature, K

P_{r2} = Total optical density at radial position r^2

V = volts

μV = microV

$\overline{M}w = M_W$ = weight-average molecular weight

Mw,r^2 = weight-average molecular weight at a particular position r^2

$\overline{v}$ = solute partial specific volume (cm^3/g)

ρ = solvent density g/cm^3)

REFERENCES

Chartier-Harlin, M-C., Crawford, F., Houlden, H., Warren, A., Hughes, D., Fidani, L., Goate, A. M., Rossor, M., Roques, P., Hardy, J., and Mullan, M. (1991) *Nature 353*, 844-846.

Crepeau, R.H., Hensley, C.P., and Edelstein, S.J. (1974) Biochemistry 13, 4860-4865.

Edalji, R.P., Pilot-Matias, T., Pratt, S.D., Egan, D.A., Severin, J., Gubbins, E.J., Petros, A., Fesik, S.W., Burres, N.J., and Holzman, T.F., (1992) *J. Prot. Chem. 11*, 213-223.

Edelstein, S.J., Rehmar, M.J., Olson, and J.S., Gibson, Q.H. (1970) *J. Biol. Chem. 245*, 4372-4381.

Egan, D.A., Logan, T.M., Liang, H., Matyoshi, E., Fesik, S.W., and Holzman, T.F. (1993) *Biochemistry 32*, 1920-1927.

Evans, D. A., Funkenstein, H. H., Albert, M. S., Scherr, P. A, Cook, N. R., Chown, M. J., Hebert, L. E., Hennekens, C. H., and Taylor, J. O. (1989) *J. Amer. Med. Assoc. 262*, 2551-2556.

Fesik, S., Gampe, R., Holzman, T.F., Egan, D.A., Edalji, R., Luly, J.R., Simmer, R., Helfrich, R., Kishore, V., and Rich, D. (1990) *Science 250*, 1406-1409.

Fesik, S.W., Gampe, R.T., Eaton, H.L., Gemmecker, G., Olejniczak, E.T., Neri, P., Holzman, T.F., Edalji, R., Simmer, R., Helfrich, R., Hochlowski, J., and Jackson, M. (1991) *Biochemistry 30*, 6574-6583.

Fujita,H. (1962) in "*Mathematical Theory of Sedimentation Analysis*" Academic Press, NY., pp 64-122.

Glenner, G. G. and Wong, C. W. (1984) *Biochem. Biophys. Res. Commun. 120*, 885-890.

Goldberg, R.J. (1953) *J. Chem. Phys. 57*, 194-202.

Goldman, R.C., Bolling, T.J., Kohlbrenner, W.E., Kim, Y., and Fox, J.L. (1986) *J. Biol. Chem. 261*, 15831-15835.

Goldgaber, D., Lerman, M. I., McBride, W. O., Saffiotti, U., and Gajdusek, C. D. (1987) *Science 235*, 877-880.

Giebeler, R. (1992) in *Analytical Ultracentrifugation in Biochemistry and Polymer Science*, (S.E. Harding, A.J. Rowe, J.C. Horton, Eds.) Roy. Soc. Chem., London, pp16-25.

Giordano, T., Pan, J.B., Monteggia, L.M., Holzman, T.F., Snyder, S.W., Krafft, G. Ghanbari, H., and Kowal, N. W. (1994) *Exptl. Neurol. 125, in press.*

Holzman, T.F., Leytus, S.P., Baldwin, T.O., Mangel, W.F. (1982) *Anal. Bioch. 119*, 62-72.

Holzman, T.F., Egan, D.A., Chung, C.C., Rittenhouse, J., and Turon, M. (1990) *Biophys. J. 57*, 378.

Holzman, T.F., Fesik, S.W., Park, C., and Kofron, J.A. (1991) in *Applications of Enzyme Biotechnology* (Kelly, J.A. and Baldwin, T.O., Eds.) Plenum, pp. 109-128.

Holzman, T.F., Egan, D.A., and Edalji, R. (1992a) *Biophys. J. 61*, 171.

Holzman, T.F., Gampe, R.T., and Fesik, S.W. (1992b) *Biophys. J. 61*, 475.

Holzman, T.F., Gampe, R.T., and Fesik, S.W. (1992c) *Biophys. J. 61*, 478.

Holzman, T.F., Kohlbrenner, W.E., Weigl, D., Rittenhouse, J., and Erickson, J. (1991) *J. Biol. Chem. 266*, 19217-19220.

Kang, J., Lemaire, H-G., Unterbeck, A., Salbaum, J. M., Masters, C. L., Grzeschik, K. H., Multhaup, G., Beyreuther, K., and Muller-Hill, B. (1987) *Nature 325*, 733-736.

Kowall, N. W., Beal, F. M., Busciglio, J., Duffy, L. K., and Yanker, B. A. (1991) *Proc. Natl. Acad. Sci. U. S. A. 88*, 7247-7251.

Meadows, R.P., Nettesheim, D.G., Xu, R.X., Olejniczak, E.T., Petros, A.M., Holzman, T.F., Severin, J., Gubbins, E.G., Smith, H., and Fesik, S.W. (1993) *Biochemistry 32*, 754-765.

Neri, P., Meadows, R., Gemmecker, G., Olejniczak, E., Nettesheim, Logan, T., Simmer, R., Helfrich, R., Holzman, T., Severin, J., and Fesik, S. (1991) *FEBS Lett. 294*, 81-88.

Petros, A.M., Gampe, R.T., Gemmecker, G., Neri, P., Holzman, T.F., Edalji, R.P., Hochlowski, J., Jackson, M., Luly, J.R., Pilot-Matias, T., Prattt, S., and Fesik, S.W. (1991) *J. Med. Chem. 34*, 2925-2928.

Pike, C.J., Burdick, D., Walencewicz, A.J., Glabe, C.G., and Cotman, C.W. (1993) *Neuroscience 13,* 1676-1687.

Propst, C.L., and Perun, T.J. (1989) in *Computer-Aided Drug Design: Methods and Applications*, (C.L. Propst and T.J Perun, *Eds.*) Marcel Dekker, New York, pp. 1-16.

Snyder, S.W., Ladror, U.S., Wang, G.T., Krafft, G.A., and Holzman, T.F. (1993a) *Biophys. J. 64*, 378.

Snyder, S.W., Ladror, U.S., Wang, G.T., Krafft, G.A., and Holzman, T.F. (1993b) *Biophys. J. 64*, 378.

Thériault, Y., Logan, T.M., Meadows, R., Yu, L., Olejniczak, E.T., Holzman, T.F., Simmer, R.L., and Fesik, S.W. (1993) *Nature 361*, 88-91.

Unger, F.M. (1981) *Adv. Carb. Chem. Biochem. 38*, 323-388.

Yanker, B. A., Duffy, L. K., and Kirschner, D. A. (1990) *Science 250*, 279-282.

THE SEDIMENTATION EQUILIBRIUM ANALYSIS OF POLYSACCHARIDES AND MUCINS:

A GUIDED TOUR OF PROBLEM SOLVING FOR DIFFICULT HETEROGENEOUS SYSTEMS

Stephen E. Harding

INTRODUCTION

For the last 12 years we have been applying hydrodynamic methods to study a particularly difficult heterogeneous class of biological macromolecule - the polysaccharides - which possibly present the ultracentrifuge with one of its biggest challenges (see, e.g. Harding, 1992). In this paper I am going to describe some of the many problems we've encountered in trying to use sedimentation equilibrium to study the size, size distribution and interactions of these molecules and, in particular, a particular class of polysaccharide-protein conjugate known as the "mucopolysaccharides" or "mucins" (Harding, 1989).

Although the vast majority of analytical ultracentrifuge users have no direct interest in these molecules in particular, they might possibly pick up a few tips or clues as to how to apply sedimentation equilibrium procedures to difficult heterogeneous macromolecules in general, of which the polysaccharides are only one small - but nontheless interesting - class.

WHY ARE POLYSACCHARIDES SO DIFFICULT?

A typical polysaccharide, for example a pectin molecule from Mrs. G. Berths laboratory at Potsdam (Fig.1a), can have a poorly defined conformation in solution with a large capacity to trap and entrain surrounding solvent molecules. This results in a high exclusion volume (sometimes >100x in excess of the anhydrous volume) and hence high thermodynamic non-ideality. Another contribution to thermodynamic non-ideality can arise from polyelectrolyte behaviour (Fig. 1b), particularly if a molecule has a high unsuppressed charge in solution, such as a pectin in a low ionic strength solvent. Many "food grade" polysaccharides for example are polyannionic. Mucins are polyanionic because of their sialic acid content. To add to the difficulty, polysaccharide preparations are usually polydisperse (Fig. 1c), that is, they contain species of different molecular weight (either in a "discrete" - viz. paucidisperse - or quasi-continuous sense). Finally, some can have the ability to self-associate in solution (Fig. 1d), such as guar. The net result is that these molecules place a considerable strain on the available ultracentrifuge methodology, and in many cases satisfactory information can only be achieved by combining results with other techniques, notably light scattering, gel permeation chromatography or electron microscopy (see, e.g. Harding et al, 1991).

BASIC INFORMATION SOUGHT

Leaving aside sedimentation velocity and analytical density gradient analyses, what is the sort of basic information we're after using sedimentation equilibrium analysis on these molecules? Clearly its in terms of molecular weight analysis in the form of either molecular weight averages (both "point" and "whole cell"), molecular weight distributions and, if the system performs in complexation phenomena or self-association reactions, in terms of stoichiometries. But there are a number of problems we have to overcome or "troubleshoot".

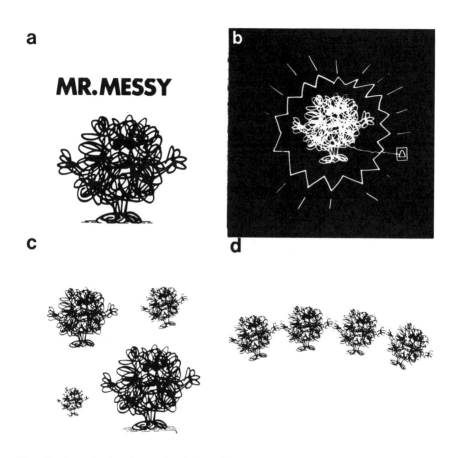

Fig. 1. A typical polysaccharide and its properties.

THE FIRST PROBLEM: LIMITED CHOICE OF OPTICAL SYSTEM

Most polysaccharides do not contain a chromophore in the visible or near (i.e. "useable") part of the ultraviolet, so in general we cannot use the absorption optical system, but must rely instead on either of two refractometric methods: schlieren optics or Rayleigh interference. The latter is preferable because of its greater sensitivity at lower concentration, but the optical record is one of solute concentration *relative* to the meniscus, rather than absolute concentration directly. For simple experiments on proteins, a popular way of avoiding this problem is to use the meniscus depletion method (Yphantis, 1964) where the ultracentrifuge is run at a high enough speed so the meniscus is effectively depleted of macromolecular solute - the optical record is then one of absolute concentration - in fringe number or weight terms - versus radius.

It is worth pointing out also that, in the future, application of the absorption optical system *may* be possible: the new XLA ultracentrifuge from Beckman instruments (Giebler, 1992) appears to have stable optics in the far ultraviolet (210-230nm). All mucins and many polysaccharides absorb significantly in this region and we are exploring the use of "far-uv" detection, with appropriate baselines, for both sedimentation equilibrium and sedimentation velocity.

THE SECOND PROBLEM: INAPPLICABILITY OF THE "HIGH SPEED" OR "MENISCUS DEPLETION" METHOD

Fig. 2 illustrates the next problem any sedimentation equilibrium analysis on these materials needs to overcome: that is we cannot use the "meniscus depletion" (see, e.g., Yphantis, 1964), method, widely used in protein biochemistry. Fig. 2 in fact shows the solution Rayleigh fringes from a low-speed sedimentation equilibrium experiment on a mucin of weight average molecular weight ~6 million, and at low loading concentration (~0.4 mg/ml), run in three separate solvents Because of the polydispersity of these materials it is generally impossible (except in cases of pseudo-non-ideality) to choose run conditions to get proper meniscus depletion without losing optical registration of the fringes near the cell base: one can observe clearly in Fig. 2 the steep rising fringes at the cell base but finite slope of the fringes near the meniscus.

There is a further problem with the high speed meniscus depletion method in terms of a speed dependent enhancement of the effective thermodynamic second virial coefficient, B_{eff}, as the following equation shows (Fujita, 1975)

$$B_{eff} = B\{1 + \frac{\lambda^2 M_z^2}{12} + \ldots\} \qquad (1)$$

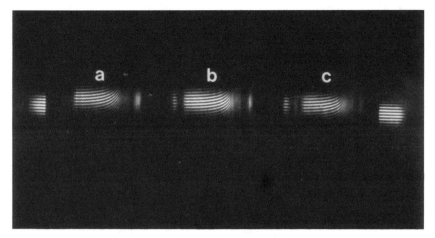

Fig. 2. Rayleigh equilibrium interference patterns for a mucin (bronchial mucin BM GRE) in three different solvents (a) a phosphate/chloride buffer containing 0.4M CsCl; (b) a phosphate/chloride buffer containing 5mg/ml fucose; (c) a phosphate/chloride buffer containing 5 mg/ml N-acetylglucosamine. The initial mucin cell loading concentration in each case was ~0.4mg/ml (30mm path length cell). The rotor speed was 1967 rev/min. From Harding (1984).

where λ is a function of the square of the rotor speed $\{\lambda = (1-v\rho_0)\omega^2(b^2-a^2)/2RT$, $\bar{v}$ being the partial specific volume, ρ_0 the solvent density, ω the angular velocity and a and b the radial positions at the cell meniscus and base respectively$\}$. The collective result is that with polysaccharides we have to use the low or intermediate speed method with the requirement of a method for evaluating the concentration at the meniscus, either in terms of g/ml or in terms of fringe numbers. The next problem is thus: how do we get meniscii concentrations out?

EVALUATION OF THE MENISCUS CONCENTRATION

My old mentor J.M. Creeth produced in the late 60's with R. Pain (Creeth & Pain, 1967) a very useful review of all the methods for getting meniscii concentrations {denoted "C_a" in weight concentra-

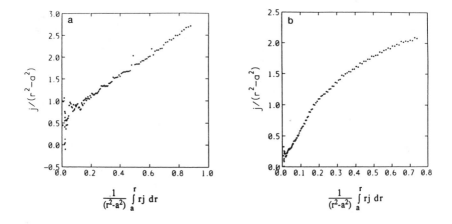

Fig. 3. Extraction of meniscus concentration using the method of ref. 9 for (a) a fairly homogeneous/ideal solution of colonic mucin "T-domains" $\{J_a \sim 0.58 \pm 0.05\}$; (b) a highly non-ideal solution of xanthan ("RD") $\{J_a \sim 0.01 \pm 0.01$, i.e. near depletion conditions$\}$. From Harding et al (1992).

tion terms or, more usually "J_a" in the equivalent fringe number terms where the subscript a represents the radial position at the meniscus} out; however, we found the most useful method was one where we get J_a by some simple graphical manipulation of the basic fringe data (Creeth & Harding, 1982a). We can get J_a usually to an accuracy of a few percent from the ratio of twice the intercept to the

limiting slope of a plot of $j/(r^2-a^2)$ against $\{1/(r^2-a^2)\}\int_a^r rjdr$ (Fig. 3),

where j is the concentration in fringe numbers ("fringe concentration") *relative to the meniscus* and r is the radial displacement. The method usually gives J_a to an accuracy of the order of 0.1 fringe. We find an adapted sliding strip procedure useful for this purpose (Harding et al, 1992), especially where the plots are strongly curved as in Fig.3b. Strong curvature is a symptom of either non-ideality or heterogeneity (the sense of the curve depends on which is the stronger effect)

MULTIPLE DATA CAPTURE AND ANALYSIS

For getting out J_a using this method multiple data collection and averaging is very important for strongly curving systems such as shown in Fig. 3b, and for this purpose the use of automatic multiple data capture and analysis is of extreme value here. We capture our data automatically but not directly *on-line,* as described by for example T. Laue (Laue, 1992) but *off-line*. That is to say we take a photograph and digitise it using a laser densitometer of the sort you can find in many Biochemistry departments (Fig. 4) - these things are normally used for scanning SDS gels, and we use a simple Fourier cosine series algorithm to average over the fringe data set to give our sedimentation equilibrium concentration distribution, to an accuracy comparable with the on-line set-up that T. Laue has described (see, e.g. Rowe et al, 1992).

EXTRACTION OF AVERAGE MOLECULAR WEIGHTS

Whole cell weight average

The next problem is in getting molecular weights out - even getting whole cell or whole distribution weight average values can be much more tricky compared with simple well behaved protein systems. Plots of the logarithm of the concentration against radial displacement squared are often strongly curved because of either heterogeneity (Fig. 5a) or non-ideality (unless by a lucky coincidence the results of the two effects cancel to give a "pseudo-ideal" profile, as in Fig. 5b) Whatever, to get the whole distribution weight average you need to estimate the concentration or the logarithm of the concentration not only at the meniscus, but also *at the cell base*, and this can be very tricky, especially in the case of strong curvature (e.g. Fig.5a) or if the base is not well defined. One way of minimising this problem is to use a function known as "M*" defined by (Creeth & Harding, 1982a)

$$M^*(r) = \frac{j}{kJ_a(r^2 - a^2) + 2k \int_a^r rj dr} \qquad (2)$$

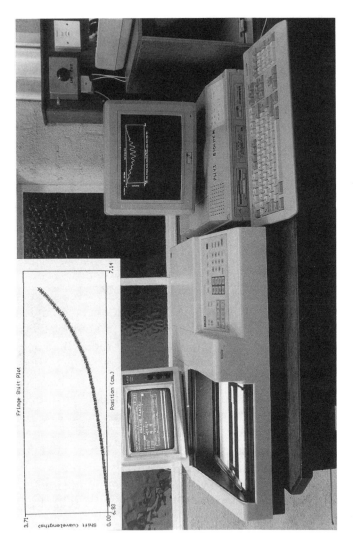

Fig. 4. LKB (Bromma, Sweden) Laser Densitometer set up at Nottingham. This is used to capture (into the Amstrad PC) automatically off-line from photographic film our Rayleigh interference data. The Fourier cosine series algorithm "ANALYSER" produces a 100-200 pt concentration versus distance dataset {as shown in the inset for a low speed sedimentation equilibrium experiment on a commercial guar sample }which is transferred to the FORTRAN programme MSTAR (Harding et al, 1992) on the mainframe IBM 3081/Q for full molecular weight analysis.

where a and b are the radial positions at the solution meniscus and cell base respectively and k is the usual constant (see e.g., Harding, 1992) in terms of rotor speed, ω, solvent density,ρ_0, (the solvent density should be used, not the solution density - see e.g., ref. 1) and partial specific volume v:

$$k = \frac{(1 - \bar{v}\rho_0)\omega^2}{2\,RT} \tag{3}$$

J.M. Creeth and myself (Creeth & Harding, 1982a) found M* defined in such a way to have some useful properties: the most important of these is that its value extrapolated to the cell base equals the weight average over the whole solute distribution "$M^o_{w,app}$", {where the "o" signifies its over the whole cell and the "app" signifies an apparent value at a finite cell loading concentration} and as we found by extensive simulations (Harding, 1992; Creeth & Harding, 1982a) for five different types of systems the M* method appeared to represent a considerable improvement for extracting $M^o_{w,app}$ compared to the conventional "average slope" or log concentration extrapolation methods (Creeth & Pain, 1967).

Point weight average molecular weights ("$M_{w,app}$")

These are relatively straightforward to produce from the fringe concentration data as $d\ln J/dr^2$ times$(1/k)$, (where $J=j + J_a$ and k is as defined in eq. (3)), but again this depends on a reasonable estimate for J_a. J (i.e. the difference in fringe concentration between the meniscus and cell base) needs to be at least 4 fringes for the $M_{w,app}$ data to be reliable without heavy smoothing. We use sliding strip procedures along the lines discussed by Teller (1965, 1973), with an 11pt sliding strip for a total data set of 100-200 radial positions.

Number and z- whole cell averages

The number point average molecular weight $M_{n,app}$ you can, in my opinion, forget about for these substances (unless, for pseudo-ideal systems, you can use the meniscus depletion method): besides J_a you also need to estimate $M_n(a)$ (Teller, 1965, 1973): the same problem applies to the whole-cell number average. The situation is not quite so bad for z-averages. If Rayleigh optics are used the point z-average is independent of errors in J_a, although it depends on the ratio of a double differential to a single differential viz, data of very high precision is necessary; the whole cell z-average requires accurate estimates of not only J_a and J_b but also the point weight averages at the meniscus and base (see e.g., Teller, 1973; Harding et al, 1992) If reliable z-averages are required then Schlieren optics should be used, which yield $M^o_{z,app}$ and $M_{z,app}$ directly via the Lamm equation (see, e.g., Creeth & Pain, 1967), and at concentrations now claimed as low (using the Fresnel fringes - Rowe et al, 1992) as can be achieved from Rayleigh fringes (say ~0.3 mg/ml): this would therefore be my method of choice. Indeed, at Nottingham, although we have two Model E's with laser light sources dedicated to Rayleigh optics, we have another dedicated to the Schlieren system producing M_z information.

THE NON-IDEALITY PROBLEM

Focussing on the problem of polysaccharide non-ideality, the first feature which I'm sure many are aware of is that the symptoms of heterogeneity, which tend to produce upward curvature in the log concentration. versus distance squared plots -as in Fig. 5a - can sometimes apparently cancel the symptoms of non-ideality which produce downward curvature to give a pseudo-ideal monodisperse linear plot - as shown in Fig. 5b for a mucin -which can be very misleading. The "pseudo-ideal" system corresponding to Fig. 5b is in fact is both very polydisprese and very non-ideal, and this feature illustrates a statement made long ago by D.Teller (1965) that a linear plot of log c versus distance squared from a single sedimentation equilibrium expt. is not by itself sufficient evidence for monodispersity or ideality.

And, as the Gilberts (G. A. Gilbert & L.M. Gilbert, 1980) and others have shown, since a single symmetric boundary from a sedimentation velocity experiment is also insufficient criterion for solute homogeneity, to this end J.M. Creeth and myself developed a simple assay for solute homogeneity using two cells of different optical path

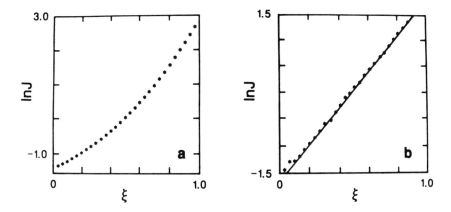

Fig. 5. Log concentration versus distance squared data evaluated from Rayleigh equilibrium optical records for 2 mucins (a) bronchial mucin (from a chronic bronchitis patient) BM GRE (b) bronchial mucin (from a cystic fibrosis patient) CF PHI. ξ is a normalised radial displacement squared parameter, normalised so it has a value of 0 at the meniscus and 1 at the cell base: ξ = {r²-a²}/{b²-a²}. Note the strong curvature of (a) and the pseudo-ideality of (b). {see, e.g. Harding et al, 1992}

length in a multihole rotor (Creeth & Harding, 1982b). The idea is to compare solution interference fringes of the same initial loading concentration if expressed on a *fringe number* basis but different loading concentrations expressed on a *weight* basis. This test can be performed in a single sedimentation equilibrium experiement if a multi-hole rotor is used together with a suitable combination of two cells (double sector, one with wedge window if the instrument does not have a multiplexing system, or multichannel). One cell or pair of interference channels has say a 30mm optical path length, the other only a 12mm path length but 2 and a half times higher weight concentration to compensate: fringe patterns, and corresponding average molecular weights will be identical *only* for a homogeneous ideal system, and not otherwise. In this way the mucopolysaccharide CFPHI, which *appeared* homogeneous and ideal from the linear log concentration versus distance squared plot of Fig. 5b can clearly be shown to be otherwise (Fig. 6). These observations clearly illustrate the dangers of inferring solute homogeneity or ideality from

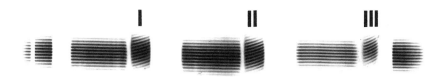

Fig. 6. Sedimentation equilibrium sample homogeneity test. Comparison of Rayleigh equilibrium patterns of the same initial loading concentration on a fringe number basis but differing initial loading concentration on a weight (mg/ml) basis, c^o. Patterns I and II: $c^o = 3.0$ mg/ml, 12mm path length multi-channel cell (inner two pairs of channels used, outer pair masked) pattern III: $c^o = 1.2$ mg/ml, 30 mm path-length multi-channel cell (outer pair of channels used, the inner two pairs masked off). Note the different curvature of III confirming the system is neither heterogeneous nor ideal. The identical nature of I and II confirms that the differing radial positions of each does not affect their concentration distribution. The solution column of II appears shorter because (i) the apparent width of the meniscus is much greater in the cell of longer optical path length and (ii) the fringes at the cell base are steeper and accordingly less intense and so are partially lost on photographic reduction (from Creeth & Harding, 1982b).

a single log concentration versus distance squared plot. Some (non-polysaccharide) systems do "pass the test" however, such as a dilute solution of the small virus TYMV (Harding & Johnson, 1985).

Moving back to the non-ideality problem, this can be very severe for polysaccharides as Table 1 shows. The interesting column is the "1 + 2BMc" one and represents the factor by which an apparent molecular weight, measured at a concentration as low as 0.2 mg/ml, underestimates the true or infinite dilution value. For many of course, the effect is not too bad - less than a few percent, but for some like alginates, serious error can result - an underestimate of over 40% for example. And for cases like these its necessary to measure the apparent molecular weight -whether it be weight or z-average, at a number of concentrations and extrapolate to zero in the standard way (see, e.g. Harding, Rowe & Creeth, 1983). In extreme cases - such

TABLE 1.   Comparative non-ideality of polysaccharides

	$10^{-6} \times M$ g/mol	$10^4 \times B$ ml.mol/g^2	BM ml/g	$^a 1+2BMc$	Ref.
Pullulan P5	0.0053	10.3	5.5	1.002	b
Pullulan P50	0.047	5.5	25.9	1.010	c
Xanthan (fraction)	0.36	2.4	86	1.035	b
β–glucan	0.17	6.1	104	1.042	d
Chitosan (KN-50-1)	0.064	1.7	109	1.044	e
Dextran T500	0.42	3.4	143	1.057	f
Pullulan P800	0.76	2.3	175	1.070	b
Chitosan (Protan 203)	0.44	5.1	224	1.090	g
Pullulan P1200	1.24	2.2	273	1.109	b
Bronchial mucin CFPHI	2.0	1.5	300	1.120	h
Chitosan (Protan SeaCure)	0.16	27.5	445.5	1.178	e
Pectin (citrus fraction)	0.045	50.0	450	1.180	i
Scleroglucan	5.7	0.50	570	1.228	j
Alginate	0.35	29.0	1015	1.406	k

a: At a loading concentration, c of 0.2 mg/ml; b: Kawahara et al (1984);   c: Sato et al (1984);   d: Woodward et al (1983);   e: Errington et al (1993); f: Edmond et al (1968); g: Muzzarelli et al (1987); h: Harding et al (1983); i: Berth et al (1990); j: Lecacheux et al (1986); k: Horton et al (1991)

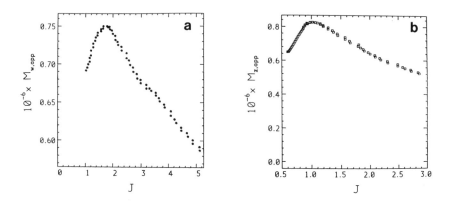

Fig. 7. Plots of (apparent) point M_w, M_z versus fringe number concentration J from a low speed sedimentation equilibrium experiment on a purified guar preparation. Rotor speed = 5200 rev/min, loading concentration $c^o \sim 0.7$ mg/ml. Rayleigh interference optics. From Jumel et al (1993).

as some alginate preparations, even under conditions of ionic strength where polyelectrolyte behaviour should be largely suppressed, the two virial coefficients (1/M and B) are not sufficient to account for the concentration behaviour, even under dilute solution conditions (Horton et al, 1991).

Non-ideality also reveals itself in *point average* representations of the data as shown in Fig. 7 for dilute solutions of a neutral but tricky polysaccharide know as guar gum. For many polysaccharides like this we observe a maximum in the point average versus fringe concentration data, both in terms of weight average (Fig. 7a) and in terms of the point z-average (Fig. 7b) and the existence of these maxima is symptomatic of a heterogeneous but highly non-ideal system, with the effects of polydispersity dominating at low radial positions and non-ideality effects dominating at the higher positions.

POLYDISPERSITY OR SELF-ASSOCIATION OR BOTH?

Fig. 8 (Creeth & Cooper, 1984) shows another point weight average versus concentration plot for a mucopolysaccharide, again with a

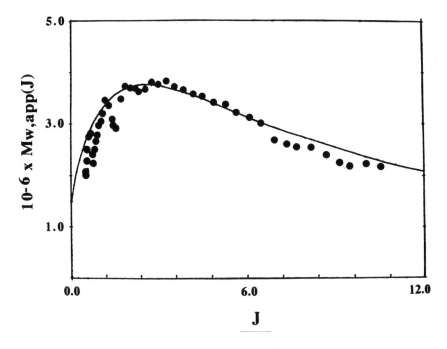

Fig. 8. Plot of (apparent) point M_w versus fringe concentration (with respect to a *12mm* optical path length cell) from a low speed sedimentation equilibrium experiment on bronchial mucin BM GRE (loading concentration, $c^o = 1.0$mg/ml, solution column length = 3mm)   The line fitted corresponds to an effective non-ideal isodesmic self-association with monomer molecular weight $M_1 = 1.5 \times 10^6$ g/mol,   second virial coefficient, B = 0.033×10^{-6} mol.ml^{-1}.fringe^{-1}, k=1.2 fringe^{-1} (see, e.g. Creeth & Cooper, 1984, where similar fits are given but with respect to a 30mm path length cell).

maximum, again characteristic of a heterogeneous but highly non-ideal system. The question of interest is what is the prime source of the heterogeneity? Is it because of the presence of components of different molecular weight that are not-interacting (i..e not in chemical equilibrium) , a phenomenon we call "polydispersity", or is self-association behaviour the main contributor?: Many polysaccharides, such as guar and mucopolysaccharides are thought to self-associate, and there is now increasing interest in the food and pharmaceutical fields of possible interactions in mixed polysaccharide systems

(Mannion et al, 1992). Both effects produce upward curvature (i.e. positive second differential) in plots of log concentration versus distance squared or point average molecular weight versus concentration. Initially when we worked on mucopolysaccharides or mucins we thought that the prime cause was self association (Harding & Creeth, 1982). Indeed, we get an excellent fit to the observed point average data if as a first approximation we ignore polydispersity and assume a self-association {Creeth & Cooper (1984) give two good examples example and another is given in Fig. 8 for the bronchial mucopolysaccharide "BM GRE"}. The fit given in Fig. 8 corresponds to a non-ideal isodesmic (i.e. each monomer is added on with constant free energy increment) self-association with plausible values for the isodesmic association constant, k, the "monomer" molecular weight, M_1 and the second virial coefficient B. We can also model the log concentration versus distance squared data directly {the line fitted in Fig. 5b for the mucin "CFPHI" corresponds to an indefinite isodesmic asociation with $M_1 = 2.15 \times 10^6$, k = 260 ml/g and B = 1.5×10^{-4} ml.mol.g^{-2} }. Despite the good fits, we know from other measurements that both BM GRE and CF PHI *are not interacting at all,* illustrating another pitfall we can easily fall into, viz. the effects of polydispersity of these types of system cannot be ignored. There are diagnostic procedures available to assay whether polydispersity effects *can* be ignored. D. Roark & D. Yphantis have shown (Roark & Yphantis, 1969) that for a purely non-ideal self-associating system plots of point weight average molecular weight versus concentration for differing cell loading concentrations should superimpose. So for a simple self-association such as lysosyme they do (Howlett et al, 1972) but for mucins for example they don't as the three examples in Fig. 9 show. In fact, further experiments comparing fringe profiles and corresponding molecular weights for these molecules in nondissociative and dissociative solvents (in the case of mucins by adding 6M GuHCl or swamping concentrations of fucose, galactose or N-acetyl glucosamine) and observing the lack of any effect of the latter (Fig. 2) (Harding, 1984, 1989) have shown that for mucopolysaccharides in dilute solution, self-association phenomena is negligible - the observed heterogeity is due virtually entirely to heterogeneity of components not in chemical equilibriium - i.e., polydispersity.

DISTRIBUTIONS OF MOLECULAR WEIGHT

The profiles in Figs. 7-9 are not molecular weight distributions, and in trying to get this sort of information, once again the main stumbling block we have to overcome is that of thermodynamic non-ideality. There are four possible routes open to us here (Table 2).

Method I. Polydispersity indices   The simplest way is by using the ratios of whole cell averages, or polydispersity indices, which is OK so long as you can measure numbers or z-averages to a reasonable precision *after* correction for non-ideality. Number averages can only be extracted with ease from Rayleigh patterns if the high speed method is used (Yphantis, 1964) - thereby effectively ruling them out for polysaccharides for the reasons given above. Z-averages can be obtained with relatively high precision using Schlieren optics and now at concentrations as low as can be obtained using

TABLE 2.   Molecular weight distribution analysis by low speed sedimentation equilibrium

Method	Type of Analyis	Ref
I	Polydispersity indices (M_Z/M_W etc)	a
II	Non-ideal-polydisperse modelling of log concentration versus distance squared data	b,c
III	Equivalent self-association fit	d
IV	Off-line coupling to gel permeation chromatography	e

a: Herdan (1949); b: Harding (1985); c: Lechner (1992); d: Creeth & Cooper, 1984; e: Harding et al (1988)

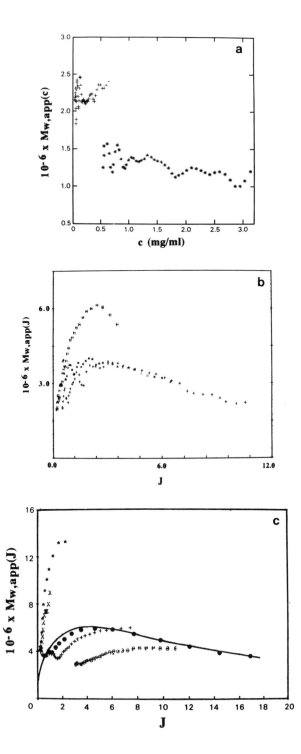

Rayleigh optics (Rowe et al, 1992) {or from Rayleigh records themselves but at much lower precision - see above}. These ratios can either be used directly, as so-called "polydispersity indices", or related to the standard deviation of a distribution (whatever form this may take) via special relations known as the "Herdan relations" (Herdan, 1949 - see also Harding, 1989 and Creeth & Pain, 1967).

Method II. Non-ideal-polydisperse modelling of log concentration versus distance squared data   The more direct way is to model directly the log concentration versus radial displacement plots by fitting the parameters of a non-ideal polydisperse system. Although this method came out in the Biophysical Journal over 8 years ago now (Harding, 1985), because of the particularly complex interdependent nature of the non-linear equations involved *for the low speed case* - largely caused by the non-ideality term, it takes a great toll on computer resources, even on the fastest computers around such as the IBM 3081 at Cambridge. So, we've presently been unable to apply it to quasi-continuous distributions of molecular weight, that are, for example, the hallmark of polysaccharides, but nontheless succesfully applied to discrete distributions of molecular weight,

---

Fig. 9. Sedimentation Equilibrium Polydispersity assay. Different loading concentration "non-overlap" plots of point weight average apparent M_w versus concentration for three mucins:

(a) Bronchial mucin CF PHI +: c^o (initial loading concentration in mg/ml) ~0.2 mg/ml, 30mm cell; *: c^o~2.0 mg/ml, 12 mm cell Both with 3mm solution columns. From Harding (1984).

(b) Bronchial mucin BMGRE ⊙:c^o~0.4 mg/ml, 30mm cell (fringe concentrations corrected to the equivalent values in a 12mm cell) *: c^o~0.7 mg/ml, 12 mm cell, +: c^o~1.0 mg/ml, 12mm cell. All three data sets correspond to 3mm solution columns. Adapted from Creeth & Cooper, 1984.

(c) Pig gastric mucin *: J^o (initial loading concentration in fringe numbers) 0.32, column length 3mm; x: J^o~0.42, 1.5mm; +: J^o~3.19, 1.5mm; ●:J^o~4.03, 3mm; : J^o~6.00, 1.5mm. All three data sets correspond to a 30mm optical path length cell. The line fitted to data set in (c) corresponds to a an effective non-ideal isodesmic association (see legend to Fig. 8), with $M_1 = 1.5$ x 10^6, B = 0.013 x 10^{-6} mol.ml^{-1}fringe^{-1} and k= 2.1 fringe^{-1} where the fringe concentration units this time refer to a 30mm path length cell. Adapted from Creeth & Cooper, 1984.

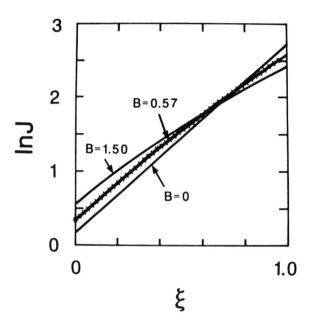

Fig. 10. Molecular weight distribution modelling of bronchial mucin CF PHI by Method II. Non-ideal three-component fit to the observed log concentration versus distance squared data for a low speed sedimentation equilibrium experiment on bronchial mucin CF PHI. The line fitted corresponds to the following parameters for each of the three components: component 1, $M_1 = 1.2 \times 10^6$ g/mol, $J^\circ_1 = 0.9$ fringe; component 2, $M_2 = 1.8 \times 10^6$, g/mol, $J^\circ_2 = 3.6$ fringe ; component 3, $M_3 = 2.4 \times 10^6$ g/mol, $J^\circ_3 = 0.9$ fringe. B has been expressed as its value x 10^4 (ml.mol.g^{-2}). The 3 component model is based on observations from platinum-shadowed electron microscopy. From Harding (1985).

which at least partially represent mucins {Harding, 1985, 1989} (Fig. 10). The problems associated with the modelling of the concentration distribution for the non-ideal case have been further examined by Lechner (1992).

Method III. Equivalent self-association fit   A much easier way, although theoretically less elegant than the previous method, is to use to our advantage the property of indistinguishability *from a single experiment* between a non-ideal polydisperse system and a non-ideal

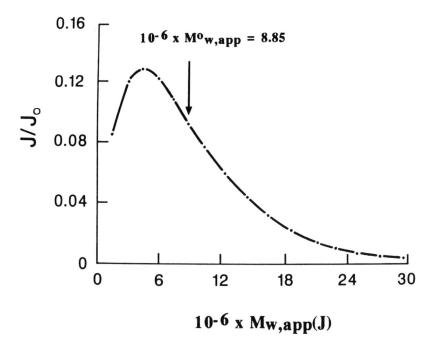

Fig. 11. Molecular weight distribution of pig gastric mucin, evaluated according to Method III. The distribution corresponds to the fit shown in Fig. 9c. The value marked by an arrow corresponds to the weight average (apparent) molecular weight for the whole solute distribution. Adapted from Creeth & Cooper, 1984 (see also Harding, 1989).

self-associating system. It is therefore possible to apply the much easier to handle equations of for example a non-ideal isodesmic association to calculate a constant which, when applied to a static system will define a distribution of molecular weight, no matter what the cause of the distribution is (Creeth & Cooper, 1984; Harding, 1989) and again, this has been successfully applied to mucins (Fig. 11).

Method IV. Off-line coupling to gel permeation chromatography. From a practical point of view, we find the best procedure is to use sedimentation equilibrium in conjunction with gel permeation chromatography (gpc) to provide an absolute calibration for the latter (Harding et al, 1988). The idea is to isolate fractions of narrow bandwidth (in terms of elution volume) from the gpc eluate, deter-

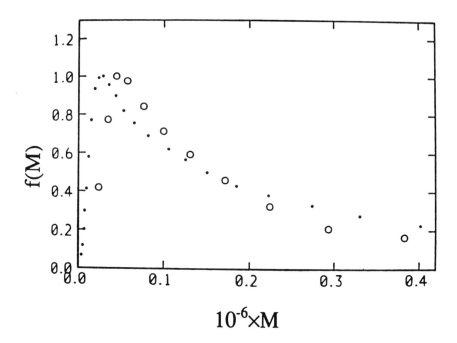

Fig. 12. Molecular weight distribution of citrus pectin, evaluated according to Method IV (open circles). The filled circles corresponds to the distribution on the same material evaluated by classical light scattering procedures coupled to gel permeation chromatography. Method IV is our method
of choice. From Harding et al, 1991.

mine their molecular weights using sedimentation equilibrium using
short solution columns (0.7mm - 1.5mm) and multi-channel cells to
speed things up, thereby giving an absolute calibration for the gpc
columns. Distributions of molecular weight found in this way have
been in remarkable agreement with similar procedures involving light
scattering as shown in Fig. 12. Method IV is in my opinion the method
of choice. It is called "off-line" because the eluate is not fed directly
into the ultracentrifuge cell whilst the ultracentrifuge is running {this
distinguishes itself from certain light scattering photometers which
are *"on-line"* - i.e. directly connected between gpc columns and a
concentration - usually refractive index based - detector}.

THE FINAL PROBLEM: SEDIMENTATION EQUILIBRIUM VERSUS LIGHT SCATTERING

The decline of sedimentation equilibrium and other ultracentrifuge techniques in the 70's and 80's in the protein biochemistry field (largely because of the advent of electrophoretic procedures and gel permeation chromatography) and followed by the revival of interest has been well documented (see e.g., Schachman, 1992). In the polysaccharide and synthetic polymer fields the decline of sedimentation equilibrium as a routine absolute molecular weight tool has been largely because of the advent of laser light scattering techniques (both "classical" and "dynamic"). Probably the bulk of the apparatus and the length of time required to reach equilibrium have contributed to its downfall, but in reality light scattering - although the apparatus is more compact and measurements themselves are a lot quicker suffers far worse disadvantages (Harding, 1988), largely through sample clarification and for this reason it is fair to say that light scattering results have a greater degree of uncertainty than sedimentation equilibrium, when used in isolation. Although sedimentation methods would be my own method of choice, we find particularly for molecular weight distribution work confirmatory measurements from both techniques (and used in conjunction with gel permeation chromatography) extremely valuable (see, e.g. Harding et al, 1991).

GLOSSARY

b   Radial position at the cell base (cm)

B   Second thermodynamic virial coefficient (ml.mol.g$_{-2}$)

B_{eff} Effective 2nd thermodynamic virial coefficient (ml.mol.g^{-2})

C   Solution concentration (g/ml)

C^o Initial loading solution concentration (g/ml)

C_a Solution concentration at the meniscus (g/ml)

j   Solution concentration in fringe numbers relative to the meniscus

J^o	Initial loading concentration (fringe numbers)
J_a	Meniscus solution concentration in fringe numbers
M	Molecular weight (g/mol)
M_1	Molecular weight of a monomer (in self-association)
M^*	Star average (operational) average molecular weight $(g.mol^{-1})$
M^o_n	Whole-cell (i.e., over all radial positions in the solution column of the cell) number average molecular weight $(g.mol^{-1})$
M^o_w	Whole-cell weight average molecular weight $(g.mol^{-1})$
M^o_z	Whole-cell z-average molecular weight $(g.mol^{-1})$
M_n	Point (i.e., at a local radial position in the cell) number average molecular weight $(g.mol^{-1})$
M_w	Point weight average molecular weight $(g.mol^{-1})$
M_z	Point z-average molecular weight $(g.mol^{-1})$
R	Gas constant (8.314×10^7 $erg.mol^{-1}.K^{-1}$)
r	Radial position (cm)
$\bar{v}$	Partial specific volume (ml/g)
T	Temperature (K)
ρ	Solution density (g/ml)
ρ_o	Solvent density (g/ml)
ω	Angular velocity of rotor $(rad.sec^{-1})$
ξ	Normalized radial displacement squared parameter (= 0 at meniscus, 1 at cell base)

REFERENCES

Berth, G., Dautzenberg, H., Lexow, D. and Rother, G. (1990) The determination of the molecular weight distribution of pectins by calibrated gel permeation chromatography Part I: Calibration by light scattering and membrane osmometry. *Carbohyd. Polym.* 12, 39-59

Creeth, J.M. and Cooper, B. (1984) Studies on the molecular weight distributions of two mucins. *Biochem. Soc. Trans.*, 12, 618-621

Creeth, J.M. and Harding, S.E. (1982a) Some observations on a new type of point average molecular weight. *J. Biochem. Biophys. Meth.* 7, 25-34

Creeth, J.M. and Harding, S.E. (1982b) A simple test for macromolecular heterogeneity in the analytical ultracentrifuge. *Biochem. J.*, 205, 639-641

Creeth, J.M. and Pain, R.H. (1967) The determination of molecular weights of biological macromolecules by ultracentrifuge methods. *Prog. Biophys. Mol. Biol.* 17, 217-287

Edmond, E., Farquhar, S., Dunstone, J.R. and Ogston, A.G. (1968) The osmotic behaviour of sephadex and its effects on chromatography. *Biochem. J.*, 108, 755

Errington, N., Harding, S.E., Vårum, K.M. and Illum, L. (1993) Hydrodynamic characterization of chitosans varying in degree of acetylation. *Int. J. Biol. Macromol.* 15, 113-117

Fujita, H. (1975) *"Foundations of Ultracentrifuge Analysis"*, Chap. 5., J. Wiley and Sons, New York

Giebler, R. (1992) The Optima XL-A: A new analytical ultracentrifuge with a novel precision absorption optical system. In *"Analytical Ultracentrifugation in Biochemistry and Polymer Science"* (Harding,. S.E., Rowe, A.J. and Horton, J.C. eds.), Chap.2, Royal Society of Chemistry, Cambridge, UK

Gilbert, G.A. and Gilbert, L.M. (1980) Determination in the ultracentrifuge of protein heterogeneity by computer modelling, illustrated by pyruvate dehydrogenase multienzyme complex. *J. Mol. Biol.* 144, 405-408

Harding, S.E. (1984) An analysis of the heterogeneity of mucins. No evidence for a self association. *Biochem, J.*, 219, 1061-1064

Harding, S.E. (1985) The representation of equilibrium solute distributions for non-ideal polydisperse systems in the analytical ultracentrifuge. Application to mucus glycoproteins. *Biophys. J.* 47, 247-250

Harding, S.E. (1988) Polysaccharide molecular weight determination: which technique? *Gums and Stabilisers for the Food Industry* 4, 15-23

Harding, S.E. (1989) The macrostructure of mucus glycoproteins in solution. *Adv. Carb. Chem. & Biochem.*, 47, 345-381

Harding, S.E. (1992) Sedimentation analysis of polysaccharides. In *"Analytical Ultracentrifugation in Biochemistry and Polymer Science"* (Harding,. S.E., Rowe, A.J. and Horton, J.C. eds.), Chap.27, Royal Society of Chemistry, Cambridge, UK

Harding, S.E., Ball, A. and Mitchell, J.R. (1988) Combined low speed sedimentation equilibrium/ gel permeaion chromatography approach to molecular weight distribution analysis. *Int. J. Biol. Macromol.*, 10, 259-264

Harding, S.E., Berth, G., Ball, A., Mitchell, J.R. and Garcia de la Torre, J. (1991) The molecular weight distribution and conformation of citrus pectins in solution studied by hydrodynamics. *Carbohyd. Polym.* 16, 1-15

Harding, S.E. and Creeth, J.M. (1982) Self-association, polydispersity and non-ideality in a cystic fibrotic glycoprotein. *IRCS (Int. Res. Commun. System) Med. Sci. Lib. Compend.*, 10, 474-475

Harding, S.E., Horton, J.C. and Morgan, P.J. (1992) MSTAR: a FORTRAN program for the model independent molecular weight analysis of macromolecules using low speed or high speed sedimentation equilibrium. In *"Analytical Ultracentrifugation in Biochemistry and Polymer Science"* (Harding,. S.E., Rowe, A.J. and Horton, J.C. eds.), Chap.15, Royal Society of Chemistry, Cambridge, UK

Harding, S.E. and Johnson, P. (1985) Physicochemical studies on turnip yellow mosaic virus: Homogeneity, molecular weights, hydrodynamic radii and concentration dependence of parameters. *Biochem. J.*, 231, 549-555

Harding, S.E., Rowe, A.J. and Creeth, J.M. (1983) Further evidence for a flexible and highly expanded spheroidal model for mucus glycoproteins in solution. *Biochem.J.*, 209, 893-896

Herdan, G. (1949) Estimation of polymer distributions on the basis of differences between the average molecular weights of different order. *Nature*, 163, 139

Horton, J.C., Harding, S.E., Mitchell, J.R. and Morton-Holmes, D.F. (1991) Thermodynamic non-ideality of dilute solutions of sodium alginate studied by sedimentation equilibrium ultracentrifugation. *Food Hydrocolloids*, 5, 125-127

Howlett, G.L., Jeffrey, P.D. and Nichol, L.W. (1972) *J. Phys. Chem.* 76, 77

Jumel, K., Mitchell, J.R. and Harding, S.E. (1993) in preparation

Kawahawa, K., Ohta, K., Miyamoto, H. and Nakamura, S. (1984) Preparation and solution properties of pullulan fractions as standard samples for water-soluble polymers. *Carbohyd. Polym.* 4, 335

Laue, T.M. (1992) On-line data acquisition and analysis from the Rayleigh interferometer. In *"Analytical Ultracentrifugation in Biochemistry and Polymer Science"* (Harding,. S.E., Rowe, A.J. and Horton, J.C. eds.), Chap.6, Royal Society of Chemistry, Cambridge, UK

Lecacheux, D., Mustiere, Y., Panaras, R. and Brigand, G. (1986) Molecular weight of scleroglucan and other extracellular microbial polysaccharides by size-exclusion chromatography and low-angle light scattering. *Carbohyd. Polym.* 6, 477-492

Lechner, M.D. (1992) Determination of molecular weight averages and molecular weight distributions from sediemtation equilibrium. In *"Analytical Ultracentrifugation in Biochemistry and Polymer Science"* (Harding,. S.E., Rowe, A.J. and Horton, J.C. eds.), Chap.16, Royal Society of Chemistry, Cambridge, UK

Mannion, R.O., Melia, C.D., Launay, B., Cuvelier, G., Hill, S.E., Harding, S.E. and Mitchell, J.R. (1992) Xanthan/ locust bean gum interactions at rom temperature. *Carbohyd. Polym.* 19, 91-97

Muzzarelli, R.A.A., Lough, C. and Emanuelli, M. (1987) The molecular weight of chitosans studied by laser light scattering. *Carbohyd. Res.* 164, 433

Roark, D. and Yphantis, D.A. (1969) *Ann. N.Y. Acad. Sci.* 164, 245-278

Rowe, A.J., Wynne-Jones, S., Thomas, D.G. and Harding, S.E. (1992). Methods for off-line anslysis of sedimentation velocity and sedimentation equilibrium patterns. In *"Analytical Ultracentrifugation in Biochemistry and Polymer Science"* (Harding,. S.E., Rowe, A.J. and Horton, J.C. eds.), Chap.5, Royal Society of Chemistry, Cambridge, UK

Sato, T., Norisuye, T and Fujita, H. (1984) Double-stranded helix of xanthan - dimensional and hydrodynamic properties in 0.1M aqueous sodium chloride. *Macromolecules*, 17, 2696-2700

Schachman, H.K. (1992) Is there a future for the Ultracentrifuge? In *"Analytical Ultracentrifugation in Biochemistry and Polymer Science"* (Harding,. S.E., Rowe, A.J. and Horton, J.C. eds.), Chap.1, Royal Society of Chemistry, Cambridge, UK

Teller, D.C. (1965) *PhD Dissertation*, University of California, Berkeley, California

Teller, D.C. (1973) Characterization of proteins by sedimentation equilibrium in the analytical ultracentrifuge. *Meth. Enzymol.*, 27, 346-441

Wedlock, D.J., Baruddin, B.A. and Phillips, G.O. (1986) Comparison of molecular weight determination of sodium alginate by sedimentation-diffusion and light scattering. *Int. J. Biol. Macromol.* 8, 57

Woodward, J.R., Phillips, D.R. and Fincher, G.B. (1983) Water soluble (1-3)(1-4)-β-D-glucans from barley (Hordeum vulgare) endosperm I. Physicochemical properties. *Carbohyd. Polym.* 1983, 3, 143

Yphantis, D.A. (1964) Equilibrium ultracentrifugation of dilute solutions. *Biochemistry*, 3, 297-317

INDEX

RELATED TITLES

❖ ❖ ❖

Tissue Culture Techniques
An Introduction
By Bernice M. Martin
1994 Approx. 250 Pages
Hardcover ISBN 0-8176-3718-4
Softcover ISBN 0-8176-3643-9

The Polymerase Chain Reaction
Kary B. Mullis, François Ferré, and Richard A. Gibbs, Editors
1994 480 Pages
Hardcover ISBN 0-8176-3607-2
Softcover ISBN 0-8176-3750-8

DNA Fingerprinting
State of the Science
S. D. J. Pena, R. Chakraborty, J. T. Epplen, and A. J. Jeffreys, Editors
1993 480 Pages
Hardcover ISBN 3-7643-2781-2
Softcover ISBN 3-7643-2906-8

Protein Analysis and Purification
Benchtop Techniques
By Ian M. Rosenberg
1994 Approx. 350 Pages
Hardcover ISBN 0-8176-3717-6
Softcover ISBN 0-8176-3665-X

A Laboratory Guide for In Vivo Studies of DNA Methylation and Protein/DNA Interactions
By H. P. Saluz and J. P. Jost
1990 286 Pages Hardcover ISBN 3-7643-2369-8

A Laboratory Guide to In Vitro Transcription
By F. Sierra
1990 148 Pages Hardcover ISBN 3-7643-2357-4

Gene Therapeutics
Methods and Applications of Direct Gene Transfer
Jon A. Wolff, Editor
1994 433 Pages Hardcover ISBN 0-8176-3650-1